Lecture Notes
in Control and Information Sciences   370

Sukhan Lee, Il Hong Suh,
Mun Sang Kim (Eds.)

Recent Progress in Robotics: Viable Robotic Service to Human

An Edition of the Selected Papers from the 13th International Conference on Advanced Robotics

 Springer

Editors

Professor Dr. Sukhan Lee

School of Information and Communication Engineering
Sungkyunkwan University
300 Chunchun-Dong
Jangan-Ku, Kyunggi-Do 440-746
Korea
Email: Lsh@ece.skku.ac.kr, Lsh1@ieee.org

Dr.-Ing. Mun Sang Kim

Center for Intelligent Robotics
Frontier 21 Program
Korea Institute of Science and Technology
L8224, 39-1, Hawolgok-dong
Sungbuk-ku, Seoul 136-791
Korea
Email: munsang@kist.re.kr

Professor Il Hong Suh

Division of Information and Communications,
Hanyang University
17 Haengdang-dong
Seongdong-gu, Seoul 133-79
Korea
Email: ihsuh@hanyang.ac.kr

ISBN 978-3-540-76728-2

e-ISBN 978-3-540-76729-9

Lecture Notes in Control and Information Sciences

ISSN 0170-8643

Library of Congress Control Number: 2007941787

Typesetting by the authors and Scientific Publishing Services Pvt. Ltd.

Printed in acid-free paper

5 4 3 2 1 0

springer.com

Preface

This volume is an edition of the papers selected from the 13th International Conference on Advanced Robotics, ICAR 2007, held in Jeju, Korea, August 22-25, 2007, with the theme: "Viable Robotics Service to Human." It is intended to deliver readers the most recent technical progress in robotics, in particular, toward the advancement of robotic service to human. To ensure its quality, this volume took only 28 papers out of the 214 papers accepted for publication for ICAR 2007. The selection was based mainly on the technical merit, but also took into consideration whether the subject represents a theme of current interest. For the final inclusion, authors of the selected papers were requested for another round of revision and expansion. In this volume, we organize the 28 contributions into three chapters. Chapter 1 covers Novel Mechanisms, Chapter 2 deals with perception guided navigation and manipulation, and Chapter 3 addresses human-robot interaction and intelligence. Chapters 1, 2 and 3 consist of 7, 13 and 8 contributions, respectively. For the sake of clarity, Chapter 2 is divided further into two parts with Part 1 for Perception Guided Navigation and Part 2 for Perception Guided Manipulation. Chapter 3 is also divided into two parts with Part 1 for Human-Robot Interaction and Part 2 for Intelligence. For the convenience of readers, a chapter summary is introduced as an overview in the beginning of each chapter. The chapter summaries were prepared by Dr. Munsang Kim for Chapter 1, Prof. Sukhan Lee for Chapter 2, and Prof. Il-Hong Suh for Chapter 3. It is the wish of the editors of this volume that readers can find this volume informative and enjoyable. We would also like to thank Springer-Verlag for undertaking the publication of this volume.

Sukhan Lee

Contents

Chapter II: Perception Guided Navigation and Manipulation

Chapter III: Human-Robot Interaction and Intelligence

Part 1: Human-Robot Interaction

Part 2: Intelligence

Chapter I

Novel Mechanisms

Summary of Novel Mechanisms

Munsang Kim

The rapid development in intelligent robotics in the 21st century makes it possible to apply robots to the areas where it was treated unrealizable before. Recently many researchers are trying to make not only humanlike robots but also diverse robot systems that can achieve missions in extreme situations of Bio, Nano and Military fields substituting human activities. In this Novel Mechanism section some efforts to make innovative robot mechanisms such that these kinds of challenges might be realized are introduced.

1. "Feasibility Study of Robust Neural Network Motion Tracking Control of Piezoelectric Actuation Systems for Micro/Nano Manipulation": This paper proposes a motion tracking control methodology for piezoelectric actuation systems in micro/nano manipulation, in which a high performance in position and velocity tracking is required: The proposed method shows a capability to overcome the problems of unknown system parameters and non-linearities.
2. "Hybrid Formation Control for Non-Holonomic Wheeled Mobile Robots": Paolo Fiorini et al propose a hybrid formation control to reduce the formation errors by adopting an additional orientation controller. This concept can handle the major formation error practically.
3. "Novel Tirpedal Mobile Robot and Considerations for Gait Planning Strategies Based on Kinematics" In this paper a unique design of 3 legged robot mechanism is proposed, which may be useful in rough terrain gaiting. By introducing a concept of a static stability margin, they can control the robot gaiting successfully.
4. "Safe Joint Mechanism based on Passive Compliance for Collision Safety", J.B. Song et al propose a novel passive mechanism in order to secure the inherently safe manipulation. They designed a safe joint mechanism, which changes the high stiffness mode to the low one by absorbing the impact force acting on the robot arm. This idea can deliver a very practical solution in designing safe manipulator to the intelligent service robots.
5. "Sensor-based Guidance Control of a Continuum Robot for Semi-Autonomous Colonoscopy": The paper by Chen et al. presents a guidance method for the ColoBot, a continuum robot for colonoscopy. Since the colon is naturally a confined space, which may change during the examination, the path for the distal end of the robot cannot be produced prior to the operation, but must be generated incrementally. Considering this, the paper proposes a sensor-based method that can generate a reference path for the distal end of ColoBot using optical sensors installed on it. The paper shows that the path of the distal end can be produced

S. Lee, I.H. Suh, and M.S. Kim (Eds.): Recent Progress in Robotics, LNCIS 370, pp. 3–4, 2008.
springerlink.com

under a simple geometric assumption, and presents a control law by formulating the inverse kinematical problem as a constrained linear optimization problem.

6. "Fully-isotropic T1R3-type Redundantly-actuated Parallel Manipulators": This paper shows that an isotropic parallel mechanism can be obtained with redundantly actuated joints. It also provides structural synthesis method to accomplish an isotropic parallel mechanism. With this paper the reader can achieve an isotropic mechanism design which provides ideal kinematic and dynamic performance.

7. "Early Reactive Grasping with Second Order 3D Feature Relations": In grasping strategy, most previous works have been based on analytical methods where the shape of objects being grasped is known a-priori. But, in this paper, the authors introduce a new method called "reflex-like" grasping strategy, where it does not require a-priori object knowledge and only predefines "spatial primitives" for grasp hypotheses. This method will make a useful progress about a cognitive concept; objects and action/grasping that can be performed on them are inseparable intertwined.

Feasibility Study of Robust Neural Network Motion Tracking Control of Piezoelectric Actuation Systems for Micro/Nano Manipulation

Hwee Choo Liaw[1], Bijan Shirinzadeh[1], Gursel Alici[2], and Julian Smith[3]

[1] Robotics and Mechatronics Research Laboratory, Department of Mechanical Engineering, Monash University, Clayton, VIC 3800, Australia
hwee-choo.liaw@eng.monash.edu.au, bijan.shirinzadeh@eng.monash.edu.au
[2] School of Mechanical, Materials, and Mechatronic Engineering, University of Wollongong, NSW 2522, Australia
gursel@uow.edu.au
[3] Monash Medical Centre, Department of Surgery, Monash University, Clayton, VIC 3800, Australia
julian.smith@med.monash.edu.au

Summary. This paper presents a robust neural network motion tracking control methodology for piezoelectric actuation systems employed in micro/nano manipulation. This control methodology is proposed for tracking desired motion trajectories in the presence of unknown system parameters, non-linearities including the hysteresis effect, and external disturbances in the control systems. In this paper, the control methodology is established including the neural networks and a sliding scheme. In particular, the radial basis function neural networks are chosen in this study for function approximations. The stability of the closed-loop systems and convergence of the position and velocity tracking errors to zero are assured by the control methodology in the presence of the aforementioned conditions. Simulation results of the control methodology for tracking of a desired motion trajectory is presented. With the capability of motion tracking, the proposed control methodology can be utilised to realise high performance piezoelectric actuated micro/nano manipulation systems.

Keywords: Micro/nano manipulation, piezoelectric actuator, motion control, neural networks, sliding control.

1 Introduction

The presence of non-linearity in the piezoelectric actuators is a major drawback in the field of micro/nano manipulation. This nonlinear effect prevents the actuators from providing the desired high-precision motion accuracy in the piezoelectric actuation systems. To resolve this problem, a considerable amount of research has been conducted. One direction of research has sought to model and compensate for the non-linearities, particularly for the hysteresis effect [1, 2, 3, 4]. However, the hysteresis effect is very complex. It is difficult to obtain an accurate model and the model parameters are difficult to quantify in practice.

S. Lee, I.H. Suh, and M.S. Kim (Eds.): Recent Progress in Robotics, LNCIS 370, pp. 5–19, 2008.
springerlink.com   © Springer-Verlag Berlin Heidelberg 2008

Alternatively, studies have been focused on the enhancement of positioning performance by proposing closed-loop control for the piezoelectric actuation systems [5, 6, 7, 8]. Nevertheless, in most of these studies, a complex feed-forward hysteresis model has been adopted in the closed-loop systems.

Neural networks provide many useful properties and capabilities. These include input-output mapping or function approximation, non-linearity, and adaptivity [9]. Furthermore, they can be trained to produce reasonable outputs even for inputs that are not available during the training process. Increasingly, neural networks have been proposed in the field of robot manipulator control for motion trajectory tracking [10, 11, 12]. This control approach can be extended to the piezoelectric actuation systems.

In this paper, a robust neural network motion tracking control methodology is established and investigated for the piezoelectric actuation systems. This work is motivated by our previous efforts in the control of the piezoelectric actuators [13, 14], and it is also inspired by the neural network control [15, 16]. The control objective is to track a specified motion trajectory in the proposed closed-loop system.

In this study, the properties of the neural networks are examined in the closed-loop control. Using the neural networks for function approximations, the proposed control methodology is established including the radial basis function neural networks and a sliding strategy. A major advantage of this control approach is that no prior knowledge of the system parameters, or of the initial estimates of the neural networks, is required in the physical realisation. The control is executed with appropriate gain settings and the neural network weights are estimated and updated on-line during the real-time control. This proposed control methodology is employed to overcome the problems of unknown system parameters, non-linearities including the hysteresis effect, and external disturbances in the piezoelectric actuation systems. The stability of the closed-loop systems is analysed in which the position and velocity tracking errors are proved to converge to zero in the tracking of a desired motion trajectory. Furthermore, a promising motion tracking performance is demonstrated in the simulation study.

This paper is organised as follows. The model of a piezoelectric actuation system is introduced in Section 2 and the radial basis function neural network is described in Section 3. The robust neural network motion tracking control methodology is established in Section 4 followed by the stability analysis in Section 5. The simulation study is detailed in Section 6 and the results are presented and discussed in Section 7. Finally, conclusions are drawn in Section 8.

2 Model of Piezoelectric Actuation System

An electromechanical model of a piezoelectric actuator has been identified based on recent studies [1, 2]. This mathematical model is divided into three stages of transformation from electrical charge to mechanical energy, and vice versa. The schematic model, as shown in Fig. 1, illustrates the transformation, which consists of a voltage-charge stage, a piezoelectric stage, and a force-displacement

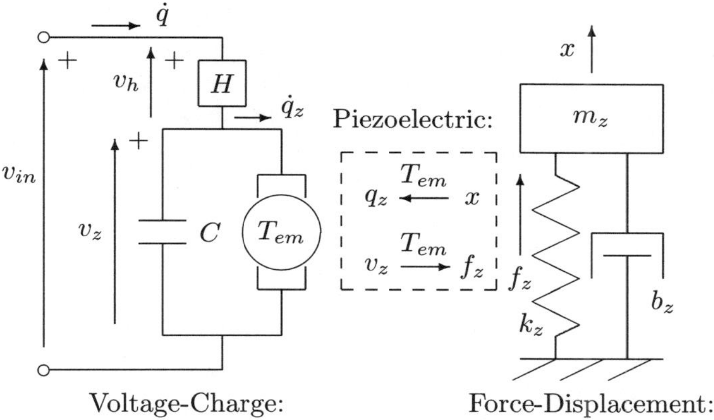

Fig. 1. Schematic model of a piezoelectric actuator

stage, and it is formulated for a voltage driven system. The dynamic equation from the electrical input to the output motion stage can be described by the following set of equations:

$$v_{in} = v_h + v_z \,, \tag{1}$$

$$v_h = H(q) \,, \tag{2}$$

$$q = C\,v_z + q_z \,, \tag{3}$$

$$q_z = T_{em}\,x \,, \tag{4}$$

$$f_z = T_{em}\,v_z \,, \tag{5}$$

$$m_z\,\ddot{x} + b_z\,\dot{x} + k_z\,x = f_z \,, \tag{6}$$

where v_{in} represents the applied (input) voltage, v_h is the voltage due to the hysteresis, v_z is the voltage related to the mechanical side of the actuator, q is the total charge in the ceramic, H is the hysteresis effect, C is the linear capacitance connected in parallel with the electromechanical transformer having a ratio of T_{em}, q_z is the piezoelectric charge related to the actuator output displacement x, and f_z is the transduced force from the electrical domain. The variables m_z, b_z, and k_z are the mass, damping, and stiffness, respectively, of the force-displacement stage. The hysteresis effect described by (2) in the piezoelectric actuator causes a highly nonlinear input/output relationship between the applied voltage and output displacement. This hysteresis effect is understood to be bounded, i.e. $|v_h| \le \delta v_h$, where δv_h is a constant number.

For control purposes, (1) and (5) are substituted into (6) to yield

$$m_z\,\ddot{x} + b_z\,\dot{x} + k_z\,x = T_{em}\,(v_{in} - v_h) \,, \tag{7}$$

and the piezoelectric actuator model is obtained by re-arranging (7),

$$m\,\ddot{x} + b\,\dot{x} + k\,x + v_h = v_{in} \,, \tag{8}$$

where $m = m_z / T_{em}$, $b = b_z / T_{em}$, and $k = k_z / T_{em}$.

8   H.C. Liaw et al.

This piezoelectric actuator model (8) can be extended to describe a piezo-electric actuation system. In reality, besides the hysteresis effect v_h described by (8), there may exist other nonlinear effects that are present in the piezoelectric actuation system. Furthermore, there are generally external disturbances in a practical dynamic system. To include these effects, the piezoelectric actuation system (8) is rewritten as

$$m\,\ddot{x} + b\,\dot{x} + k\,x + v_n + v_d = v_{in}\,, \tag{9}$$

where v_n and v_d represent all the nonlinear effects and external disturbances, respectively, encountered in the motion system. It must be noted that the non-linear effects and external disturbances are bounded and there exits an upper bound δv_{nd} such that

$$|\,v_n\,| + |\,v_d\,| \leq \delta v_{nd}\,, \tag{10}$$

where δv_{nd} is a positive constant number.

3   Radial Basis Function Neural Network

Many different models of neural networks have been established. They are mostly designed for specific objectives. Among these models, the radial basis function (RBF) neural networks are found to be well suited for the purpose of modelling uncertain or nonlinear functions [9, 16]. A typical RBF neural network is shown in Fig. 2. It is a two-layer network comprising a hidden layer and a output layer. The hidden layer consists of an array of functions, i.e. RBFs, and the output layer is merely a linear combination of the hidden layer functions. With this simple structure, the RBF neural network permits a more effective weight updating procedure compared to other complex multi-layer networks.

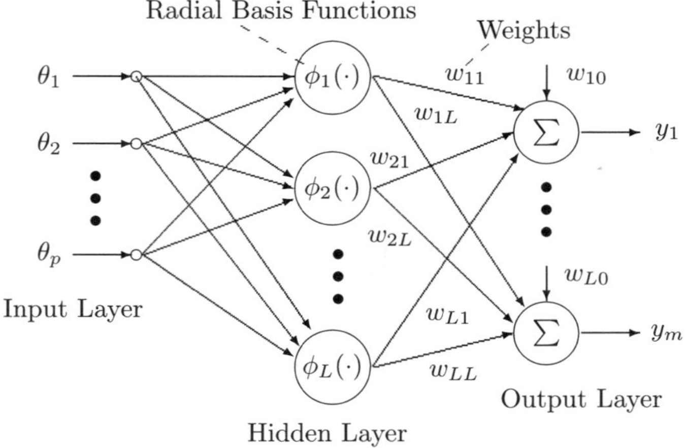

Fig. 2. Structure of a two-layer radial basis function neural network

In this study, the RBF neural network is employed for function approximation. It is assumed that a smooth function $\boldsymbol{f}(\boldsymbol{\theta}) : \Re^p \to \Re^m$ is expressed in terms of the RBF neural network as [15]

$$\boldsymbol{f}(\boldsymbol{\theta}) = \boldsymbol{W}^T \boldsymbol{\phi}(\boldsymbol{\theta}) + \boldsymbol{\varepsilon}(\boldsymbol{\theta}), \tag{11}$$

$$\boldsymbol{W}^T = \begin{bmatrix} w_{10} & w_{11} & w_{12} & \cdots & w_{1L} \\ w_{20} & w_{21} & w_{22} & \cdots & w_{2L} \\ \vdots & \vdots & \vdots & \ddots & \vdots \\ w_{m0} & w_{m1} & w_{m2} & \cdots & w_{mL} \end{bmatrix}, \tag{12}$$

$$\boldsymbol{\phi}(\boldsymbol{\theta}) = \begin{bmatrix} 1 & \phi_1(\boldsymbol{\theta}) & \phi_2(\boldsymbol{\theta}) & \cdots & \phi_L(\boldsymbol{\theta}) \end{bmatrix}^T, \tag{13}$$

where $\boldsymbol{\theta} \in \Re^p$ is the input vector, $\boldsymbol{W} \in \Re^{(L+1)\times m}$ are the ideal thresholds and weights of the RBF neural network, $\boldsymbol{\phi}(\boldsymbol{\theta}) \in \Re^{(L+1)}$ is the activation vector comprising the RBFs, and $\boldsymbol{\varepsilon}(\boldsymbol{\theta}) \in \Re^m$ are the neural network function approximation errors. Using the RBF neural network, as shown in Fig. 2, with a sufficiently large number L of RBF in the hidden layer to approximate the smooth function $\boldsymbol{f}(\boldsymbol{\theta})$ described by (11), there exists a positive number $\delta\varepsilon$ such that

$$\| \boldsymbol{\varepsilon}(\boldsymbol{\theta}) \| \leq \delta\varepsilon \quad \forall \boldsymbol{\theta}. \tag{14}$$

A suitable RBF can be selected from a large class of functions for the activation vector described by (13). Commonly, the Gaussian function [16, 9] is chosen and the RBF $\phi_i(\boldsymbol{\theta})$ for $i = 1, ..., L$ is given by

$$\phi_i(\boldsymbol{\theta}) = \exp\left[-\frac{1}{2\,\eta_i^2} \| \boldsymbol{\theta} - \boldsymbol{\mu}_i \|^2 \right], \tag{15}$$

where $\boldsymbol{\mu}_i$ is the mean or centre of the function, and $\boldsymbol{\eta}_i$ denotes its width.

4 Robust Neural Network Motion Tracking Control Methodology

For the piezoelectric actuation system described by (9), a robust neural network control methodology can be formulated for the purpose of tracking a desired motion trajectory $x_d(t)$. Under the proposed control approach, the physical system parameters in (9) are assumed to be unknown. Furthermore, there exist bounded nonlinear effects and external disturbances within the system. The motion trajectory $x_d(t)$ is assumed to be at least twice continuously differentiable and both $\dot{x}_d(t)$ and $\ddot{x}_d(t)$ are bounded and uniformly continuous in $t \in [0, \infty)$. The control methodology is established such that the closed-loop system will follow the required motion trajectory. In the formulation, the position tracking error $e_p(t)$ is defined as

$$e_p(t) = x(t) - x_d(t), \tag{16}$$

and a tracking or switching function σ is defined as

$$\sigma = \dot{e}_p + \alpha\, s(e_p)\,, \tag{17}$$

where α is a strictly positive scalar, and $s(e_p)$ is the saturation error function. This saturation function is derived from a special positive definite function $\rho(e_p)$, which is defined as [13]

$$\rho(e_p) = \sqrt{c^2 + e_p{}^2} - |c|\,, \tag{18}$$

where c is an arbitrary constant with its absolute value $|c| > 0$. The saturation error function $s(e_p)$ described by (17) is expressed by using (18) as

$$s(e_p) = \frac{\mathrm{d}\rho(e_p)}{\mathrm{d}e_p} = \frac{e_p}{\sqrt{c^2 + e_p{}^2}}\,, \tag{19}$$

and the time derivative of the positive definite function (18) is given by

$$\dot{\rho}(e_p) = s(e_p)\,\dot{e}_p\,. \tag{20}$$

The time derivative of the tracking function (17) is given by

$$\dot{\sigma} = \ddot{e}_p + \alpha\,\dot{s}(e_p)\,. \tag{21}$$

The term $\dot{s}(e_p)$ described by (21) is derived from (19) and is given by

$$\dot{s}(e_p) = \frac{\mathrm{d}s(e_p)}{\mathrm{d}e_p}\,\dot{e}_p = \frac{c^2\,\dot{e}_p}{\sqrt{(c^2 + e_p{}^2)^3}}\,. \tag{22}$$

Rearranging (16), (17), and (21), yields

$$x = e_p + x_d\,, \tag{23}$$
$$\dot{x} = \sigma + \dot{x}_d - \alpha\, s(e_p)\,, \tag{24}$$
$$\ddot{x} = \dot{\sigma} + \ddot{x}_d - \alpha\,\dot{s}(e_p)\,. \tag{25}$$

Substituting (23), (24), and (25) into (9) to express the dynamic model in terms of the tracking errors, yields

$$m\,\dot{\sigma} = v_{in} - b\,\sigma - k\,e_p - v_n - v_d - f\,, \tag{26}$$

where the function f is assumed to be unknown and it is given by

$$f = m\,[\ddot{x}_d - \alpha\,\dot{s}(e_p)] + b\,[\dot{x}_d - \alpha\, s(e_p)] + k\,x_d\,. \tag{27}$$

It must be noted that the RBF neural network described by (11) is employed for the function approximation of the unknown function f, and it is assumed that

$$f(\boldsymbol{\theta}) \equiv f\,. \tag{28}$$

The vector $\boldsymbol{\theta}$ described by (28) is defined according to (27) as

$$\boldsymbol{\theta} \equiv \left[\, x_d \;\; [\dot{x}_d - \alpha\, s(e_p)] \;\; [\ddot{x}_d - \alpha\, \dot{s}(e_p)] \,\right]^T . \tag{29}$$

The function $\boldsymbol{f}(\boldsymbol{\theta})$ given by (11) for (28) is redefined as $f(\boldsymbol{\theta}) : \Re^3 \to \Re^1$, and it is rewritten as

$$f(\boldsymbol{\theta}) = \boldsymbol{w}^T \boldsymbol{\phi}(\boldsymbol{\theta}) + \varepsilon(\boldsymbol{\theta}) , \tag{30}$$

where $\boldsymbol{w} \in \Re^{(L+1)}$ are the ideal threshold and weights.

For the piezoelectric actuation system described by (9), a robust neural network motion tracking control methodology is proposed and described by the following equations:

$$v_{in} = \hat{f}(\boldsymbol{\theta}) - k_p\, e_p - k_v\, \dot{e}_p - k_s\, \sigma - d\, \frac{\sigma}{|\sigma|} , \tag{31}$$

$$d \geq \delta v_{nd} + \delta\varepsilon + \epsilon , \tag{32}$$

where $\hat{f}(\boldsymbol{\theta})$ is an estimate of $f(\boldsymbol{\theta})$; k_p, k_v, and k_s are the control gains; and ϵ is a strictly positive scalar. The estimated function $\hat{f}(\boldsymbol{\theta})$ is established as

$$\hat{f}(\boldsymbol{\theta}) = \hat{\boldsymbol{w}}^T \boldsymbol{\phi}(\boldsymbol{\theta}) , \tag{33}$$

where $\hat{\boldsymbol{w}} \in \Re^{(L+1)}$ are the estimated threshold and weights of the neural network. They are provided by the tuning algorithm, which is designed as

$$\dot{\hat{\boldsymbol{w}}} = -\boldsymbol{L}^{-1}\, \boldsymbol{\phi}(\boldsymbol{\theta})\, \sigma , \tag{34}$$

where $\boldsymbol{L} \in \Re^{(L+1)\times(L+1)}$ is a symmetric positive definite tuning matrix. With the estimated function $\hat{f}(\boldsymbol{\theta})$ given by (33), the control law (31) becomes

$$v_{in} = \hat{\boldsymbol{w}}^T \boldsymbol{\phi}(\boldsymbol{\theta}) - k_p\, e_p - k_v\, \dot{e}_p - k_s\, \sigma - d\, \frac{\sigma}{|\sigma|} , \tag{35}$$

It must be noted that the estimated threshold and weights of the neural network are adjusted on-line by the tuning algorithm (34) while the closed-loop system is tracking the desired motion trajectory. No prior off-line learning procedure is needed and the control law (35) can be executed with any appropriate initial estimated threshold and weight values.

5 Stability Analysis

The closed-loop behaviour of the proposed control methodology must be carefully examined in the system stability study. For this purpose, the closed-loop dynamics is derived and analysed. Substituting the control input (31) into the piezoelectric actuation system described by (26), yields

$$m\,\dot{\sigma} = -(k_p + k)\, e_p - k_v\, \dot{e}_p - (k_s + b)\, \sigma - d\, \frac{\sigma}{|\sigma|} - v_n - v_d + \tilde{f}(\boldsymbol{\theta}) , \tag{36}$$

12   H.C. Liaw et al.

where $\tilde{f}(\boldsymbol{\theta}) = \hat{f}(\boldsymbol{\theta}) - f(\boldsymbol{\theta})$ is the function estimation error. This estimation error is expressed according to (33) and (30) as

$$\tilde{f}(\boldsymbol{\theta}) = \tilde{\boldsymbol{w}}^T \boldsymbol{\phi}(\boldsymbol{\theta}) - \varepsilon(\boldsymbol{\theta}) \,, \tag{37}$$

where $\tilde{\boldsymbol{w}}$ are the threshold and weight estimation errors, which are defined as

$$\tilde{\boldsymbol{w}} = \hat{\boldsymbol{w}} - \boldsymbol{w} \,. \tag{38}$$

Multiplying both sides of (36) by the tracking function σ and applying (37), the closed-loop dynamics described by (36) is rewritten as

$$\sigma \, m \, \dot{\sigma} = -y - d\,|\sigma| - \sigma\,[\,v_n + v_d + \varepsilon(\boldsymbol{\theta})\,] + \sigma \, \tilde{\boldsymbol{w}}^T \boldsymbol{\phi}(\boldsymbol{\theta}) \,, \tag{39}$$

where the term y is a function given by

$$y = \sigma \, [\, (k_p + k) \, e_p + k_v \, \dot{e}_p + (k_s + b) \, \sigma \,] \,. \tag{40}$$

Expanding the right-hand side of (40) and replacing σ by (17), yields

$$\begin{aligned}
y = {} &(k_p + k)\, e_p\, \dot{e}_p + (k_v + k_s + b)\, \dot{e}_p{}^2 + \alpha \, (k_p + k)\, e_p\, s(e_p) \\
&+ \alpha\,[\,k_v + 2\,(k_s + b)\,]\, s(e_p)\, \dot{e}_p + \alpha^2 \, (k_s + b)\, s^2(e_p) \,.
\end{aligned} \tag{41}$$

The function y can be expressed in terms of two positive definite functions u_1 and v, and it is given by

$$y = \dot{u}_1 + v \,, \tag{42}$$

and

$$u_1 = \frac{1}{2}\,(k_p + k)\, e_p{}^2 + \alpha\,[\,k_v + 2\,(k_s + b)\,]\, \rho(e_p) \,, \tag{43}$$

$$v = (k_v + k_s + b)\, \dot{e}_p{}^2 + \alpha\,(k_p + k)\, e_p\, s(e_p) + \alpha^2\,(k_s + b)\, s^2(e_p) \,, \tag{44}$$

where the term $s(e_p)\, \dot{e}_p$ in (41) is related to $\rho(e_p)$ in (43) described by (20). By closely examining the terms $\rho(e_p)$ and $s(e_p)$ given by (18) and (19), respectively, the functions u_1 and v given by (43) and (44), respectively, are positive definite for the positive chosen control gains k_p, k_v, and k_s.

Theorem 1. *For the piezoelectric actuation system described by (9), the robust neural network motion tracking control law (35) assures the convergence of the motion trajectory tracking with $e_p(t) \to 0$ and $\dot{e}_p(t) \to 0$ as $t \to \infty$.*

Proof. It must be noted that for the system described by (9) with the proposed control law (35), the functions u_1 and v from (43) and (44), respectively, are positive definite in all non-zero values of $e_p(t)$ and $\dot{e}_p(t)$.

A Lyapunov function u_2 is proposed for the closed-loop system,

$$u_2 = \frac{1}{2}\, m\, \sigma^2 + \frac{1}{2}\, \tilde{\boldsymbol{w}}^T \boldsymbol{L}\, \tilde{\boldsymbol{w}} \,, \tag{45}$$

Differentiating u_2 with respect to time, yields

$$\dot{u}_2 = \sigma\, m\, \dot{\sigma} + \tilde{\boldsymbol{w}}^T \boldsymbol{L}\, \dot{\tilde{\boldsymbol{w}}}. \tag{46}$$

Substituting the closed-loop dynamics (39) into (46), yields

$$\dot{u}_2 = -y - d\,|\sigma| - \sigma\,[\,v_n + v_d + \varepsilon(\boldsymbol{\theta})\,] + \tilde{\boldsymbol{w}}^T[\,\boldsymbol{\phi}(\boldsymbol{\theta})\,\sigma + \boldsymbol{L}\,\dot{\tilde{\boldsymbol{w}}}\,]. \tag{47}$$

From (38), the term $\dot{\tilde{\boldsymbol{w}}} = \dot{\hat{\boldsymbol{w}}}$ as $\boldsymbol{w}$ is a constant. Substituting the tuning algorithm (34) into (47), yields

$$\dot{u}_2 = -y - d\,|\sigma| - \sigma\,[\,v_n + v_d + \varepsilon(\boldsymbol{\theta})\,]. \tag{48}$$

Replacing the function y in (48) by using (42), and considering (10), (14), and (32), yields

$$\begin{aligned}
\dot{u}_2 &= -\dot{u}_1 - v - d\,|\sigma| - \sigma\,[\,v_n + v_d + \varepsilon(\boldsymbol{\theta})\,], \\
\dot{u} &= -v - d\,|\sigma| - \sigma\,[\,v_n + v_d + \varepsilon(\boldsymbol{\theta})\,], \\
&\leq -v - d\,|\sigma| + |\sigma|\,[\,|v_n| + |v_d| + |\varepsilon(\boldsymbol{\theta})|\,], \\
&\leq -v - d\,|\sigma| + |\sigma|\,[\,\delta v_{nd} + \delta\varepsilon\,], \\
&\leq -v - \epsilon\,|\sigma|,
\end{aligned} \tag{49}$$

where $u = u_1 + u_2$ is a Lyapunov function. This shows that $u \to 0$ and implies $e_p(t) \to 0$ and $\dot{e}_p(t) \to 0$ as $t \to \infty$. Both the system stability and tracking convergence are guaranteed by the control law (35) driving the system (9) closely tracking a desired motion trajectory. $\qquad\square$

Remark 1. In the realisation of the control law (35), the discontinuous function $\frac{\sigma}{\|\sigma\|}$ will give rise to control chattering due to imperfect switching in the computer control. This is undesirable, as un-modelled high frequency dynamics might be excited. To eliminate this effect, the concept of boundary layer technique [17] is applied to smooth the control signal. In a small neighbourhood of the sliding surface ($\sigma = 0$), the discontinuous function is replaced by a boundary saturation function $\mathrm{sat}(\frac{\sigma}{\Delta})$ defined as

$$\mathrm{sat}(\frac{\sigma}{\Delta}) = \begin{cases} -1 & : & \sigma < -\Delta, \\ \sigma/\Delta & : & -\Delta \leq \sigma \leq \Delta, \\ +1 & : & \sigma > \Delta, \end{cases} \tag{50}$$

where Δ is the specified boundary layer thickness, and the robust neural network motion tracking control law (35) becomes

$$v_{in} = \hat{\boldsymbol{w}}^T\boldsymbol{\phi}(\boldsymbol{\theta}) - k_p\,e_p - k_v\,\dot{e}_p - k_s\,\sigma - d\,\mathrm{sat}(\frac{\sigma}{\Delta}). \tag{51}$$

6 Simulation Study

Computer simulation is essential for the development of a piezoelectric actuation system for micro/nano manipulation. Several objectives can be achieved from the simulation. They include the realisation of the control algorithms, tuning of the controller, and investigation of the behaviour of the proposed closed-loop system.

In this study, the unknown function f described by (27) is divided into three sub-functions:

$$f_m = m\left[\ddot{x}_d - \alpha\,\dot{s}(e_p)\right],\tag{52}$$

$$f_b = b\left[\dot{x}_d - \alpha\,s(e_p)\right],\tag{53}$$

$$f_k = k\,x_d\,.\tag{54}$$

To cater for these sub-functions, three separate neural networks are utilised to approximate the unknown functions f_m, f_b, and f_k described by (52), (53), and (54), respectively. The estimated function described by (33) becomes

$$\hat{f}_m = \hat{\boldsymbol{w}}_m^T\,\boldsymbol{\phi}_m(\theta_m)\,,\tag{55}$$

$$\hat{f}_b = \hat{\boldsymbol{w}}_b^T\,\boldsymbol{\phi}_b(\theta_b)\,,\tag{56}$$

$$\hat{f}_k = \hat{\boldsymbol{w}}_k^T\,\boldsymbol{\phi}_k(\theta_k)\,,\tag{57}$$

where $\theta_m = \ddot{x}_d - \alpha\,\dot{s}(e_p)$, $\theta_b = \dot{x}_d - \alpha\,s(e_p)$, and $\theta_k = x_d$. The terms $\hat{\boldsymbol{w}}$ and $\boldsymbol{\phi}(\boldsymbol{\theta})$ of the control law (51) are modified as

$$\hat{\boldsymbol{w}} = \left[\,\hat{\boldsymbol{w}}_m^T\ \ \hat{\boldsymbol{w}}_b^T\ \ \hat{\boldsymbol{w}}_k^T\,\right]^T,\tag{58}$$

and

$$\boldsymbol{\phi}(\boldsymbol{\theta}) = \left[\,\boldsymbol{\phi}_m^T(\theta_m)\ \ \boldsymbol{\phi}_b^T(\theta_b)\ \ \boldsymbol{\phi}_k^T(\theta_k)\,\right]^T.\tag{59}$$

The matrix $\boldsymbol{L}$ of tuning algorithm (34) becomes

$$\boldsymbol{L}^{-1} = diag\left\{\,\boldsymbol{L}_m^{-1}\ \ \boldsymbol{L}_b^{-1}\ \ \boldsymbol{L}_k^{-1}\,\right\}.\tag{60}$$

The simulation model is constructed by using the piezoelectric actuation system described by (9) and the robust neural network motion tracking control law given by (51). In the simulation, a desired motion trajectory is employed as shown in Fig. 3 for position, velocity, and acceleration. The desired motion trajectory is formed by segments of quintic polynomials [18] for the investigation of the tracking performance of the proposed control methodology.

The neural networks are designed with each activation function $\boldsymbol{\phi}_m$, $\boldsymbol{\phi}_b$, and $\boldsymbol{\phi}_k$ in (59) having 30 RBFs, i.e. $L = 30$. The centres and widths described by (15) for the activation functions are chosen to be evenly spaced between $[\,0,\,7\times10^{-2}\,(m/s^2)\,]$ for $\boldsymbol{\phi}_m$, $[\,0,\,1.1\times10^{-3}\,(m/s)\,]$ for $\boldsymbol{\phi}_b$, and $[\,0,\,3\times10^{-5}\,(m)\,]$ for $\boldsymbol{\phi}_k$. The initial estimated threshold and weights $\hat{\boldsymbol{w}}_m$, $\hat{\boldsymbol{w}}_b$, and $\hat{\boldsymbol{w}}_k$ in (58) are chosen to be zero. The tuning gains in (60) are selected as $\boldsymbol{L}_m^{-1} = \boldsymbol{L}_b^{-1} = \boldsymbol{L}_k^{-1} = 8\times10^4\,diag\{1\times10^{-4}, 1, 1, ..., 1\}$.

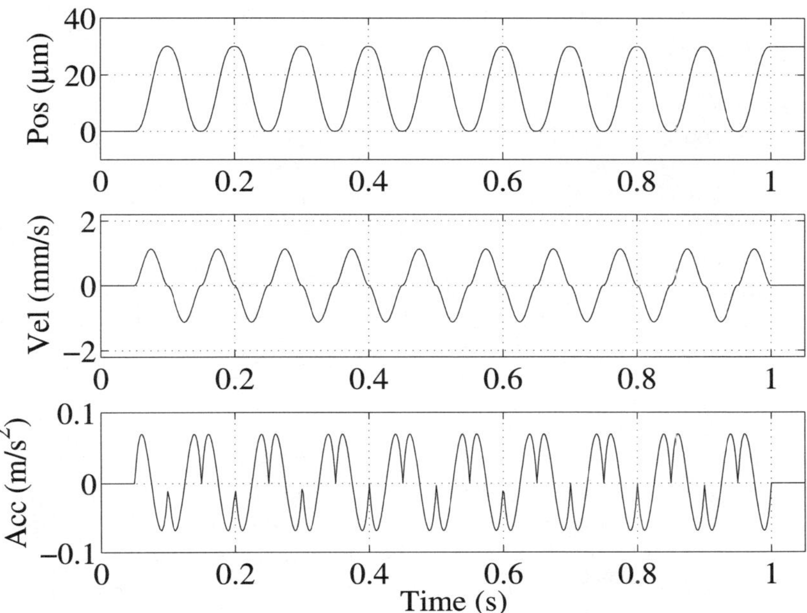

Fig. 3. Desired motion trajectory

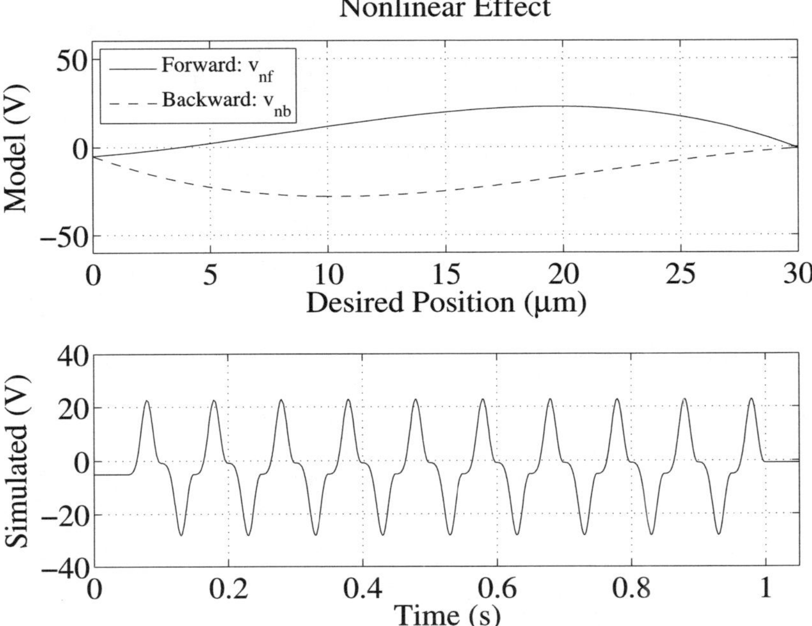

Fig. 4. Nonlinear effect in simulation

The piezoelectric actuation system described by (9) is simulated to posses the dynamic coefficients of $m = 1\,(Vs^2/m)$, $b = 2.5 \times 10^3\,(Vs/m)$, and $k = 2 \times 10^6\,(V/m)$. The control gains in (51) are chosen as $k_p = 2 \times 10^4\,(V/m)$, $k_v = 500\,(Vs/m)$, and $k_s = 200\,(Vs/m)$. The total value of bounds in (32) for (51) is specified as $d = 35\,(V)$. The boundary layer thickness in (50) for (51) is chosen as $\Delta = 0.005\,(m/s)$. The saturation error function $s(e_p)$ in (17) and its time derivative $\dot{s}(e_p)$ in (21) are realised by using (19) and (22), respectively. The arbitrary constant c in (19) and positive scalar α in (17) are selected as $1 \times 10^{-6}\,(m)$ and $1.0 \times 10^{-2}\,(m/s)$, respectively.

For a more realistic study, a model of nonlinear effect v_n described by (9) is introduced in the simulation. This model and its simulated effect are shown in Fig. 4.

7 Results and Discussion

Tracking the desired motion trajectory, as shown in Fig. 3, the resulting positions and velocities are shown in Fig. 5. Despite unknown system parameters and together with the introduction of nonlinear effect in the simulation, the proposed robust neural network motion tracking control law (51) showed a promising tracking ability.

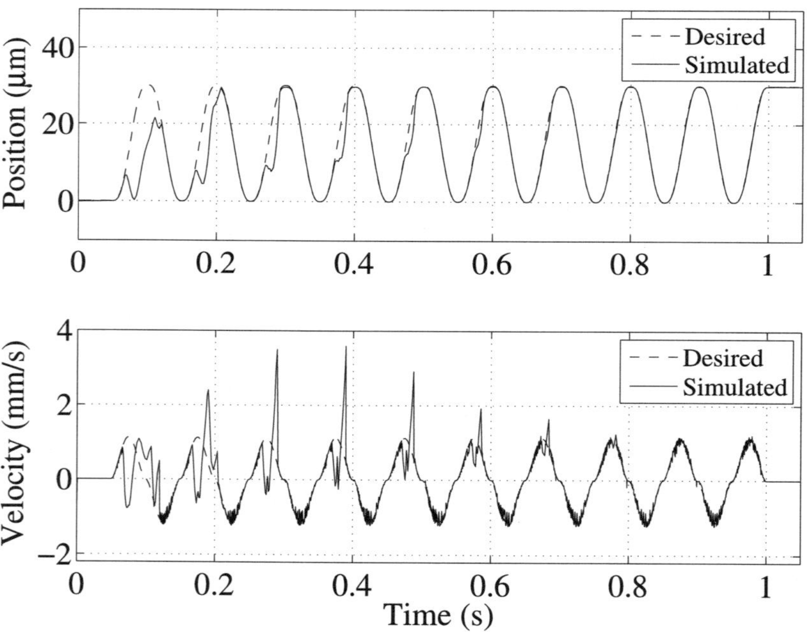

Fig. 5. Simulated positions and velocities compared to their desired values

Estimated threshold and first ten weights

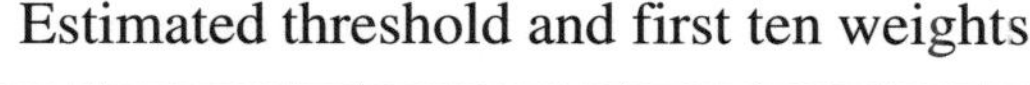

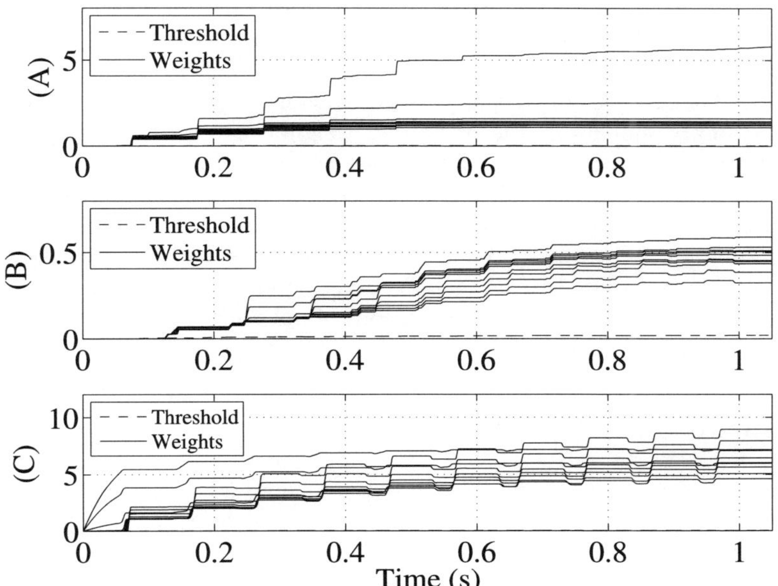

Fig. 6. Estimated threshold and weights: (A) $\hat{\boldsymbol{w}}_m$, (B) $\hat{\boldsymbol{w}}_b$, and (C) $\hat{\boldsymbol{w}}_k$

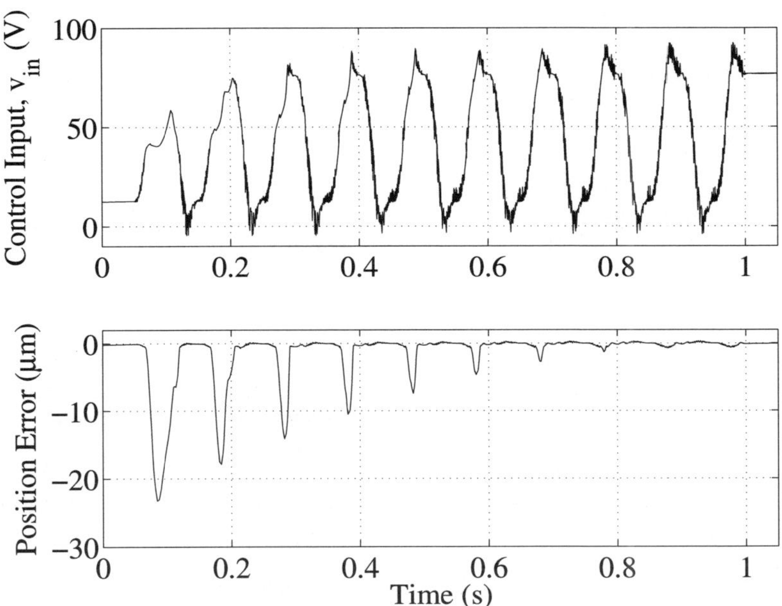

Fig. 7. Control input and position tracking errors

The estimated threshold and the first ten estimated weights for the neural networks $\hat{f}_m$, $\hat{f}_b$, and $\hat{f}_k$ described by (55), (56), and (57), respectively, are shown in Fig. 6. With the appropriate tuning gains, the estimated weights were gradually settled to their final values during the simulation.

The control input v_{in} and the position tracking errors are shown in Fig. 7. With the improvements in the estimated weights, as presented in Fig. 6, the position tracking errors reduced significantly from -23.22 (μm) to -0.65 (μm) during the dynamic motion. This demonstrated the feasibility and effectiveness of the proposed control methodology for the piezoelectric actuation system.

In summary, the proposed robust neural network motion tracking control methodology is simulated to be stable, robust, and capable of tracking the desired motion trajectory under the conditions of unknown system parameters and nonlinear effect.

8   Conclusions

A robust neural network motion tracking control methodology has been proposed and investigated for the tracking of a desired motion trajectory in a piezoelectric actuation system.

This proposed control methodology is formulated to accommodate unknown system parameters, non-linearities including the hysteresis effect, and external disturbances in the motion system.

The stability of the proposed closed-loop system has been analysed and the convergence of the position and velocity tracking errors to zero is guaranteed by the control methodology. Furthermore, a promising tracking ability has been demonstrated in the simulation study.

Acknowledgment

This work is supported by an Australian Research Council (ARC) Linkage Infrastructure, Equipment and Facilities (LIEF) grant and an ARC Discovery grant.

References

1. Goldfarb, M., Celanovic, N.: Modeling piezoelectric stack actuators for control of micromanipulation. IEEE Control Systems Magazine 17(3), 69–79 (1997)
2. Adriaens, H.J.M.T.A., de Koning, W.L., Banning, R.: Modeling piezoelectric actuators. IEEE/ASME Transactions on Mechatronics 5(4), 331–341 (2000)
3. Oh, J., Bernstein, D.S.: Semilinear Duhem model for rate-independent and rate-dependent hysteresis. IEEE Transactions on Automatic Control 50(5), 631–645 (2005)
4. Ge, P., Jouaneh, M.: Generalized Preisach model for hysteresis nonlinearity of piezoceramic actuators. Precision Engineering 20(2), 99–111 (1997)
5. Chang, T., Sun, X.: Analysis and control of monolithic piezoelectric nano-actuator. IEEE Transactions on Control Systems Technology 9(1), 69–75 (2001)

6. Shieh, H.-J., Lin, F.-J., Huang, P.-K., Teng, L.-T.: Adaptive tracking control solely using displacement feedback for a piezo-positioning mechanism. IEE Proceedings of Control Theory and Applications 151(5), 653–660 (2004)
7. Song, G., Zhao, J., Zhou, X., Abreu-García, J.A.D.: Tracking control of a piezo-ceramic actuator with hysteresis compensation using inverse Preisach model. IEEE/ASME Transactions on Mechatronics 10(2), 198–209 (2005)
8. Ru, C., Sun, L.: Improving positioning accuracy of piezoelectric actuators by feed-forward hysteresis compensation based on a new mathematical model. Review of Scientific Instruments 76(9), 095 111-1-8 (2005)
9. Haykin, S.: Neural networks: a comprehensive foundation, 2nd edn. Prentice-Hall, Upper Saddle River, New Jersey (1999)
10. Lewis, F.L.: Neural network control of robot manipulators. IEEE Expert 11(3), 64–75 (1996)
11. Sanner, R.M., Slotine, J.-J.E.: Structurally dynamic wavelet networks for adaptive control of robotic systems. International Journal of Control 70(3), 405–421 (1998)
12. Kim, Y.H., Lewis, F.L.: Neural network output feedback control of robot manipulators. IEEE Transactions on Robotics and Automation 15(2), 301–309 (1999)
13. Liaw, H.C., Oetomo, D., Alici, G., Shirinzadeh, B.: Special class of positive definite functions for formulating adaptive micro/nano manipulator control. In: Proceedings of 9th IEEE International Workshop on Advanced Motion Control, Istanbul, Turkey, March 27-29, 2006, pp. 517–522 (2006)
14. Liaw, H.C., Oetomo, D., Shirinzadeh, B., Alici, G.: Robust motion tracking control of piezoelectric actuation systems. In: Proceedings of 2006 IEEE International Conference on Robotics and Automation, Orlando, Florida, May 15-19, 2006, pp. 1414–1419 (2006)
15. Lewis, F.L., Jagannathan, S., Yeşildirek, A.: Neural network control of robot manipulators and nonlinear systems. Taylor & Francis, London (1999)
16. Huang, S., Tan, K.K., Tang, K.Z.: Neural network control: theory and applications. Research Studies Press, Baldock, Hertfordshire, England (2004)
17. Slotine, J.-J.E., Li, W.: Applied nonlinear control. Prentice-Hall, Englewood Cliffs, New Jersey (1991)
18. Craig, J.J.: Introduction to robotics: mechanics and control. Addison-Wesley, Reading, MA (1989)

Hybrid Formation Control for Non-Holonomic Wheeled Mobile Robots

Juan Marcos Toibero[1], Flavio Roberti[1], Paolo Fiorini[2], and Ricardo Carelli[1]

[1] Instituto de Automatica. Universidad Nacional de San Juan. Argentina
{mtoibero,froberti,rcarelli}@inaut.unsj.edu.ar
[2] Department of Computer Science. University of Verona. Italy
paolo.fiorini@univr.it

Summary. This paper presents a hybrid formation controller approach for non-holonomic mobile robots. This approach is based on the stable switching between a leader-following formation controller and an orientation controller. The switching attempts to maintain low values of formation errors during specific leader movements that otherwise will produce a significant increment on such errors. Experimental results on commercial unicycle-like mobile robots are provided to show the feasibility and performance of the proposed control strategy.

Keywords: Non-Holonomic Wheeled Mobile Robots, Formation Control, Lyapunov Stability.

1 Introduction

Many cooperative tasks in real world environments, such as exploring, surveillance, search and rescue, transporting large objects and capturing a prey, need the robots to maintain some desired formations when moving. The hybrid control strategies developed along this paper involves mobile robot formation control when considering obstacles. By formation control, it is referred the problem of controlling the relative position and orientations of robots in a group, while allowing the group to move as a whole. Problems in formation control that have been investigated include assignment of feasible formations, moving into formation, maintenance of formation shape [15] and switching between formations ([4] and [3]). The work of [16] is a very good example of the state of the art in robot formation control, in which it is presented a complete framework to achieve a stable formation for car-like and unicycle-like mobile robots.

In this paper, an extension of the previous work [5] for multiple mobile robots formation is presented, where it was also considered a leader-follower controller [13] and the work of [7] in order to deal with the non-holonomic constraint of the unicycle-like mobile robots. In the last paper, the authors state that complicated coordinated tasks can be interpreted in terms of simpler coordinate tasks that are to be manipulated sequentially, which is one of the key points of this paper. The considered formation has a leader robot -see e.g. [8] for a review on the leader-following method- which has an omnidirectional camera,

S. Lee, I.H. Suh, and M.S. Kim (Eds.): Recent Progress in Robotics, LNCIS 370, pp. 21–34, 2008.
springerlink.com

a laser range-finder and odometry sensors; whereas the followers have odometry and collision (sonar) sensors. The omnidirectional camera is used to identify the relative follower positions. Hence, follower postures are determined relative to the leader coordinate system. This assumption is not a constraint since the leader could get access to these positions using another absolute position sensor such as, for instance, a GPSs or odometry. This centralized control architecture, where the control actions for all the followers are generated by the leader (which has the main visual and laser sensors) could be decentralized by allowing the followers to estimate the leader movements (angular and translational velocities) and performing a minimal communication between the robots. Due only to sensory limitation on our robot team we exposed experimental results that were obtained by using a centralized approach. Nevertheless, a decentralized control scheme could be supported by this strategy. Using hybrid control systems in formation control is not a novelty; in fact, several papers can be found in the literature using hybrid control systems: including a discrete event system at the supervisory level and continuous controllers to give the control actions [12], [8], [10], [11] and [4]. Indeed, the main contribution of this paper is not in the formation control framework, but in the way the non-holonomic mobile robots are coordinated in order to maintain the formation following the leader motion while guaranteeing the asymptotic stability of the hybrid formation control system. The developed control system aims at satisfying two major objectives: i) get the robots at desired initial positions in the given formation. This problem is the so-called static formation problem [9]; and ii) avoid unknown obstacles while maintaining the formation geometry, instead of changing the formation geometry as in [16]. In this case, it is considered the obstacle contour-following strategy for the leader robot as presented in [6]. Regarding the follower robots, it is proposed a hybrid approach based on a formation controller.

The remainder of this paper is organized as follows: Section 2 presents a stable leader-based formation controller. Then, in Section 3 it is described the hybrid control system including simulations results and stability considerations. Finally, in Section 4 several experimental results are reported to state conclusions and future work in Section 5.

2 Stable Formation Control

The kinematics model employed in this paper considers formation errors with respect to a Cartesian mobile coordinate system over the leader robot, which Y-axle coincides with the heading of this robot (Fig. 1.) The movement of each robot in the world coordinate system (with upper index w) is ruled by the well-known unicycle-like mobile robot kinematics: for the leader in (1) and for the i-th follower in (2)

$$^{w}\dot{x} = \nu \cos(^{w}\theta); \quad ^{w}\dot{y} = \nu \sin(^{w}\theta); \quad ^{w}\dot{\theta} = \omega \tag{1}$$

$$^{w}\dot{x}_i = \nu_i \cos(^{w}\theta_i); \quad ^{w}\dot{y}_i = \nu_i \sin(^{w}\theta_i); \quad ^{w}\dot{\theta}_i = \omega_i \tag{2}$$

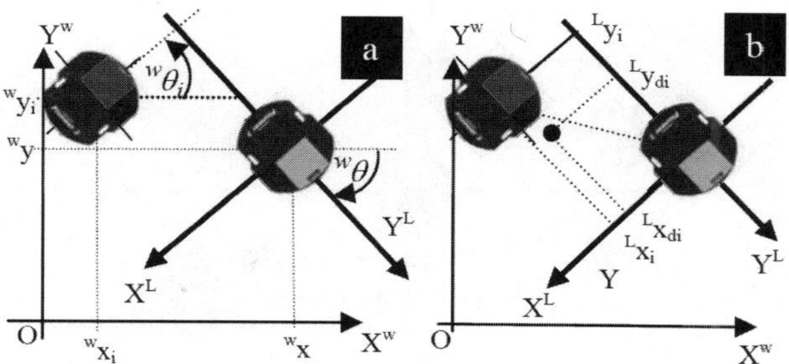

Fig. 1. a) Reference systems: world (absolute) reference coordinate system (OX^wY^w), and a second coordinate system attached on the (blue) leader robot $(0X^LY^L)$ where the desired positions for each of the followers are defined. b) i-th follower robot positioned at coordinates $(^Lx_i, {}^Ly_i)$ on the leader Cartesian reference where the reference position is given by $(^Lx_{di}, {}^Ly_{di})$.

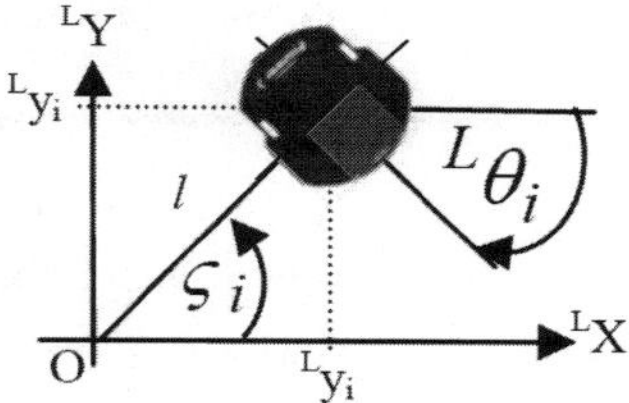

Fig. 2. i-th follower in the leader coordinate system

Leader movement is controlled by selecting its absolute velocities: ν and ω. The formation controller objective is to find the values of the velocities ν_i and ω_i for the follower robots in such a way that the formation errors decay asymptotically to zero. A third kinematics model must be considered in order to obtain the i-th follower coordinates relative to the leader 0^LX^LY coordinate system which moves at a linear velocity ν and an angular velocity ω

$$^L\dot{x}_i = \nu_i \cos(^L\theta_i) + \omega l \sin(\zeta_i) \tag{3}$$

$$^L\dot{y}_i = \nu_i \sin(^L\theta_i) - \omega l \cos(\zeta_i) - \nu \tag{4}$$

$$^L\dot{\theta}_i = \omega_i - \omega \tag{5}$$

here, l is the distance between the robot center and the origin of the mobile coordinate system and ζ_i is the angle between the ^LX axis and l (Fig. 2.) Note that for a static leader, these equations become

$$^L\dot{x}_i = \nu_i \cos(^L\theta_i); \quad {}^L\dot{y}_i = \nu_i \sin(^L\theta_i); \quad {}^L\dot{\theta}_i = \omega_i \tag{6}$$

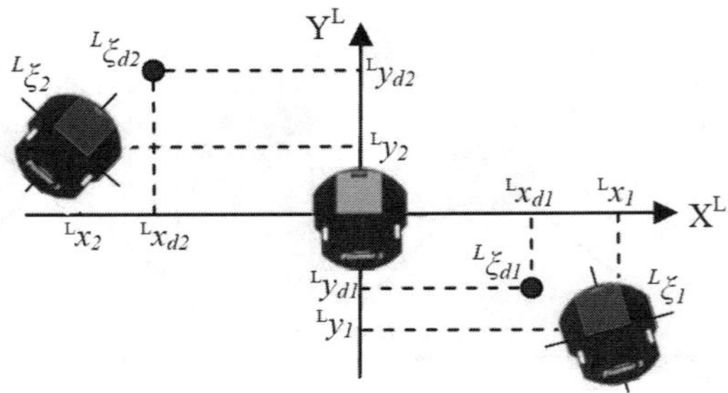

Fig. 3. Representation of the vectors $^L\boldsymbol{\xi}_i$ and $^L\boldsymbol{\xi}_{di}$

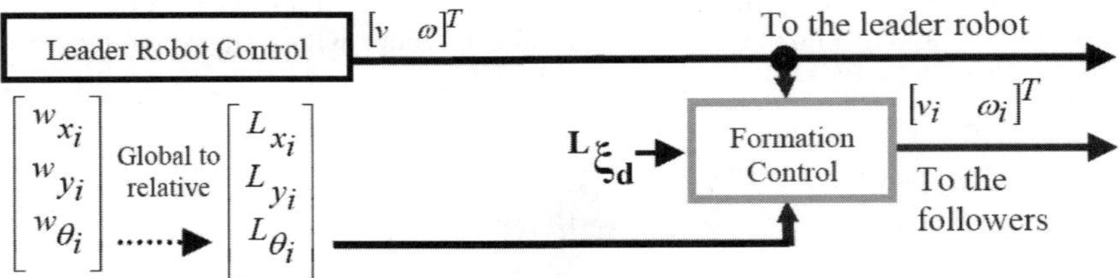

Fig. 4. Formation control block diagram

which describe the i-th follower movement on the leader coordinate system. The consideration of the postures of the followers relative to the leader coordinate system allows managing the entire formation without knowing the absolute positions of the followers. This relative posture can be obtained using a sensor system (e.g. a catadioptric vision system) mounted on one of the robots (Fig. 3). However, if the absolute postures information is available, it can be easily converted to the leader coordinate system with the transformation

$$^L x_i = (^w x_i - {}^w x)\sin(^w\theta) - (^w y_i - {}^w y)\cos(^w\theta) \tag{7}$$

$$^L y_i = (^w x_i - {}^w x)\cos(^w\theta) + (^w y_i - {}^w y)\sin(^w\theta) \tag{8}$$

$$^L\theta_i = {}^w\theta_i - {}^L\theta + \pi/2 \tag{9}$$

which gives the relation between the absolute $(^w x_i, {}^w y_i)$ and the relative $(^L x_i, {}^L y_i)$ coordinates. On the other hand, the i-th follower absolute heading angle can be transformed to the leader coordinate system with (9) In Fig. 4 it can be seen the block diagram of the proposed controller. From this figure, it must be noted the formation controller independence on the leader motion generation.

In order to calculate an error indicator between the current and the desired positions of the robots in the formation, let consider that (10) is the position vector of the i-th follower robot, and that (11) denotes the i-th follower desired position, with i=1,2,..,n.

$$^{L}\boldsymbol{\xi}_i = [^{L}x_i \quad ^{L}y_i]^T \tag{10}$$

$$^{L}\boldsymbol{\xi}_{di} = [^{L}x_{di} \quad ^{L}y_{di}]^T \tag{11}$$

both vectors are defined on the framework attached to the leader robot (Fig. 3.) The n position vectors (10) and (11) can be arranged in the global position vectors:

$$^{L}\boldsymbol{\xi} = [^{L}\xi_1^T \quad ^{L}\xi_2^T \quad \cdots \quad ^{L}\xi_n^T]^T \tag{12}$$

$$^{L}\boldsymbol{\xi}_d = [^{L}\xi_{d1}^T \quad ^{L}\xi_{d2}^T \quad \cdots \quad ^{L}\xi_{dn}^T]^T \tag{13}$$

The difference between the actual and the desired robot position is (14) and the formation error is defined as follows [14] in (15):

$$^{L}\tilde{\boldsymbol{\xi}} = {}^{L}\boldsymbol{\xi}_d - {}^{L}\boldsymbol{\xi} \tag{14}$$

$$\tilde{\mathbf{h}} = \mathbf{h}_d - \mathbf{h} = \mathbf{h}(^{L}\boldsymbol{\xi}_d) - \mathbf{h}(^{L}\boldsymbol{\xi}) \tag{15}$$

$$\mathbf{h} = \mathbf{h}(^{L}\xi) = \mathbf{h}(^{L}\boldsymbol{\xi}_d - {}^{L}\tilde{\boldsymbol{\xi}}) \tag{16}$$

where $\mathbf{h}$ is a suitable selected output variable representing the formation parameters, which captures information about the current conditions of the group of robots; $\mathbf{h}_d$ represents the desired output variable. For instance, $\mathbf{h}$ can be selected as the xy-position of each follower robot. Function $\mathbf{h}(^{L}\boldsymbol{\xi})$ must be defined in such a way that it be continuous and differentiable, and the Jacobian matrix $\mathbf{J}$ has full rank.

$$\dot{\mathbf{h}} = \mathbf{J}(\xi)^{L}\dot{\xi} = \frac{\partial \mathbf{h}(^{L}\boldsymbol{\xi})}{\partial^{L}\boldsymbol{\xi}}^{L}\dot{\xi} \tag{17}$$

Vector $^{L}\dot{\boldsymbol{\xi}}$ (robot translational velocities in the leader reference system: $^{L}\dot{x}_i$ and $^{L}\dot{y}_i$) has two different components,

$$^{L}\dot{\boldsymbol{\xi}} = {}^{L}\dot{\boldsymbol{\xi}}_s - {}^{L}\dot{\boldsymbol{\xi}}_l \tag{18}$$

where $^{L}\dot{\boldsymbol{\xi}}_s$ is the time variation of $^{L}\boldsymbol{\xi}$ produced by the velocities of the follower robots $\|^{L}\dot{\boldsymbol{\xi}}_{si}\| = \nu_i$; and $^{L}\dot{\boldsymbol{\xi}}_l$ is the time variation of $^{L}\boldsymbol{\xi}$ produced by the velocities of the leader robot. Now, (17) can be written as:

$$\dot{\mathbf{h}} = \mathbf{J}(^{L}\xi)(^{L}\dot{\boldsymbol{\xi}}_s - {}^{L}\dot{\boldsymbol{\xi}}_l) \tag{19}$$

The control objective is to guarantee that the mobile robots will asymptotically achieve the desired formation: (defined by $\mathbf{h}_d$,) or

$$\lim_{t \to \infty} \tilde{\mathbf{h}}(t) = 0 \tag{20}$$

to this aim, it is first defined a reference velocities vector as:

$$^{L}\dot{\boldsymbol{\xi}}_{r} = \mathbf{J}^{-1}(^{L}\boldsymbol{\xi})\{\dot{\mathbf{h}}_{d} + \mathbf{f}_{\tilde{\mathbf{h}}}(\tilde{\mathbf{h}})\} +^{L} \dot{\boldsymbol{\xi}}_{l} \tag{21}$$

where $\mathbf{f}_{\tilde{\mathbf{h}}}(\tilde{\mathbf{h}}) = \tanh(\tilde{\mathbf{h}})$ is a saturation function applied to the output error, such that $\mathbf{x}^{T}\mathbf{f}_{\tilde{\mathbf{h}}}(\mathbf{x}) > \mathbf{0} \ \forall \mathbf{x} \neq 0$, This function could be considered e.g. as $f(\mathbf{x}) = \tanh(\mathbf{x})$. $^{L}\dot{\boldsymbol{\xi}}_{r}$ represents the velocities of the followers robots on the framework attached to the leader robot that allow them to reach (and to maintain) the desired formation while following the leader. Assuming perfect velocity servoing

$$^{L}\dot{\boldsymbol{\xi}}_{r} =^{L} \dot{\boldsymbol{\xi}}_{r} \tag{22}$$

then from (21),(19) and (15) the following closed loop equation can be obtained:

$$\dot{\tilde{\mathbf{h}}} + \mathbf{f}_{\tilde{\mathbf{h}}}(\tilde{\mathbf{h}}) = 0 \tag{23}$$

Now, in order to consider the formation errors analysis under the perfect servoing assumption the following Lyapunov candidate [2] function (24) is introduced with its time-derivative along system trajectories (25)

$$V = \tilde{\mathbf{h}}^{T} \tilde{\mathbf{h}}/2 \tag{24}$$

$$\dot{V} = \tilde{\mathbf{h}}^{T} \dot{\tilde{\mathbf{h}}} = -\tilde{\mathbf{h}}^{T} \mathbf{f}_{\tilde{\mathbf{h}}}(\tilde{\mathbf{h}}) \tag{25}$$

it is clear that if $^{L}\dot{\boldsymbol{\xi}}_{s} \equiv^{L} \dot{\boldsymbol{\xi}}_{r} \Rightarrow \tilde{\mathbf{h}} \to 0$ asymptotically. *Remark.* This condition is verified for the ideal case in which the robots follow exactly the reference velocity (22). However, for a real controller this velocity equality will eventually be reached asymptotically. The convergence of the control error to zero under this real condition will be analyzed at the end of this section. Vector $^{L}\dot{\boldsymbol{\xi}}_{l}$ (29) is computed using the knowledge of linear and angular velocities of the leader robot, and the relative positions of the follower robots:

$$r_{1} = \nu/\omega \tag{26}$$

$$r_{2i} = \sqrt{(r_{1} + x_{i})^{2} + y_{i}^{2}} \tag{27}$$

$$\beta_{i} = \arctan(y_{i}/(r_{1} + x_{i})) \tag{28}$$

$$\|^{L}\dot{\boldsymbol{\xi}}_{li}\| = \omega \, r_{2i} \qquad \begin{bmatrix} ^{L}\dot{\boldsymbol{\xi}}_{lxi} \\ ^{L}\dot{\boldsymbol{\xi}}_{lyi} \end{bmatrix} = \begin{bmatrix} -\|^{L}\dot{\boldsymbol{\xi}}_{li}\| \sin(\beta_{i}) \\ \|^{L}\dot{\boldsymbol{\xi}}_{li}\| \cos(\beta_{i}) \end{bmatrix} \tag{29}$$

where r_{1} and r_{2i} are virtual turning radius (Fig. 5) and subscript i denotes the i-th follower. The commands for the linear and angular velocities of each robot are computed in order to secure that the robots will reach asymptotically the velocity reference ($^{L}\boldsymbol{\xi}_{s} \to^{L} \boldsymbol{\xi}_{\mathbf{r}}$.)

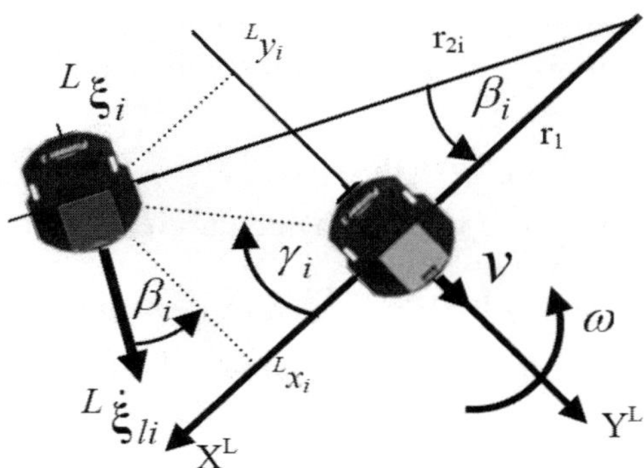

Fig. 5. ξ_{li} velocity computation

The proposed control law for heading control is:

$$\omega_i = k_{\omega_i}\, f(^L\tilde{\theta})_i +^L \dot{\theta}_{ri} + \omega \tag{30}$$

where $^L\tilde{\theta}_i =^L \theta_{ri} -^L \theta_i$ is the angular error between the i-th follower robot heading $^L\theta_i$ and the angle of its reference velocity $^L\theta_{ri} = \angle^L\dot{\xi}_{ri}$; consequently $^L\dot{\theta}_{ri}$ is the time derivative of this reference velocity heading for the i-th robot; ω is the angular velocity of the mobile framework attached to the leader robot; and $f(^L\tilde{\theta})_i$ is a saturation function applied to the angular error, such as in (21); and k_{ω_i} is a positive constant. Next, by equating (30) and (5), the following closed-loop equation can be obtained:

$$^L\dot{\tilde{\theta}}_i + k_{\omega i}\, f(^L\tilde{\theta}_i) \tag{31}$$

Now, in order to analyze the stability for the orientation error $^L\tilde{\theta}_i$ it is introduced the following Lyapunov candidate (32) and its time derivative (33):

$$V =^L \tilde{\theta}_i^{\,2}/2 \tag{32}$$

$$\dot{V} =^L \tilde{\theta}_i\, ^L\dot{\tilde{\theta}}_i = -k_{\omega i}\, ^L\tilde{\theta}_i\, f(^L\tilde{\theta}_i) < 0 \tag{33}$$

which implies that $^L\tilde{\theta}_i \to 0$ as $t \to \infty$. That is, the robot orientation on the leader Cartesian coordinate system tends asymptotically to the desired reference orientation, which guarantees maintaining the desired formation. Once it was proved that $^L\theta_i(t) \to^L\theta_{ri}(t)$, it must now be proved that the same occurs for $\|^L\dot{\xi}_{si}\| = \nu_i \to \|^L\dot{\xi}_{ri}\|$. To this aim, the following control law for the linear velocity is proposed:

$$\nu_i = \|^L\dot{\xi}_{ri}\|\, \cos(^L\tilde{\theta}_i) \tag{34}$$

which obviously produces that $\nu_i \to \|^L\dot{\xi}_{ri}\|$, since it has been proved that $^L\tilde{\theta}_i \to 0$. The factor $\cos(^L\tilde{\theta}_i)$ has been added to prevent high control actions when a large angular error exists. Now, we have proved that $^L\dot{\xi}_r -^L \dot{\xi}_s = \rho$ with

$\rho \to 0$, which is a more realistic assumption than (22). Then, formation errors are considered again in order to analyze its stability under the new condition. So, (23) can be written as:

$$\dot{\tilde{\mathbf{h}}} + \mathbf{f}_{\tilde{\mathbf{h}}}(\tilde{\mathbf{h}}) = \mathbf{J}\,\rho \tag{35}$$

Let us consider the same Lyapunov candidate (24) but now with its time derivative:

$$\dot{V} = \tilde{\mathbf{h}}^T\,\dot{\tilde{\mathbf{h}}} = -\tilde{\mathbf{h}}^T\,\mathbf{f}_{\tilde{\mathbf{h}}}(\tilde{\mathbf{h}}) - \mathbf{J}\,\rho \tag{36}$$

A sufficient condition for (36) to be negative definite is $\|\mathbf{f}_{\tilde{\mathbf{h}}}(\tilde{\mathbf{h}})\| > \|\mathbf{J}\|\,\|\rho\|$. Since if $\rho(t) \to 0 \Rightarrow \|\tilde{\mathbf{h}}(t)\| \to 0$ for $t \to \infty$.

3 Hybrid Formation Control

The interaction between the leader and the follower controllers must be in such a way that the followers always maintain their desired positions independently of the leader manoeuvres. This allows preserving the formation and therefore the involved robots can perform a cooperative task. Our approach is based on the detection of leader movements that will significantly increase formation errors. In Fig. 6 we present a hybrid formation control strategy where it can be appreciated the inclusion of a supervisor which generates switching signals at both levels: leader (σ_L) and followers (σ_{Fi}) based on: i) the follower posture, ii) the leader absolute posture and iii) the leader control actions. Besides, it was also included a new orientation controller, that corrects the followers heading accordingly to a given logic (next, in Fig.9.) The main idea is to detect leader movements that will immediately produce formation errors. These errors will arise due to the nonholonomic constraint of the unicycle-like wheeled mobile robots (mostly due to different robot headings.) Hence, the headings of the followers are set to values that prevent these initial errors and only after this correction is done, the leader is allowed to continue with its planned movement. These leader movements (which are detected directly from the leader control commands) are namely: i) "stop and go" (a step in the forward velocity) and ii) "only-heading movements" (a step in the angular velocity command with null forward velocity.) The leader motion control is based on the results of [5] and gives the robot the capability to get a desired posture in the world coordinate system $[{}^w x_d\ {}^w y_d\ {}^w \theta_d]$ avoiding obstacles. This motion could only be stopped by the supervisor ("leader stopped" in the block diagram.) This strategy allows separating completely the control analysis into the leader motion control analysis and the follower motion control analysis. For the leader this analysis is trivial since its motion control is asymptotically stable, then the new control system which is assumed to include the possibility to stop the leader during a finite time, will also be asymptotically stable. Now, regarding the follower robots, the inclusion of the orientation controller must be considered into the stability analysis (Section 3.2). Note the existence of a switching signal σ_{Fi} for each follower, and consequently, an orientation controller available for each follower robot.

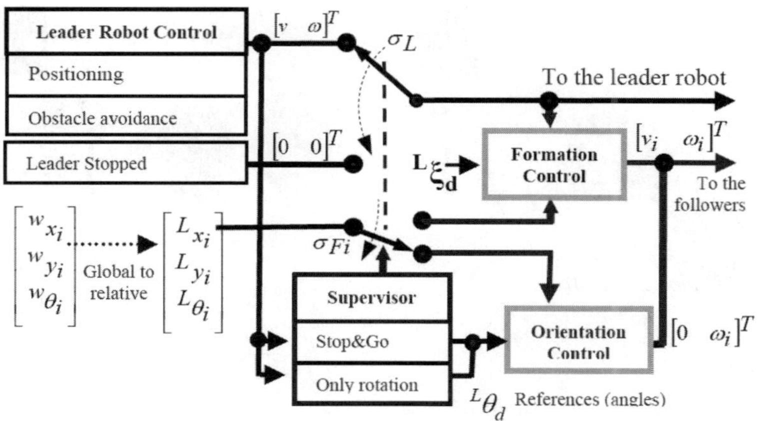

Fig. 6. Hybrid formation control block diagram

3.1 Follower Robots: Heading Control

In this section it is introduced a proportional only-bearing controller that allows the follower robots to set their headings to desired computed values that will provide good initial heading conditions for the future formation evolution. The position of each follower robot in the formation is defined by its coordinates $(^L x_{id}, ^L y_{id})$ regardless of its orientation. Taking advantage of the unicycle kinematics, it will be assumed from here on that the robots can rotate without distorting the formation (allowing change the "formation heading.") In other words, if e.g. the robots are transporting an object, they must be able to turn freely over its own centers without changing the transported object orientation. It is proposed a proportional controller for the heading error

$$\tilde{\theta}_i = {}^L\theta_d - {}^L\theta_i \tag{37}$$

where the desired value $^L\theta_d$ is computed according to Section 3.2. Then, proposing the following Lyapunov function (38) with the control action (39) the asymptotic stability of this control system could be immediately proved by (40)

$$V_{or_i} = \tilde{\theta}_i^2/2 \tag{38}$$

$$\omega_i = -k_{\omega i}\tanh(\tilde{\theta}_i) \tag{39}$$

$$\dot{V}_{or_i} = \tilde{\theta}_i\,\dot{\tilde{\theta}}_i = -k_{\omega i}\tilde{\theta}_i\tanh(\tilde{\theta}_i) \tag{40}$$

The importance of introducing the heading controller can be appreciated from Fig. 7 starting with a null error formation (Fig. 7a), the leader develops a pure-rotational evolution (Fig. 7b), and the follower try to keep the formation with

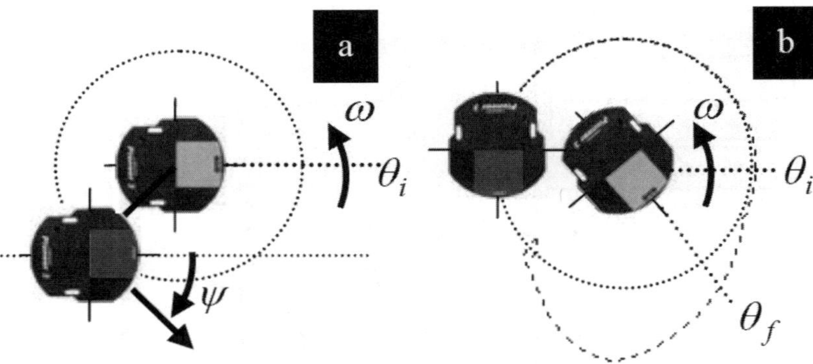

Fig. 7. Formation control without orientation control for a leader (blue) and a follower (red): a) initial configuration; b) the path described by the follower (dotted line)

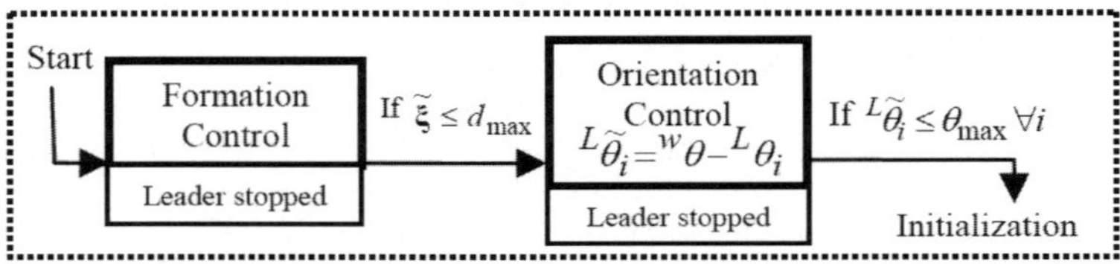

Fig. 8. Initialization logic: static formation

a significant transition error. It is clear that this error could be avoided if the starting orientation of the follower robot is set to ψ before the leader starts its rotation. This angle is computed depending of the sign of the leader angular velocity according to:

$$\psi = \gamma_i + sgn(\omega)\,\pi/2 \tag{41}$$

where the angle γ_i is given by

$$\gamma_i = \tan^{-1}\left(\frac{^L y_{id}}{^L x_{id}}\right) \tag{42}$$

Moreover, depending on the leader angular velocity, formation errors could be greater or even produce follower backwards movements. This ψ-angle correction avoids transitory formation errors improving the whole control system performance. The same analysis could be done for the leader "stop&go" movement with heading errors on the follower robots. In this case, the leader attempts to start its translational motion and it is easy to see that the robot configuration that will present minimal formation error at this transitory will be the formation in which all robots have the same heading angle.

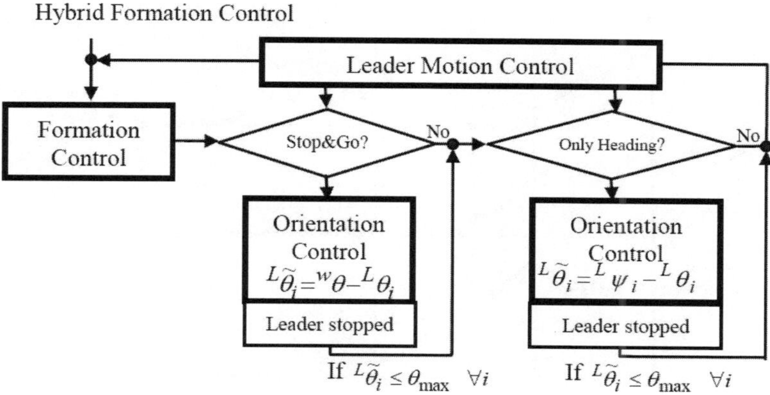

Fig. 9. General Hybrid formation logic

3.2 Stability Analysis

The supervisor logics were divided into two cases: an initialization case (or static formation case of Fig. 8) that corrects followers' initial postures to a new posture with null formation error and with the same leader heading; and the general case that allows keeping the this formation geometry (Fig. 9) when the leader starts moving. The orientation control is used in two situations: one related to the leader only-bearing movement that corrects the ψ-angle for each follower; and the other related to the leader "stop&go" movement, that equals all the followers headings to the leader heading. In both cases the objective is to minimize the heading errors before starting the leader movement. Accordingly to the exposed logics, it is considered a switching between the formation controller of Section II and a orientation controller of Section 3.1 for each follower which stability at switching times must be analyzed. This is done by considering Multiple Lyapunov Functions [1]: It must be guaranteed that the sequence associated to the discontinuous Lyapunov Functions (when are active) be decreasing for all the controllers involved and furthermore, it must be guaranteed also that the switching is not arbitrarily fast. In Fig. 10 it is depicted a typical switching instant (at $<t1>$) for a three robot formation. At this point, the leader is stopped, and the orientation controllers compute their references (note the existence of different values for each robot); then, at $<t2>$ follower 1 has achieved the maximum acceptable error θ_{max}, however the formation controller will not start with its movement after instant $<t3>$ when the second follower has achieved its maximum heading error. In consequence the logics secure: i) that the switching from the orientation control back to the formation control is enough slow to allow the followers to achieve its desired postures avoiding the undesirable *chattering* effect; ii) that the value of (24) is the same before and after the switching since do not depend on the follower' headings. This fact can be seen in Fig.10 where $V(t_1)=V(t_3)$. This way, the asymptotic stability proved for the formation controller (and its performance) will not be affected by the proposed switching.

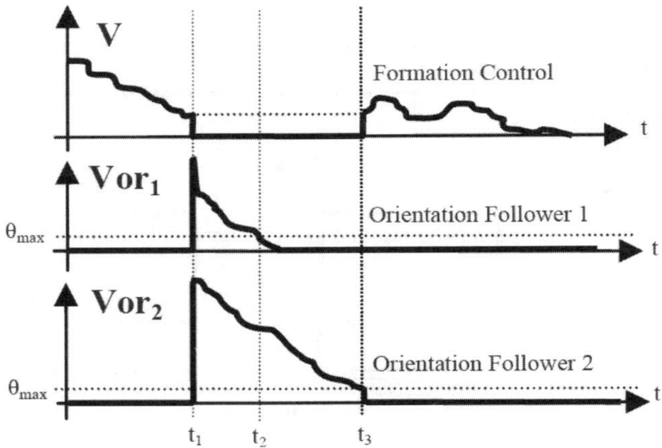

Fig. 10. Multiple Lyapunov Functions approach

4   Experimental Results

Experimental results were performed by using two Pioneer robots with onboard PCs and wireless internet connection. This way, each sample time ($Ts = 100ms$) the leader robot asks for the follower absolute position and after computing the control commands, sends them back to the follower. It was considered a reference translational velocity of $150mm/s$ and a maximum angular velocity of $50/s$ for the leader. In Fig. 11 it can be appreciated the formation evolution within a room without obstacles. In the first part, it is considered the static formation problem for a follower initial posture of $[^L x_1, {}^L y_1, {}^L \theta_1]^T = [1200, -1000, 180]^T$ and a desired formation at $(^L x_{d1}, {}^L y_{d1}) = (600, -600)$. It can be appreciated the formation

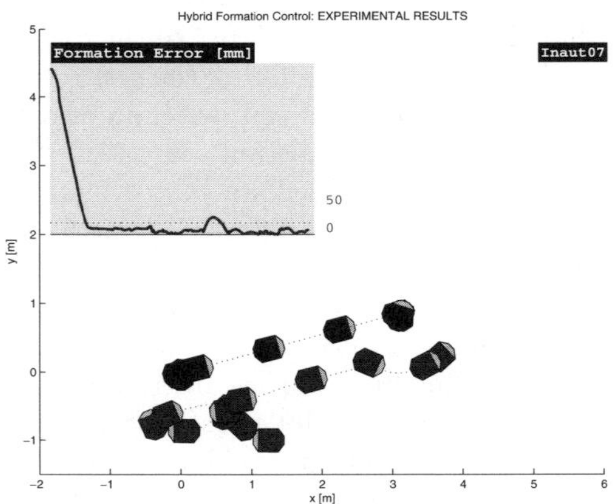

Fig. 11. Experimental results: formation control without obstacles including initial formation error

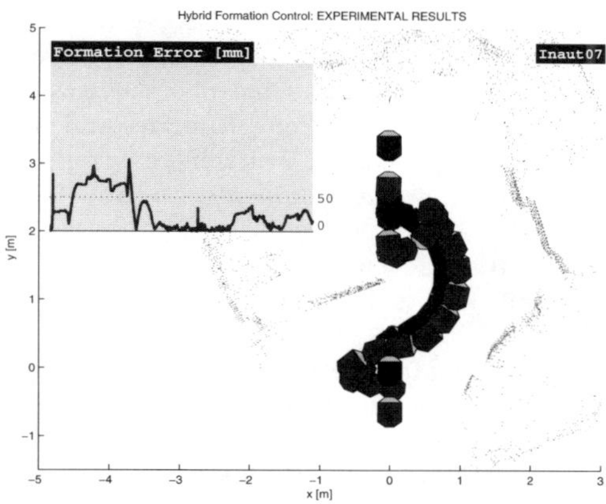

Fig. 12. Experimental results: formation control with obstacles

error correction according to the initialization logic of Fig. 9. After the formation geometry is achieved, the leader robot is allowed to start with its motion towards the goal point in $[^{w}x_{DES}, {}^{w}y_{DES}, {}^{w}\theta_{DES}]^{T} = [3100, 850, 90]^{T}$ from the initial posture at $[0, 0, 90]^{T}$. Finally, Fig. 12 shows the robot trajectories for a similar experiment but considering an obstacle which is detected with the leader laser range finder. In this case the desired formation point was set to $(^{L}x_{d1}, {}^{L}y_{d1}) = (0, -600)$ and the maximum formation error was 105mm while the mean value was of 28mm (that could be compared with the error values for the previous experiment: maximum value of 720mm and mean value of 22mm.) From these two previous plots, it can be concluded the low formation error values achieved for the robot formations.

5 Conclusions

In this paper it has been considered a hybrid formation control approach. A continuous (leader-based) formation controller has been complemented with an orientation controller for each follower, allowing a considerable reduction of formation errors during leader manoeuvres. The good performance of this approach has been shown through experimental results.

Acknowledgments

The authors gratefully acknowledge SEPCIT (FONCYT) and CONICET of Argentina for partially funding this research.

References

1. Liberzon, D.: Switching in Systems and Control. Birkhauser (2003)
2. Slotine, J., Li, W.: Applied Non Linear Control. Prentice-Hall, Englewood Cliffs (1991)

3. Fierro, R., Song, P., Das, A.K., Kumar, V.: Cooperative control of robot formations. In: Cooperative control and cooperation, Ch. 5, Dordrecht (2002)
4. Desai, J.P., Kumar, V., Ostrowski, J.P.: Control of changes in formation for a team of mobile robots. In: Proc. IEEE Int. Conf. Robotics and Automation (1999)
5. Toibero, J.M., Carelli, R., Kuchen, B.: Switching control of mobile robots for autonomous navigation in unknown environments. In: Proc. IEEE Int. Conf. Robotics and Automation (2007)
6. Toibero, J.M., Carelli, R., Kuchen, B.: Stable Switching Contour-Following controller for wheeled mobile robots. In: Proc. IEEE Int. Conf. Robotics and Automation (2006)
7. Gulec, N., Unel, M.: Coordinated motion of autonomous mobile robots using nonholonomic reference trajectories. In: Proc. IEEE IECON (2005)
8. Shao, J., Xie, G., Yu, J., Wang, L.: Leader-following formation control of multiple mobile robots. In: Proc. IEEE Int. Symp. Intelligent Control (2005)
9. Antonelli, G., Arrichiello, F., Chiaverini, S.: Experiments of formation control with collision avoidance using the null-space-based behavioral control. In: Proc. IEEE Int. Conf. Intelligent Robots and Systems (2006)
10. Ogren, P., Leonard, N.E.: Obstacle avoidance in formation. In: Proc. IEEE Int. Conf. Robotics and Automation (2003)
11. Ogren, P.: Split and join of vehicle formations doing obstacle avoidance. In: Proc. IEEE Int. Conf. Robotics and Automation (2004)
12. Chio, T., Tarn, T.: Rules and control strategies of multi-robot team moving in hierarchical formation. In: Proc. IEEE Int. Conf. Robotics and Automation (2003)
13. Carelli, R., Roberti, F., Vasallo, R., Bastos, T.: Estrategia de control de formacin estable para robots mviles. In: Proc. AADECA (2006)
14. Kelly, R., Carelli, R., Ibarra, J.M., Monroy, C.: Control de una pandilla de robots mviles para el seguimiento de una constelacin de puntos objetivo. In: Proc. Mexic. Conf. on Robotics (2004)
15. Desai, J.P., Ostrowski, J.P., Kumar, V.: Robotics and Automation 17, 905–908 (2001)
16. Das, A.K., Fierro, R., Kumar, V., Ostrowski, J.P., Spletzer, J., Taylor, J.C.: Robotics and Automation 18, 813–825 (2002)

Novel Tripedal Mobile Robot and Considerations for Gait Planning Strategies Based on Kinematics

Ivette Morazzani, Dennis Hong, Derek Lahr, and Ping Ren

RoMeLa: Robotics and Mechanisms Laboratory, Virginia Tech, Blacksburg, VA, USA
dhong@vt.edu

Abstract. This paper presents a novel tripedal mobile robot STriDER (Self-excited Tripedal Dynamic Experimental Robot) and considerations for gait planning strategies based on kinematics. To initiate a step, two of the robot's legs are oriented to push the center of gravity outside the support triangle formed by the three foot contact points, utilizing a unique abductor joint mechanism. As the robot begins to fall forward, the middle leg or swing leg, swings in between the two stance legs and catches the fall. Simultaneously, the body rotates 180 degrees around a *body pivot line* preventing the legs from tangling up. In the first version of STriDER the concept of passive dynamic locomotion was emphasized; however for the new version, STriDER 2.0, all joints are actively controlled for robustness. Several kinematic constraints are discussed as the robot takes a step including; stability, dynamics, body height, body twisting motion, and the swing leg's path. These guidelines will lay the foundation for future gait generation developments utilizing both the kinematics and dynamics of the system.

1 Introduction

STriDER (Self-excited Tripedal Dynamic Experimental Robot) is a novel three-legged walking robot that utilizes a unique tripedal gait to walk [1, 2, 3, 4, 5]. To initiate a step, two of its legs are oriented to push the center of gravity outside a support triangle formed by the three foot contact points, using a unique abductor joint mechanism. As the robot begins to fall forward, the middle leg or swing leg, swings in between the two stance legs and catches the fall. Simultaneously, the body rotates 180 degrees preventing the legs from tangling up.

The first version of STriDER [1, 3, 4] emphasizes on the passive dynamic nature of its gaits. Passive dynamics locomotion utilizes the natural built in dynamics of the robots body and limbs to create the most efficient walking and natural motion [6, 7]. In the new version, STriDER 2.0, all of its joints are actuated for robustness. The inverse and forward displacement analysis is preformed by treating the robot as a parallel manipulator when all three feet are on the ground [5]. STriDER is developed for deploying sensors rather than task manipulations. The robot's tall stance is ideal for surveillance and setting cameras at high positions [1]. The current research focuses on posturing, gait synthesis, and trajectory planning for which the concept of passive dynamics is not emphasized. Since STriDER is a

S. Lee, I.H. Suh, and M.S. Kim (Eds.): Recent Progress in Robotics, LNCIS 370, pp. 35–48, 2008.
springerlink.com   © Springer-Verlag Berlin Heidelberg 2008

non-linear, under-actuated mechanical system in nature (there can not be an actuator between the foot and the ground), the dynamics is a key factor in the planning of gait. Recent research on the optimization of bipedal gait with dynamic constraints includes [8, 9]. The technical approaches intensively discussed in those works can be utilized as the source of reference for the novel tripedal gait in this study. In this paper, we present considerations for gait planning strategies based on kinematics and lay out the foundation and guidelines for future work on a single step gait generation based on both kinematics and dynamics.

2   Background

In this section, the concept of the tripedal gait, locomotion strategies, turning ability, mechanical design, kinematic configuration, and inverse and forward displacement analysis of STriDER are discussed.

2.1   STriDER: Self-excited Tripedal Dynamic Experimental Robot

The design and locomotion strategies of robots are often inspired by nature; however, STriDER utilizes an innovative tripedal gait not seen in nature. Unlike common bipeds, quadrupeds, and hexapods, STriDER, shown in Fig. 1, is an innovative three-legged walking machine that can incorporate the concept of actuated passive dynamic locomotion. Thus, the proper mechanical design of a robot can provide energy efficient locomotion without sophisticated control methods [10]. However, STriDER is inherently stable with its tripod stance and can easily change directions. This makes it uniquely capable to handle rugged terrain where the path planning, turning, and positioning strategies studied here are crucial.

Fig. 1. STriDER 2.0 prototype on right of its predecessor, STriDER

The novel tripedal gait (patent pending) is implemented, as shown in Fig. 2 for a single step. During a step, two legs act as stance legs while the other acts as a swing leg. STriDER begins with a stable tripod stance (Fig. 2(a)), then the hip links are oriented to push the center of gravity forward by aligning the stance legs' pelvis links (Fig. 2(b)). As the body of the robot falls forward (Fig. 2(c)), the swing leg naturally swings in between the two stance legs (Fig. 2(d)) and catches the fall (Fig. 2(e)). As the robot takes a step, the body needs to rotate 180 degree to prevent the legs from tangling up. Once all three legs are in contact with the ground, the robot regains its stability and the posture of the robot is reset in preparation for the next step (Fig. 2(f)) [1, 3].

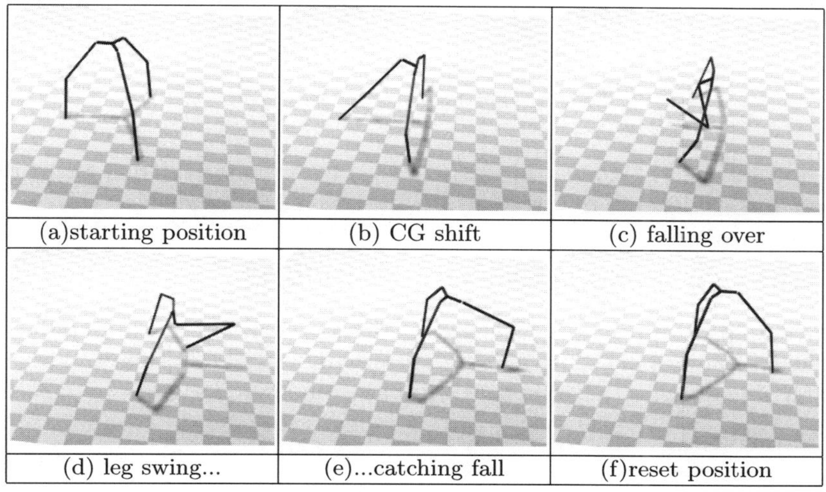

<table>
<tr><td>(a)starting position</td><td>(b) CG shift</td><td>(c) falling over</td></tr>
<tr><td>(d) leg swing...</td><td>(e)...catching fall</td><td>(f)reset position</td></tr>
</table>

Fig. 2. The motion of a single step [1]

Gaits for changing directions can be implemented in several ways, one of which is illustrated in Fig. 3. By changing the sequence of choice of the swing leg, the tripedal gait can move the robot in 60 degree interval directions for each step [4]. Alternatively, the step direction can be modified such that the stance momentarily changes to an iscoceles or scalene triangle as opposed to an equilateral. This will then change the orientation of the following stance legs from the customary 60 degree angle and therefore the direction of the robot's travel as well. This method is of particular interest because of the inherent flexibility which is more conducive to rugged environments [1].

The design of the first prototype with optimized design parameters for a smooth dynamic gait, and the resulting simple experiments for a single step tripedal gait are presented in [3]. Dynamic modeling, simulation, and motion generation strategies using the concept of self-excitation are presented in [1]. A second prototype, STriDER 2.0, has been fabricated as shown to the right of STriDER in Fig. 1.

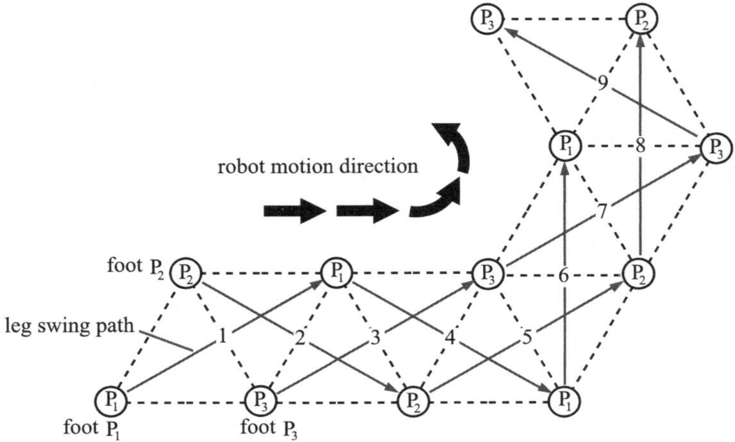

Fig. 3. Gait for changing direction

These models will be used in future experiments to examine STriDER's transitions between gaits, adaptation to various terrains, and stability analysis.

2.2   Kinematic Configuration of STriDER 2.0

The definition of coordinate systems for each leg is shown in Fig. 4. Details of the coordinates frames and link parameters are presented in [5]. The subscript i denotes the leg number (i.e. i=1, 2, 3) in the coordinate frames, links, and joint labels.

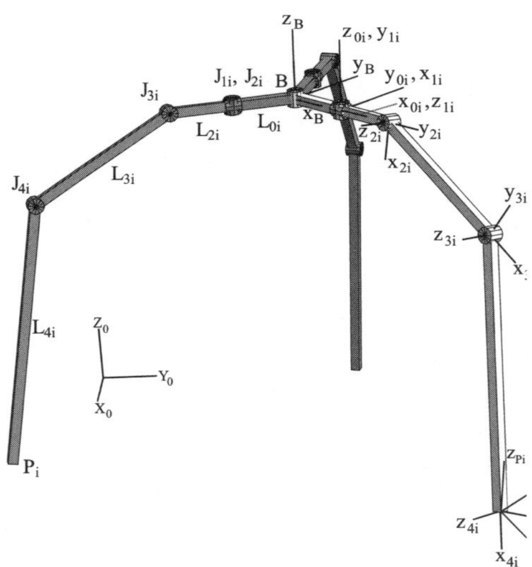

Fig. 4. Coordinate frame and joint definitions [5]

Table 1. Nomenclature

i	Leg number (i=1,2,3)
$\{X_0, Y_0, Z_0\}$	Global fixed coordinate system
$\{x_B, y_B, z_B\}$	Body center coordinate system
J_{1i}	Hip abductor joint for leg i
J_{2i}	Hip rotator joint for leg i
J_{3i}	Hip flexure joint for leg i
J_{4i}	Knee joint for leg i
P_i	Foot contact point for leg i
L_{0i}	Body link for leg i
L_{1i}	Hip link for leg i (length=0)
L_{2i}	Pelvis link for leg i
L_{3i}	Thigh link for leg i
L_{4i}	Shank link for leg i

Table 1 lists the nomenclature used to define the coordinate frames, joint and links. A global coordinate system, $\{X_0, Y_0, Z_0\}$, is established and used as the reference for positions and orientations where the negative Z_0 vector is in the same direction as gravity. Each leg includes four actuated joints, J_{1i}, J_{2i}, J_{3i}, and J_{4i}. Because the three abductor joints are actuated together in STriDER 2.0 [2], as described in the following section, J_{1i} is not treated as an active joint in this paper.

STriDER can be considered as a three-branch in-parallel manipulator when all three foot contact points are fixed on the ground. Then the ground is modeled as "the base" of a parallel manipulator, with the body as "the moving platform". The foot can be treated as a passive spherical joint connecting each leg to the ground with the no slip condition assumption. Given the fact that the knee joints, hip flexure joints and hip rotator joints are all revolute joints and each of the three legs mainly has two segments i.e. thigh and shank link, STriDER belongs to the class of in-parallel manipulators with 3 - SRRR (Spherical-Revolute-Revolute-Revolute) configuration. Detailed discussion and the development of the solutions for the inverse and forward kinematics mentioned can be found in [5].

2.3 Mechanical Design of STriDER and STriDER 2.0

STriDER stands roughly 1.8[m] tall with a base that is approximately 15[cm] wide. As stated earlier the leg lengths were determined through an optimization process with consideration for passive dynamic motion. As of now STriDER 2.0 stands only .9[m] tall but this height was chosen somewhat arbitrarily and may change as this version will be used primarily for the investigation of its kinematics as opposed to the dynamics. The body of STriDER 2.0 was designed 18[cm] wide at its base. Both robots are actuated using DC servo motors through distributed control with position feedback.

Because of the continuous inverting motion inherent to the locomotion strategy of this robot, slip rings were built into each of the three rotator joints [1]. It is necessary then to remove the actuator away from the rotation axis of the joint such that wires could be routed through the rotator shaft. In both STriDER and STriDER 2.0 this is accomplished using a spur gear pair [2].

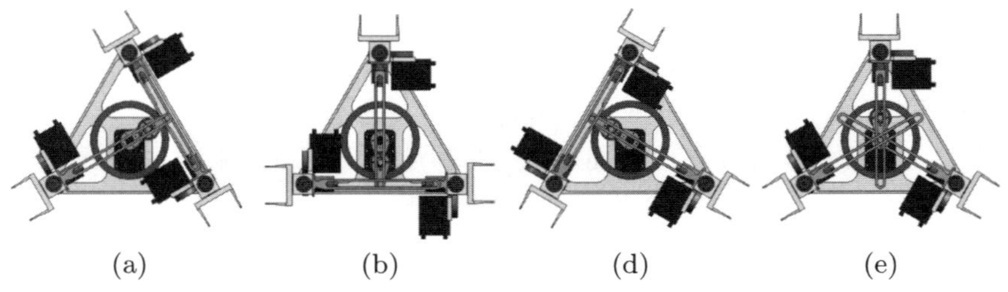

(a)   (b)   (d)   (e)

Fig. 5. The four positions of the rotator joint aligning mechanism with internal gear set

The tripedal gait requires the entire body of STriDER to rotate about the two hip rotator joints of the stance legs as the swing leg swings between them. Since any one of the three legs can be chosen as the swing leg, any two of the three hip rotator joints need to be able to align to each other. The hip abductor joints perform this motion by changing the angle of the hip rotator joints so that the axis of one hip rotator joint can be aligned to another while the third is set to be perpendicular to this axis. In addition to the three orientations in which a pair of rotator joints is aligned, it is also desirable that all rotator axes intersect in the center of the body. In the first prototype of STriDER the three hip abductor joints were independently actuated and controlled with three separate DC motors. While this approach worked, the size and weight of the two additional motors made the design undesirable, as it essentially requires only a single degree of freedom motion to successfully aligning the rotator joints in the four desired configurations.

In [2], a new abductor joint mechanism is presented which aligns the rotator joints using only one actuator which can replace the three motors of STriDER's abductors. This mechanism uses an internal gearset to generate a special trifolium curve with a pin which guides the hip rotator joints via slotted arms through the four specific positions shown in Fig. 5.

3 Gait Planning Constraints for a Single Step

Many factors and constraints contribute to the development of STriDER 2.0's path planning strategies and gait generation. To correctly generate a gait both kinematics and dynamics must be considered. Although dynamics plays a major role in gait generation, the following sections discuss possible considerations for gait planning strategies solely based on kinematics.

3.1   Stability

The robot's static stability is important during a step, as the novel tripedal gait requires the robot to become statically unstable forcing the robot to fall forward and swing its middle leg in between the stance legs and catch the fall. However, when all three feet are touching the ground, the robot must be statically stable by keeping the projected center of gravity point in the support triangle, formed by the three foot positions. Thus, the location of the projected center of gravity point plays an important role in the generation of a gait. A detailed discussion of a quantitative static stability margin is discussed in Sections 4.

3.2   Dynamics

Dynamics plays a key role in producing the gait for walking robots. STriDER can be modeled as a planar four-link invert pendulum in the sagittal plane by treating the two stance legs as a single link connected to the ground by a revolute joint, as shown in Fig. 6 [1]. In this figure, the angle between the link representing the stance legs and the ground is called the tilting angle. Since there is no active actuator between the foot and the ground, STriDER is inherently an under-actuated mechanical system. Assuming no slipping on the ground, the tilting angle during a gait is affected by the coupled dynamics of the other links in the system. The rotation of the body or any of the other actuated links will drive the unactuated links. In [7], self-excited control is utilized to enable a three-link planar robot to walk naturally on level ground. Utilizing this concept of self-excitation, STriDERs passive dynamic gait was produced in [1, 3]. [9] proved the existence of limit-cycle motion of multi-link planar robots by using differential flatness and dynamic-based optimization. This methodology will be utilized in generating the gait for STriDER 2.0 in future research where all of the joints of the robot are actively controlled

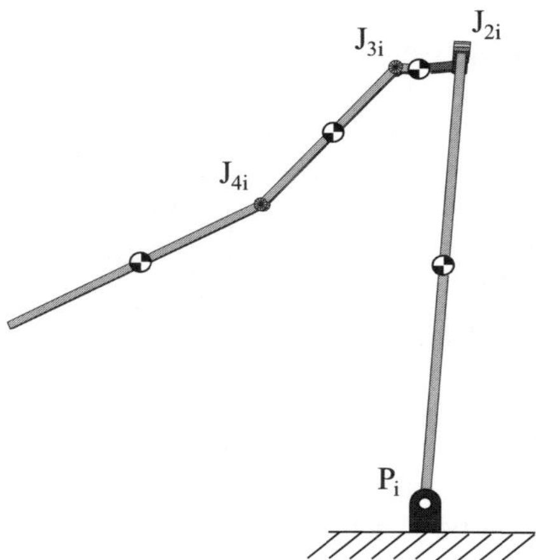

Fig. 6. Inverted four link pendulum [[1, 3]]

to control the unactuated tilting angle of the robot. In this paper, all joint angles of STriDER are calculated based on kinematics only to illustrate the concept of a single-step gait and to emphasize the importance of the kinematic constraints for the system. Future research will address the dynamics of the system together with kinematics considerations developed in this paper.

3.3 Height of the Body

The height of the body must also be considered when taking a step which is defined as the distance from the center of the body (point B in Fig. 4) to the ground in the negative Z_0 direction. The body's maximum height depends on the geometry of the support triangle. Thus, the height of the body when all links of the stance legs are aligned from the center of the body to the stance leg foot position is the maximum height during that step with that specific support triangle's geometry. However, the maximum possible height for any geometry is the total length of the thigh and shank link. The minimum height must allow the swing leg to swing underneath the body as the body rotates 180 degrees without scuffing the ground. The height of the body also affects the speed of the fall. The higher the body the slower the fall of the robot, and the faster the body position is slower the fall of the robot.

3.4 Body Twisting Motion During a Step

During a step, two pivot lines must be considered; one is the pivot line formed by aligning the stance legs hip abductor joints that allows the body to rotate 180 degrees called the *body pivot line*, while the other is the pivot line formed by the two stance leg's foot contact point that allows the entire robot to pivot called the *stance leg pivot line*. When the *body pivot line* and *stance leg pivot line* are parallel while the robot takes a step, the kinematic analysis is greatly simplified and collision between the swing leg and stance legs is prevented. However, for uneven terrains it might be beneficial for the pivot lines to be skewed, as it may aid the swing leg in avoiding obstacles.

STriDER 2.0 has to align two of its rotator joints to prepare for each step. A top view of the support triangle formed by the foot contact points, P_1, P_2, and P_3 is shown in Fig. 7. $\overline{P_2P_3}$ is the *stance leg pivot line* and P_1 is the initial location of the swing leg foot contact point. Line f is formed by points P_1 and P_2 and line e is formed by points P_1 and P_3. Region I is the boundary created between line f, line e and $\overline{P_2P_3}$. For the case presented here, it is assumed that initially, the *body pivot line* is parallel to the *stance leg pivot line* and point P_{12} is the final swing leg foot contact position which must lie in Region I. Since P_1 and P_{12} form a straight line going through Region I, the body has to twist its facing angle and make its projected pivot line perpendicular to $\overline{P_1P_{12}}$. The twisting motion of the body is controlled with the stance legs and during the twisting the plane of the body is parallel with the ground. The twisting angle θ_{TW}, as shown in Fig. 7, is defined as the rotation of the *body pivot line* about its midpoint in $\pm Z_B$ directions, where Z_B is the z-axis of the body coordinate system shown in Fig. 4. θ_{TW} can be determined from the coordinates of P_1, P_2, P_3 and P_{12}, and satisfies the following constraints:

$$-\theta_C < \theta_{TW} < \theta_B \tag{1}$$

$$\theta_A = \theta_B + \theta_C \tag{2}$$

$$\theta_B = ArcTan\left(\frac{\overline{P_3H}}{\overline{HP_1}}\right) \tag{3}$$

$$\theta_C = ArcTan\left(\frac{\overline{P_2H}}{\overline{HP_1}}\right) \tag{4}$$

Note that, θ_B and θ_C are two extreme cases when the final foot position P_{12} lies on line e or f.

The twisting angle of the body is an important factor for the turning strategy of STriDER on various terrains. A large turning angle per step can increase the mobility of STriDER in complicated environments [11].

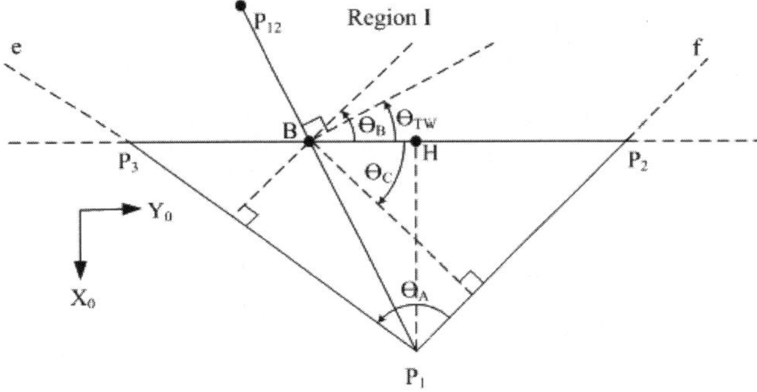

Fig. 7. Top view of the support triangle

3.5   Swing Leg's Clearance and Landing Position

The swing leg's foot path is also an important variable to consider as the robot takes a step. The swing leg's foot should not scuff the ground during the swing portion of the gait, thus the knee must be bent at certain angles to prevent the foot from touching the ground. Also, when considering a single step an allowable region for the subsequent swings leg's foot contact position must be constrained, as mentioned in Section 3.4.

4   Static Stability Margin

A specific quantitative static stability margin (SSM) was developed to assess the stability of STriDER. First, the CG_P point, shown in Fig. 8, is the center of gravity point projected in the negative Z_0 direction to the triangular plane formed by the robot's three foot contact points in 3D space. When the CG_P lies inside the support triangle, the SSM is calculated for a stable condition as shown in Equation (5),

$$SMM = Min \left[\frac{d_1}{r}, \frac{d_2}{r}, \frac{d_3}{r} \right] \tag{5}$$

where d_1, d_2, and d_3 is the distance from point CG_P to each side of the support triangle and r is the radius of the support triangle's incircle, as shown in Fig. 8. The center of the support triangle, labeled I in Fig. 8, was chosen as the center of the incircle of the support triangle since it is the point that represents the maximum equal distance from each side of the triangle.

If the point CG_P lies outside the support triangle the robot is statically unstable, as shown in Fig. 9 . In this case, the static stability margin depends upon the region, defined by the lines connecting point I to the three foot positions, P_1, P_2, and P_3, in which CG_P lies, as shown in Fig. 10.

Therefore the angles, θ_{CG}, θ_2, and θ_3, are defined as that between lines $\overline{IP_1}$ and $\overline{ICG_P}$ and $\overline{IP_1}$ and $\overline{IP_2}$ respectively as in Fig. 10. The static stability margin is then given as Equation (6),

$$SMM = \begin{cases} -\frac{d_3}{r} & 0 \leq \theta_{CG} < \theta_2 \\[2mm] -\frac{d_1}{r} & \theta_2 \leq \theta_{CG} < \theta_3 \\[2mm] -\frac{d_2}{r} & \theta_3 \leq \theta_{CG} < 2\pi \end{cases} \tag{6}$$

where r, d_1, d_2, and d_3 are defined as before. When the projected center of gravity point, CG_P, lies on any of the support triangle's sides it is marginally stable and the SSM is equal to 0. Table 2 shows the SSM range for these three cases.

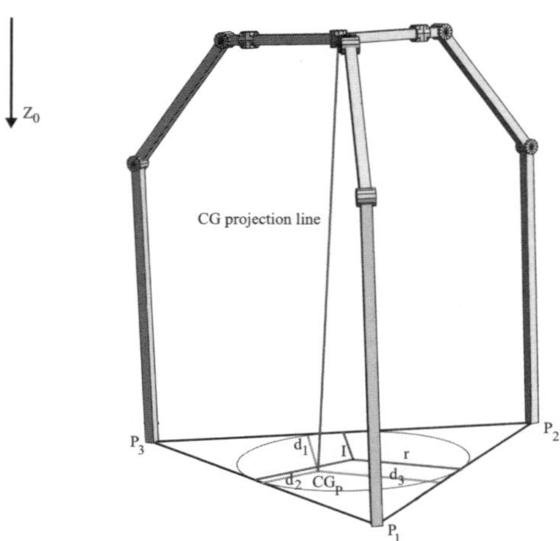

Fig. 8. Stable configuration with SM=0.555

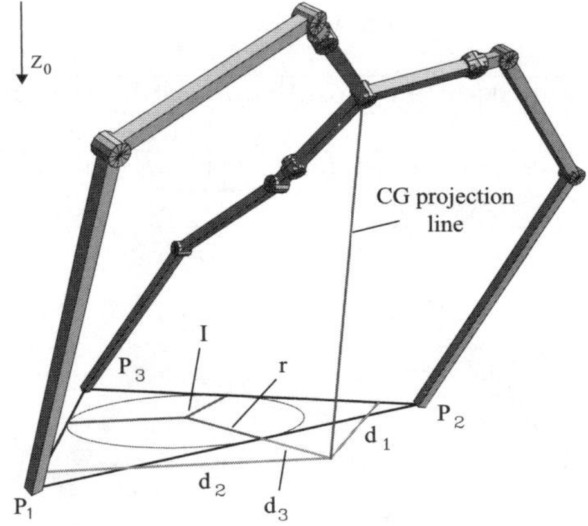

Fig. 9. Unstable configuration with a SM=-0.723

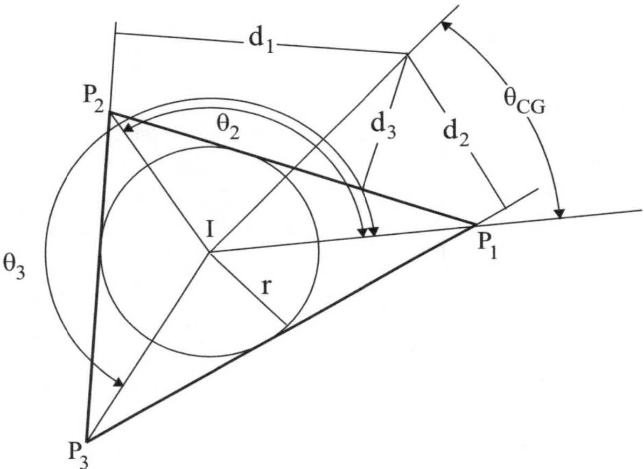

Fig. 10. SSM definition when CG_P lies outside the support triangle

Note, the robot is most stable when the projected center of gravity point lies on point I, thus the SSM is equal to 1. As the point CG_P moves closer to the sides of the triangle the SSM decreases and once CG_P lies any of the sides, the SM is equal to 0. As the CG_P point continues to move further outside the support triangle the SSM increases in magnitude in the negative direction. Fig. 8 and 9 show a stable and unstable case with their corresponding SSM values, respectively.

5   Foundations for a Single Step Gait Generation

This section lays out the foundation and guidelines for future work on a single step gait generation based on both kinematics and dynamics. Several of the

Table 2. SSM Range

Static Stability Condition	SSM Range
Stable	$1 > SSM > 0$
Marginally Stable	$SSM = 0$
Unstable	$-\infty > SSM < 0$

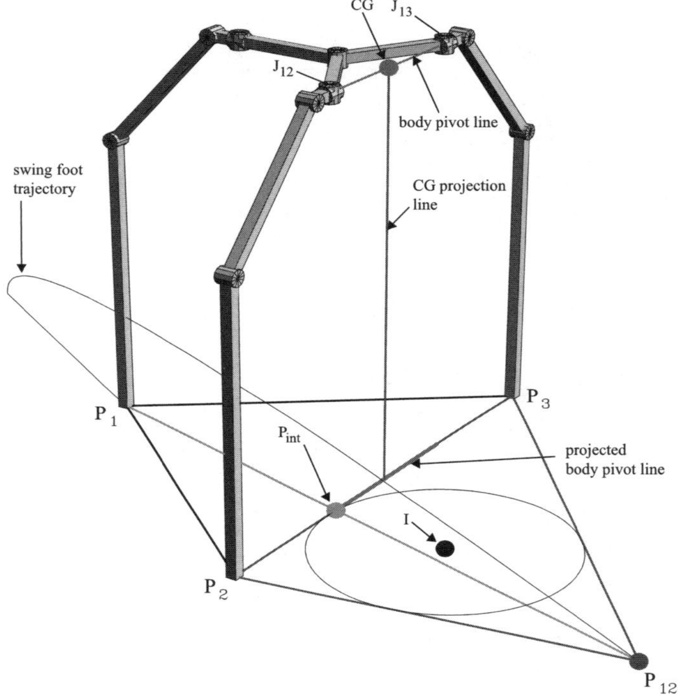

Fig. 11. Gait simulation labels

constraints addressed in Sections 3 should be considered when taking a single step. The objective is to achieve a single step from an initial swing leg foot position, P_1, to a desired final swing leg foot position P_{12} (within Region I), on an even ground, as shown in Fig. 7.

In Fig. 11, the center of gravity can be assumed to be located in the midpoint of the *body pivot line* formed by global positions of the hip abductor joints J_{12} and J_{13}. The swing foot projected path line, $\overline{P_1 P_{12}}$, is formulated from an initial swing leg foot position, P_1, to a final foot position, P_{12}. The *stance leg pivot line*, $\overline{P_2 P_3}$, is defined as the line connecting the stance leg's foot contact points, P_2 and P_3. P_{int}, is the intersection point of lines $\overline{P_1 P_{12}}$ and $\overline{P_2 P_3}$.

First, the robot may begin its gait at marginally stable state, where the projected center of gravity point lies on the *stance leg pivot line*, $\overline{P_2 P_3}$, as shown in Fig. 11 and discussed in Section 4. The robot must then shift so the projected center of gravity point, CG_P, coincides with P_{int}, the intersection of lines $\overline{P_1 P_{12}}$ and $\overline{P_2 P_3}$. Then, as mentioned in Section 3.4, the body must twist so the projected *body pivot line* is perpendicular to $\overline{P_1 P_{12}}$. The robot is now in position to fall forward and reach its desired final foot location. The rotation of the body or any other actuated links will force the robot to fall forward to initiate the swing portion of the step. Also, the body should be set at a height below the maximum height but high enough so the swing leg would have adequate room to swing in-between the stance legs.

6   Conclusions and Future Research

As an initial investigation, the gait planning strategies for STriDER were studied by discussing several kinematic constraints as the robot takes a step, without dynamic considerations. A static stability margin criterion was developed to quantify the static stability of the posture. Finally, the foundations for a single step gait were presented. Trajectory planning strategies and the generation of optimal gait will be conducted based on both kinematics and dynamics.

References

1. Heaston, J.: Design of a novel tripedal locomotion robot and simulation of a dynamic gait for a single step. Ma, Virginia Polytechnic and State University (2006)
2. Hong, D.W., Lahr, D.F.: Synthesis of the body swing rotator joint aligning mechanism for the abductor joint of a novel tripedal locomotion robot. In: 31st ASME Mechanisms and Robotics Conference, Las Vegas, Nevada (September 2007)
3. Heaston, J., Hong, D.W.: Design optimization of a novel tripedal locomotion robot through simulation and experiments for a single step dynamic gait. In: 31st ASME Mechanisms and Robotics Conference, Las Vegas, Nevada (September 2007)
4. Hong, D.W.: Biologically inspired locomotion strategies: Novel ground mobile robots at romela. In: URAI International Conference on Ubiquitous Robots and Ambient Intelligence, Seoul, S. Korea (October 2006)
5. Ren, P., Morazzani, I., Hong, D.W.: Forward and inverse displacement analysis of a novel three-legged mobile robot based on the kinematics of in-parallel manipulators. In: 31st ASME Mechanisms and Robotics Conference, Las Vegas, Nevada (September 2007)
6. McGeer, T.: Passive dynamic walking. International Journal of Robotics Research 9(2), 62–82 (1990)
7. Takahashi, R., Ono, K., Shimada, T.: Self-excited walking of a biped mechanism. International Journal of Robotics Research 20(12), 953–966 (2001)
8. Agrawal, S.K., Sangwan, V.: Differentially flat design of bipeds ensuring limit-cycles. In: Proceedings of IEEE International Conference on Robotics and Automation, Rome, Italy (April 2007)

9. Sangwan, V., Agrawal, S.K.: Design of under-actuated open-chain planar robots for repetitive cyclic motions. In: Proceedings of IDETC/CIE, ASME International Design Engineering Technical Conferences and Computers and Information in Engineering Conference, Philadelphia, Pennsylvania, USA (September 2006)
10. Spong, M.W., Bhatia, G.: Further results on control of the compass gait biped. In: International Conference on Intelligent Robots and Systems, Las Vegas, Nevada (October 2003)
11. Worley, M.W., Ren, P., Sandu, C., Hong, D.W.: The development of an assessment tool for the mobility of lightweight autonomous vehicles on coastal terrain. In: SPIE Defense and Security Symposium,Orlando, Florida (April 2007)

Safe Joint Mechanism Based on Passive Compliance for Collision Safety

Jung-Jun Park[1], Jae-Bok Song[1], and Hong-Seok Kim[2]

[1] Department of Mechanical Engineering, Korea University, Seoul, Korea
{hantiboy, jbsong}@korea.ac.kr
[2] Center for Intelligent Robot, Korea Institute of Industrial Technology
hskim@kitech.re.kr

Summary. A safe robot arm can be achieved by either a passive or active compliance system. A passive compliance system composed of purely mechanical elements often provide faster and more reliable responses for dynamic collision than an active one involving sensors and actuators. Since both positioning accuracy and collision safety are important, a robot arm should exhibit very low stiffness when subjected to a collision force greater than the one causing human injury, but maintain very high stiffness otherwise. To implement these requirements, a novel safe joint mechanism (SJM), which consists of linear springs and a slider-crank mechanism, is proposed in this research. The SJM has the advantages of variable stiffness which can be achieved only by passive mechanical elements. Various experiments of static and dynamic collisions showed the high stiffness of the SJM against an external force of less than the critical impact force, but an abrupt drop in the stiffness when the external force exceeds the critical force, thus guaranteeing collision safety. Furthermore, the critical impact force can be set to any value depending on the application.

1 Introduction

For industrial robots, safe human-robot coexistence is not as important as the fast and precise manipulation. However, service robots often interact directly with humans for various tasks. For this reason, safety has become one of the most important issues in service robotics. Therefore, several types of compliant joints and flexible links of a manipulator have been proposed for safety.

A safe robot arm can be achieved by either a passive or active compliance system. In the actively compliant arm, collision is detected by various types of sensors, and the stiffness of the arm is properly controlled. The active compliance-based approach suffers from the relatively low bandwidth because it involves sensing and actuation in a response to dynamic collision. This rather slow response can be improved slightly when non-contact sensors such as proximate sensors are employed. Furthermore, the installation of the sensor and actuator in the robot arm often leads to high cost, an increase in system size and weight, possible sensor noise, and actuator malfunction.

On the other hand, the robot arm based on passive compliance is usually composed of the mechanical components such as a spring and a damper, which absorb the excessive collision force. Since this approach does not utilize any

S. Lee, I.H. Suh, and M.S. Kim (Eds.): Recent Progress in Robotics, LNCIS 370, pp. 49–61, 2008.
springerlink.com

sensor or actuator, it can provide fast and reliable responses even for dynamic collision. Various safety mechanisms based on passive compliance have been suggested so far. The programmable passive impedance component using an antagonistic nonlinear spring and a binary damper was proposed to mimic the human muscles [1]. The mechanical impedance adjuster with a variable spring and an electromagnetic brake was developed [2]. The programmable, passive compliance-based shoulder mechanism using an elastic link was proposed [3]. A passive compliance joint with rotary springs and a MR damper was suggested for the safe arm of a service robot [4]. A variable stiffness actuator with the nonlinear torque transmitting system composed of a spring and a belt was developed [5].

Most passive compliance-based devices use linear springs. However, one drawback to the use of a linear spring is positioning inaccuracy due to the continual operation of the spring even for small external forces that do not require any shock absorption and due to undesirable oscillations caused by the elastic behavior of the spring. To cope with this problem, some systems adopt the active compliance approach by incorporating extra sensors and actuators such as electric dampers or brakes, which significantly impair the advantages of a passive system. In this research, therefore, a novel passive compliance-based safety mechanism that can overcome the above problems is proposed.

Some tradeoffs are required between positioning accuracy and safety in the design of a manipulator because high stiffness is beneficial to positioning accuracy whereas low stiffness is advantageous to collision safety performance. Therefore, the manipulator should exhibit very low stiffness when subjected to collision force greater than the one that causes injury to humans, but should maintain very high stiffness otherwise. Of course, this ideal feature can be achieved by the active compliance approach, but this approach often causes the several shortcomings mentioned above.

In the previous research, this ideal feature was realized by a novel design of the safe link mechanism (SLM) which was based on the passive compliance [6]. However, a safe mechanism, which is simpler and more lightweight than SLM, is desirable for a service robot arm. To implement these requirements, the safe joint mechanism (SJM) which possesses the same characteristics as SLM is proposed in this research. The SJM is composed of the passive mechanical elements such as linear springs and a slider-crank mechanism. The springs are used to absorb the high collision force for safety, while the slider-crank mechanism determines the safety or non-safety of the external force so that the SJM operates only in case of an emergency. The main contribution of this proposed device is the variable stiffness capability implemented only by use of passive mechanical elements. Without compromising positioning accuracy for safety, both features can be achieved simultaneously with the SJM.

The rest of the paper is organized as follows. The operating principle of the SJM is discussed in detail in section 2. Section 3 presents further explanation about its operation based on simulations. Various experimental results for both static and dynamic collisions are provided in section 4. Finally, section 5 presents conclusions and future work.

2   Operation Principle of Safe Joint Mechanism

The passive safety mechanism proposed in this research is composed of a spring and a slider-crank mechanism. This chapter presents the concept of the transmission angle of the slider-crank mechanism and the characteristics of the slider-crank mechanism in combination with the spring.

Springs have been widely used for a variety of safety mechanisms because of their excellent shock absorbing property. Since the displacement of a linear spring is proportional to the external force, the robot arm exhibits deflection due to its own weight and/or payloads when a spring is installed at the manipulator joint. This characteristic is beneficial to a safe robot arm, but has an adverse effect on positioning accuracy. To cope with this problem, it is desirable to develop a spring whose stiffness remains very high when an external force acting on the end-effector is within the range of the normal operation, but becomes very low when it exceeds a certain level of force due to collision with the object. However, no such springs with this ideal feature exist. In this research, the power transmission characteristics of the 4-bar linkage are exploited to achieve this nonlinear spring feature.

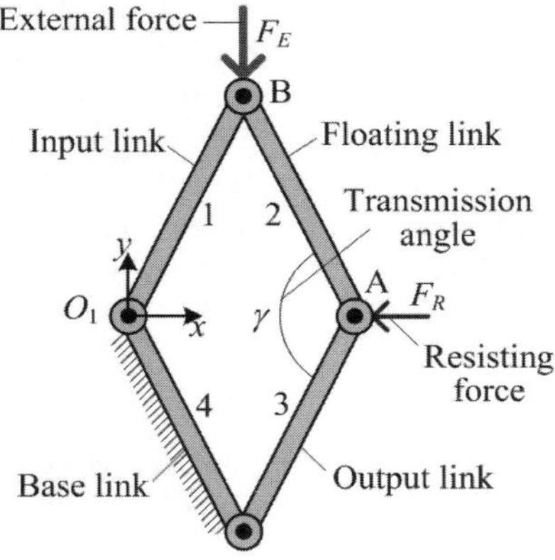

Fig. 1. 4-bar linkage

Consider a 4-bar linkage mechanism shown in Fig. 1. When an external force F_E is exerted on point B of the input link in the y-axis direction, an appropriate resisting force F_R acting in the x-axis direction can prevent the movement of the output link. In the 4-bar linkage, the transmission angle is defined as the angle between the floating and the output link. The power transmission efficiency from input to output varies depending on this transmission angle. If the transmission angle γ is less than 45° or greater than 135°, a large force is required at the input link to move the output link. That is, only a small F_R is sufficient to prevent the output link from moving for a given F_E in this case. However, as the transmission angle approaches 90°, the power transmission efficiency improves, thus leading to

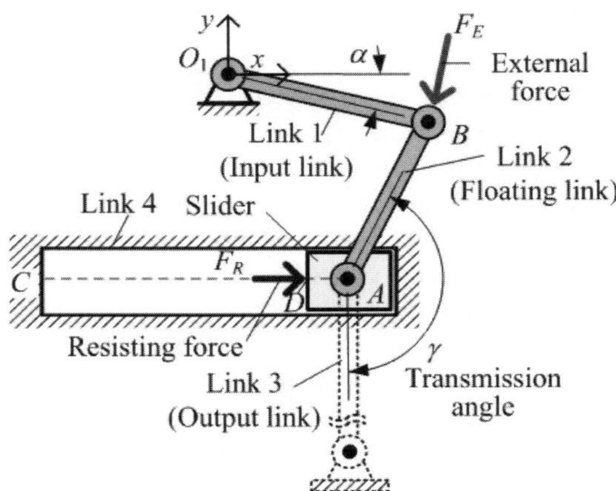

Fig. 2. Slider-crank mechanism

easy movement of the output link of a 4-bar linkage [7]. Therefore, a large F_R is required to prevent the output link from moving for a given F_E.

The slider-crank linkage can be regarded as the 4-bar linkage if the slider is replaced by an infinitely long link perpendicular to the sliding path as shown Fig. 2. Therefore, revolute joint A between link 2 and link 3 can move rectilinearly only in the x-axis direction. Note that the transmission angle of a slider-crank mechanism can be also defined as the angle between the floating link (link 2) and the output link (link 3). The force balance of the forces acting on the slider and the input link can be given by

$$F_E = -F_E \frac{\sin \gamma}{\cos(\gamma + \alpha)} \tag{1}$$

where α is the inclined angle of link 1. Note that the value of $\sin\gamma/\cos(\gamma + \alpha)$ is always negative because γ is in the range of 90° to 180° in Fig. 2, which requires the minus sign in Eq. (1). In Eq. (1), for the same external force, the resisting force changes as a function of γ.

If the pre-compressed spring is installed between points C and D in Fig. 3, the spring force F_S can offer the resisting force F_R, which resists the movement of the slider caused by the external force F_E. When the external force is balanced against the spring force, the external force can be described in terms of the transmission angle and the other geometric parameters as follows:

$$F_E = -k \left(s_o - c + d + l_2 \sin \gamma\right) \frac{\cos(\gamma + \alpha)}{\sin \gamma} \tag{2}$$

where k is the spring constant, s_o the initial length of the spring, l_2 the length of link 2 and x the displacement of the slider. Although x does not explicitly appear in Eq. (2), it is directly related to γ by the relation of $x = c - d - l_2 \cos(\gamma - 90)$. For example, when $k = 0.8\text{kN/m}$, $l_1 = l_2 = 19\text{mm}$, $s_o = 34\text{mm}$,

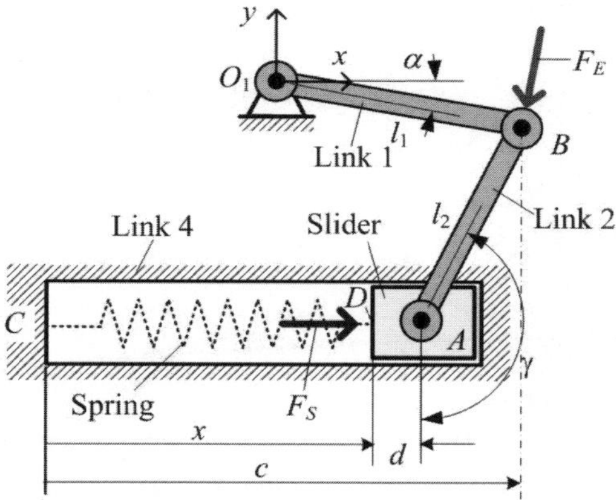

Fig. 3. Slider-crank mechanism combined with spring

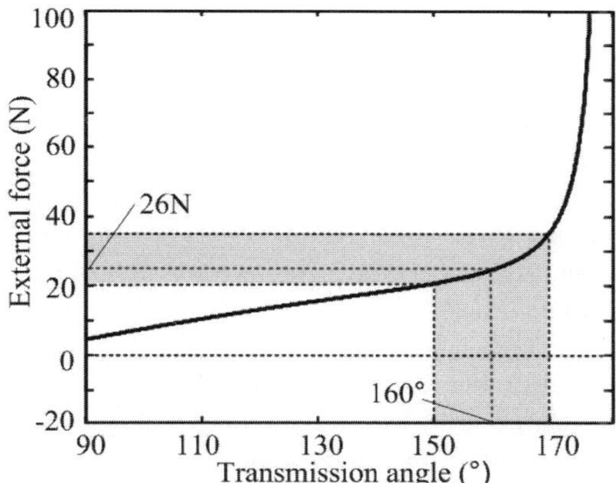

Fig. 4. External force as a function of transmission angle

$c = 36$mm, $d = 6.5$mm and $\alpha = 20°$, the external force for the static force balance can be plotted as a function of γ in Fig. 4. The spring force does not need to be specified for static balance because it is automatically determined for a given γ. As shown in the figure, the external force diverges rapidly to positive infinity as γ approaches 180°, so even a very small spring force can make this mechanism statically balanced against a very large external force. In this research, the transmission angle in the range of 150° to 170° is mainly used in consideration of the mechanical strength of the mechanism.

In this proposed mechanism, the external force required to balance with the spring force is defined as the critical impact force. For a given γ, a static balance is maintained when the external force equals the critical impact force, as shown in Fig. 4, but the spring is rapidly compressed once the external force greater

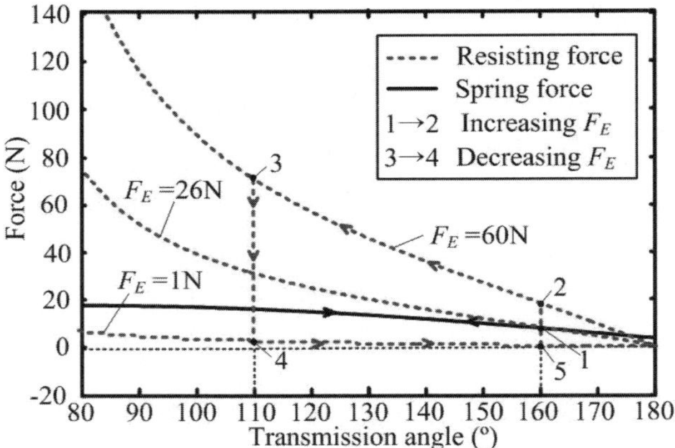

Fig. 5. Plots of resisting force and spring force versus transmission angle for high and low critical impact forces

than this critical value acts on this mechanism. The detailed explanation about this phenomenon is given below. Figure 5 shows the resisting force curves for the three given external forces ($F_E = 1, 26, 60$N) as a function of the transmission angle γ, which is computed by Eq. (1). The spring force as a function of γ is also plotted in Fig. 5. Note that the variation of the spring force is much smaller than that of the resisting force throughout the wide range of γ. Since the spring force provides the resisting force, when the two forces are equal, the mechanism becomes statically balanced, as shown in Fig. 3.

Suppose the critical impact force is set to 26N. Then the transmission angle for the static equilibrium becomes 160° from Eq. (2) with $F_E = 26$N. This corresponds to equilibrium point 1, which is the intersection of the resisting force curve of $F_E = 26$N and the spring force curve. Now suppose the external force abruptly increases to 60N, which is larger than the critical impact force (1→2), then γ reduces as the slider moves to the left in Fig. 3. As γ decreases, the resisting force rapidly increases (2→3), and the spring force also slightly increases, as shown in Fig. 5. Since the resisting force required for the static equilibrium becomes much larger than the spring force, the static equilibrium cannot be maintained, and thus the slider moves left rapidly. When the external force is reduced to 1N which is less than the critical impact force (3→4), the spring force becomes larger than the resisting force required and γ increases because the slider is pushed right (4→5) by the spring force.

3   Safe Joint Mechanism Model

3.1   Protorype Modeling

The mechanisms introduced conceptually in the previous section are now integrated into the safe joint mechanism (SJM), which suggests a new concept of a safe

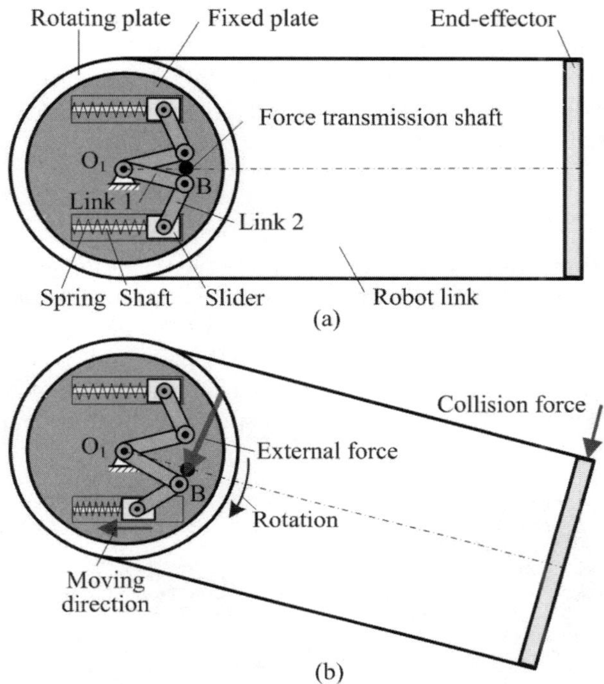

Fig. 6. Operation of SJM; (a) before collision, and (b) after collision

robot arm. The SJM consists of a slider-crank mechanism and a linear spring. As shown in Fig. 6, the slider-crank mechanism is installed at the fixed plate and the robot link is connected to the rotating plate. The rotation centers of both plates are identical and the collision force can be transmitted to the slider-crank mechanism by means of the force transmission shaft fixed at the rotating plate. The slider-crank mechanisms are arranged symmetrically so that they can absorb the collision force applied in both directions. In this prototype, the collision force acting on the end-effector is amplified according to the ratio of the rotation radius of the force transmission shaft to that of the end-effector, and is transmitted to point B of the input link by the force transmission shaft. Therefore, the external force exerted on the SJM is proportional to the collision force.

If the external force exceeding the critical impact force is applied to the input link of the slider-crank mechanism by the force transmission shaft connected to the rotating plate, then the input link is rotated around point O_1, as shown in Fig. 6(b). Then, the slider connected to link 2 is forced to move left on the guide shaft to compress the spring. This movement of the slider reduces the transmission angle, so maintaining the static balance requires a greater resisting force for the same external force. However, the increased spring force due to its compression is not large enough to sustain the balance. This unbalanced state causes the slider to rapidly slide left. As a result, the force transmission shaft fixed at the rotating plate is rotated and the robot link is also rotated, which absorbs the collision force. However, if the external force amplified from the collision force is less than the critical impact force, the end-effector does not move at all, and the slider-crank mechanisms maintain the static equilibrium, thus providing high stiffness to the SJM.

3.2   Simulation of Prototype

Various simulations were conducted to evaluate the performance of the proposed SJM. As shown in Fig. 7, the components of the mechanism were modeled by Solidworks and its dynamics was analyzed by Visual Nastran 4D. For simplicity of simulation, only one slider-crank mechanism of the SJM was modeled by assuming that the external force directly acted on the end-effector of the robot link. The slider moving on the guide shaft was modeled as a spring-damper system. The damper

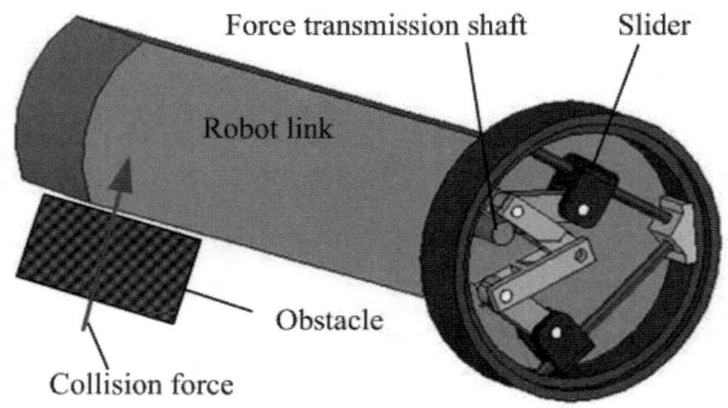

Fig. 7. Modeling of safety mechanism

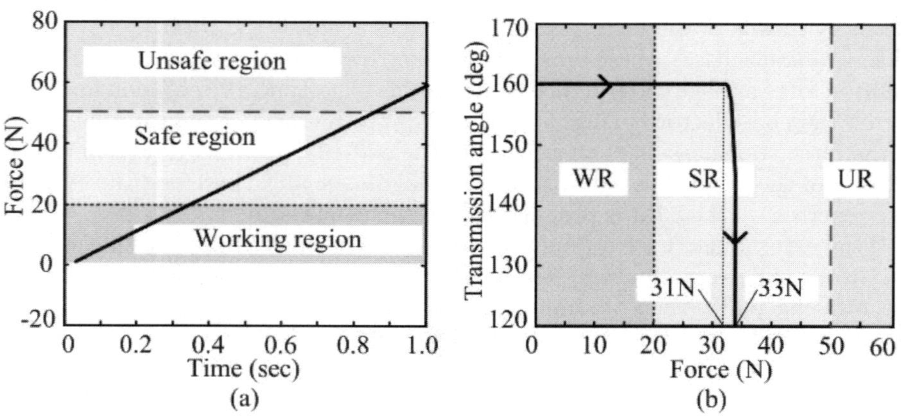

Fig. 8. Simulation results for static collision for initial transmission angle of 160°; (a) external force versus time, and (b) transmission angle versus external force

was modeled to represent the friction between the slider and the shaft, although a damper was not used for the real system.

Figure 8 shows the simulation results for a static collision. As the external force acting on the end-effector increases linearly up to 60N during 1sec, the transmission angle is drastically changed, as shown in Fig. 8(b). In this simulation, the damping coefficient is set to $c = 1\mathrm{kg/s}$, the spring constant to $k = 8\mathrm{kN/m}$ and the initial transmission angle to 160°. As shown in Fig. 8(b), the transmission angle does not change for the external force less than the critical impact force (in this simulation, 31N). However, once the external force exceeds this critical impact force, the transmission angle sharply decreases. In summary, the SJM stiffness remains very high like a rigid joint while the external force is below 31N. In the range of 31 to 33N, the transmission angle and thus the stiffness decrease. As the collision force approaches 33N, the stiffness abruptly diminishes, and consequently, the SJM behaves as a compliant joint.

4   Experiments for Safe Joint Mechanism

4.1   Protorype of SJM

The prototype of the SJM shown in Fig. 9 was constructed to conduct various experiments related to the performance of the SJM. Most components are made of duralumin and polyoxymethylene which can endure the shock exerted on the SJM. The slider can slide and the spring can be compressed by means of the linear bushing guides.

4.2   Safety Criterion

The safety criterion can be divided into static and dynamic collisions. The static collision means that the collision speed of the robot arm relative to a human is

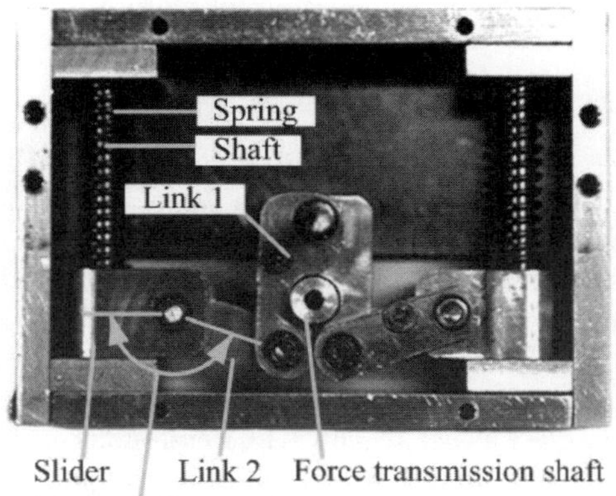

Fig. 9. Prototype of SJM

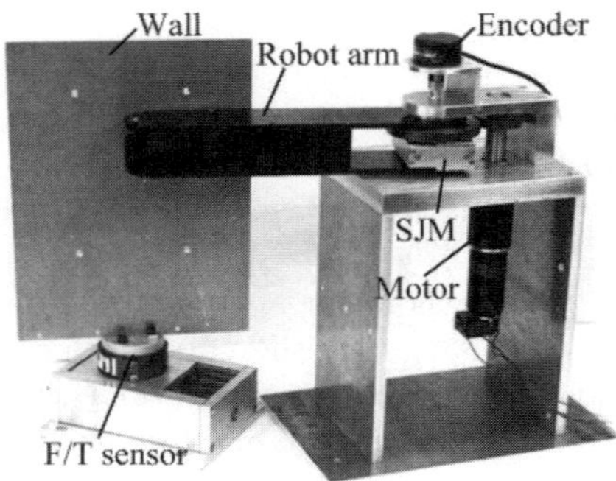

Fig. 10. Experimental setup for robot arm with SJM

very low (e.g., below 0.6m/s). The human pain tolerance for static collision can be expressed by

$$F \leq F_{\text{limit}} \tag{3}$$

where F_{limit} is the injury criterion value which has been suggested as 50N by several experimental researches [8].

In the case of dynamic collision, both the collision force and the collision speed are important. To represent human safety associated with the dynamic collision of the SJM, the head injury criterion (HIC), which is used to quantitatively measure head injury risk in car crash situations, is adopted in this research [9].

$$HIC = T \left(\frac{1}{T} \int_0^T a(t)dt \right)^{2.5} \tag{4}$$

where T is the final time of impact and $a(t)$ is the acceleration in the unit of gravitational acceleration g. An HIC value of 1,000 or greater is typically associated with extremely severe head injury, and a value of 100 can be considered suitable to normal operation of a machine physically interacting with humans.

4.3   Experimental Results

Figure 10 shows an experimental setup in which the SJM is installed at the 1-DOF robot arm. The fixed plate of the SJM in Fig. 6 is attached to the motor and the force transmitting shaft is connected to the robot arm. Therefore, the torque of a motor can be transmitted to the robot link via the SJM. A force/torque sensor is installed at the end of the wall to measure the collision force. The displacement of the SJM is measured by an encoder attached to the SJM.

In the experiment for static collision, the spring constant was 8kN and the initial transmission angle was set to 160°. The end-effector of the robot arm was initially placed to barely touch a fixed wall, and its joint torque provided by the motor was increased slowly. The static collision force between the robot link and the wall was measured by the 6 axis force/torque sensor. Experiments were conducted for the robotic arms with and without the SJM.

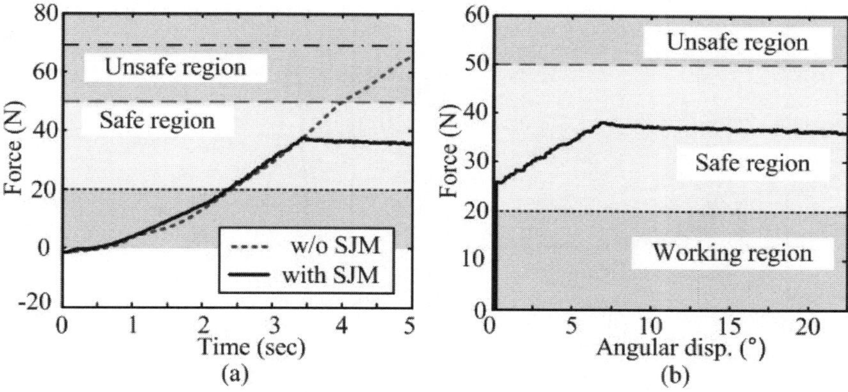

Fig. 11. Experimental results for static collision for robot arm; (a) collision force versus time without and with SJM, and (b) collision force versus angular displacement of SJM

The robot arm without the SJM delivered a contact force that increased up to 70N to the wall, as shown in Fig. 11(a). However, the contact force of only up to 38N was transmitted to the wall for the robot arm with the SJM, as shown in Fig. 11(a). In other words, the contact force above the pain tolerance does not occur because the excessive force is absorbed by the SJM. In Fig. 11(b), virtually no angular displacement of the SJM attached to the robot arm occurs when the contact force is below the critical impact force of 26N. Therefore, the robot arm with the SJM can accurately handle a payload up to approximately 2kg as if it were a very rigid joint. As the contact force rises above the critical impact force, the SJM stiffness quickly diminishes and the angular displacement occurs, thus maintaining the robot arm in the safe region. In summary, the SJM provides high positioning accuracy of the robot arm in the working region, and guarantees safe human-robot contact by absorbing the contact force above 50N in the unsafe region.

Next, some experiments on dynamic collision were conducted for the robot arm equipped with the SJM. The experimental conditions including the spring constant and the initial transmission angle were set to the same values as those of static collision experiments. For dynamic collision, a plastic ball of 1.5kg moving at a velocity of 3m/s was forced to collide with the end-effector of the robot arm. The acceleration of the ball was measured by the accelerometer mounted at the ball.

The experimental results are shown in Fig. 12. At the instant the ball contacts the end-effector, the acceleration of the ball reached a peak value of 80g, but immediately after collision, the collision force delivered to the ball dropped rapidly because of the operation of the SJM. The dynamic collision safety of the robot arm with the SJM can be verified in terms of HIC defined by Eq. (4). The HIC value was computed as 50, which is far less than 100. Therefore, the safe human-robot contact can be achieved even for this harsh dynamic collision. Figure 12(b)

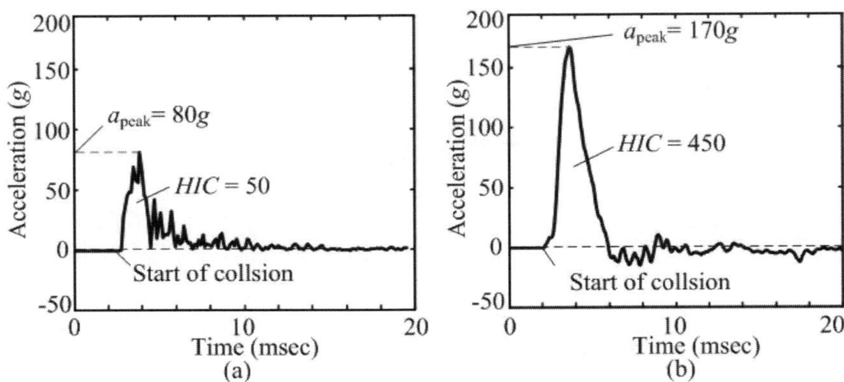

Fig. 12. Experimental results for dynamic collision of robot arm; acceleration versus time (a) with SJM, and (b) without SJM

shows the experimental results for the dynamic collision of the robot arm without the SJM. The peak value of the acceleration is almost twice that of the robot arm with the SJM, and the HIC value reached as high as 450, which indicates a high risk of injury to a human. Therefore, the robot arm with the SJM provides much greater safety for human-robot contact than that without the SJM.

5 Conclusion

In this research, the safe joint mechanism (SJM) was proposed. The SJM maintains very high stiffness up to the pre-determined critical impact force, but provides very low stiffness above this critical value, at which point the SJM absorbs the impact acting on the robot arm. From the analysis and experiments, the following conclusions are drawn:

1) The SJM has very high stiffness like a rigid joint when the external force acting on it is less than the critical impact force. Therefore, high positioning accuracy of the robot arm can be achieved in normal operation.

2) When the external force exceeds the critical impact force, the stiffness of the SJM abruptly drops. As a result, the robot arm acts as a compliant joint with high compliance. Therefore, human-robot collision safety can be attained even for a high-speed dynamic collision.

3) The critical impact force of the SJM can be set accurately by adjusting the initial transmission angle of the slider-crank mechanism, the spring constant and the initial spring length.

4) The proposed SJM is based on passive compliance, so it shows faster response and higher reliability than that based on the active compliance having sensors and actuators.

Currently, to apply the SJMs to two or more joints of the robot arm, the simpler and lightweight model is under development.

Acknowledgements. This research was supported by the Personal Robot Development Project funded by the Ministry of Commerce, Industry and Energy of Korea. fdgdfgfdgfd

References

1. Laurin-Kovitz, K., Colgate, J.: Design of components for programmable passive impedance. In: Proceedings of the IEEE International Conference on Robotics and Automation, pp. 1476–1481 (1991)
2. Morita, T., Sugano, S.: Development of one-D.O.F. robot arm equipped with mechanical impedance adjuster. In: Proceedings of the IEEE/RSJ International Conference on Intelligent Robots and System, pp. 407–412 (1995)
3. Okada, M., Nakamura, Y., Ban, S.: Design of programmable passive compliance shoulder mechanism. In: Proceedings of the IEEE International Conference on Robotics and Automation, pp. 348–353 (2001)
4. Kang, S., Kim, M.: Safe Arm Design for Service Robot. In: Proceedings of the International Conference on Control, Automation and System, pp. 88–95 (2002)
5. Tonietti, G., Schiavi, R., Bicchi, A.: Design and Control of a Variable Stiffness Actuator for Safe and Fast Physical Human/Robot Interaction. In: Proceedings of the IEEE International Conference on Robotics and Automation, pp. 528–533 (2000)
6. Park, J., Kim, B., Song, J.: Safe Link Mechanism based on Passive Compliance for Safe Human-Robot Collision. In: Proceedings of the IEEE International Conference on Robotics and Automation, pp. 1152–1157 (2007)
7. Wilson, C., Sadler, J.: Kinematics and Dynamics of Machinery, 2nd edn. Haper-Collins, New York (1993)
8. Yamada, Y., Hirasawa, Y., Huang, S., Umetani, Y.: Fail-safe human/robot contact in the safety space. In: Proceedings of the IEEE International Workshop on Robot and Human Communication, pp. 59–64 (1996)
9. Versace, J.: A review of the severity index. In: Proceedings of the 15th Stapp Car Crash Conference, pp. 771–779 (1971)

A Guidance Control Strategy for Semi-autonomous Colonoscopy Using a Continuum Robot

Gang Chen[1], Minh Tu Pham[2], and Tanneguy Redarce[2]

[1] Unilever R&D Port Sunlight, United Kingdom
   gang.chen@unilever.com
[2] Laboratoire Ampère
   UMR CNRS 5005 INSA Lyon, Villeurbanne, France
   minh-tu.pham@insa-lyon.fr

Summary. Due to their compliance and high dexterity, biologically-inspired continuum robots have attracted much interest for applications such as medical surgery, urban search and rescue, de-mining etc. In this paper, we will present an application to medical surgery-colonoscopy by designing a pneumatic-driven flexible robotic manipulator, called ColoBot. The focus of this paper lies in the sensor-based guidance control of the ColoBot in a tubular, compliant and slippery environment of human colon. The kinematic model related the position and orientation of distal end of ColoBot to the actuator inputs is firstly developed and formulated for orientation control of ColoBot. For the autonomous guidance inside the colon, a method based on a circumscribed circle is utilized to calculate the real-time reference paths from the measurement of three sensors for orientation control of the Colobot. This proposed approach can be extended to the control of continuum robots in the conditions of a dynamically confined space. The experimental results on an emulation platform will be presented in order to validate the proposed control strategy.

1 Introduction

Biologically-inspired continuum robots [17] have attracted much interest from robotics researchers during the last decades. These kinds of systems are characterized by the fact that their mechanical components do not have rigid links and discret joints in contrast with traditional industry robots. The design of these robots are inspired by movements of natural animals such as tongues, elephant trunks and tentacles etc. The unusual compliance and redundant degrees of freedom of these robots provide strong potential to achieve delicate tasks successfully even in cluttered and/or unstructured environments such as undersea operations [13], urban search and rescue, wasted materials handling [9], Minimally Invasive Surgery [1, 6, 15, 18].

Although continuum robots presented potential advantages for many applications, they also present a more complex problem on kinematics than rigid-link robots due to its lack of joints and rigid links. Hirose developed kinematics in 2-D for snake-like robots by introducing the serpenoid curve which closely

S. Lee, I.H. Suh, and M.S. Kim (Eds.): Recent Progress in Robotics, LNCIS 370, pp. 63–78, 2008.
springerlink.com   © Springer-Verlag Berlin Heidelberg 2008

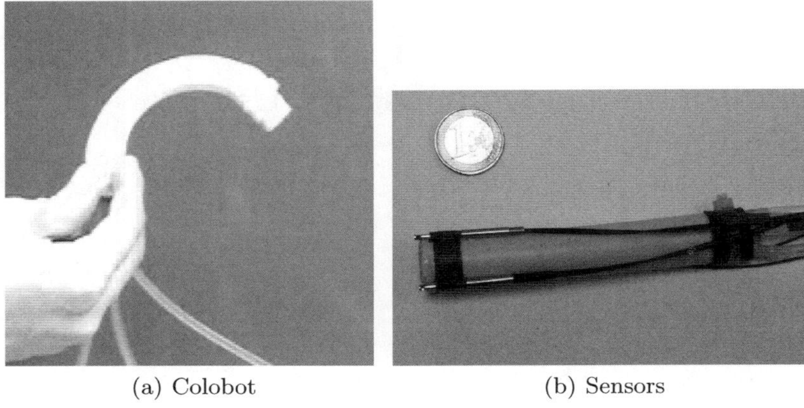

(a) Colobot   (b) Sensors

Fig. 1. Colobot and distance sensors

matches the shape of a snake [7]. In contrast, Chirikjian [5] developed kinematics of hyber-redundant robots by using a continuous backbone curve. But few continuum robots match the proposed curve which limits the application of this approach. Recently, researchers have developed kinematic models [3, 10, 18] and implemented kinematic control [11, 12] of continuum robots.

This paper deals with an application of a continuum robot for semi-autonomous colonoscopy. The designed robot called ColoBot [2] is a silicon-based micro continuum robot with 3 Degrees Of Freedrom (DOF) (Fig. 1(a)). The robotic manipulator is a unique unit with 3 active chambers regularly disposed 120 degrees apart. The outer diameter of the tip is 17 mm that is lesser than the average diameter of the colon. The diameter of the inner hole is 8mm, which is used in order to place the camera or other lighting tools. The weight of the prototype is 20 grams. The internal pressure of each chamber is independently controlled by using pneumatic jet-pipe servovalves. The promising result obtained from the preliminary experiments showed that this tip could bend up to 120°. During the operation of colonoscopy, this robotic tip will be mounted to the distal-end of an actual colonoscope to automatically guide the exploration of Colobot inside the human colon.

In order to achieve an automatic guidance, motion control and safe motion planning are two important problems to be solved. As mentioned earlier, some studies have been done on motion control [11, 12] based on a new developed kinematic model [3, 4, 10, 18]. However, these works did not deal with the autonomous manipulation. The desired motions for controlling continuum robots are obtained through either predefined trajectory or tele-operation by using a joystick in these control strategies.

In our case, motion planning is a very important element for autonomomous guidance. A contact-free (minimal contact) trajectory for the distal end of the ColoBot should be planned in order to guide the colonoscope deeply into the colon. However, the tubular, compliant and slippery environment of human colon brings out two main constraints in generating a trajectory. They are:

- The colon is a three-dimensional confined space and there is very little knowledge and very little works on the model of the colon for the purpose of motion planning.
- the colonoscopy procedure is performed in a dynamic environment due to the breathing activities of the patient and the movement induced by the insertion of the colonoscope body.

The main contribution of this paper is the sensor-based reactive planning in a 2-D plane and guidance control in a 3-D space for the semi-autonomous colonoscopy. Three special-designed optical distance sensors [16] disposing at 120 degrees are used to measure the relative position between the top-end of the Colobot and the colon wall. Each sensor is placed in front of each active pneumatic chamber (Fig. 1(b)). For more simplicity but without loss of generality, it is assumed that a colon is a cylindrical tube and its cross section is an ellipse at the sensor plane. A method based on a circumscribed circle is utilized to calculate the reference position of the ColoBot inside the colon. Kinematic-based position control will use these reference paths to adjust the position of ColoBot inside the colon in order to achieve guidance. The paper will touch upon the following topics: Section 2 will deal with the kinematics and characterization of pneumatic actuators. The implementation of sensor-based planning algorithm and kinematic control will be detailed in Section 3 and 4 respectively. Experimental results will be presented in Section 5, followed by the conclusions.

2   Kinematic Analysis of the Robotic Tip

The formulation of kinematic model which relates the position and orientation of the distal end of the Colobot in a cartesian frame to the actuator inputs is composed of three steps. Firstly, the relationship between the stretch length of each chamber of the silicone-based actuator and the applied pressure in each chamber is determined experimentally. Secondly, the robot bending shape relating to the actuator inputs (length of each chamber or applied pressure in each chamber) is determined through geometric relationships. Finally, a kinematic model relates the task space frame (distal end position inside the environment) to the robot deformation shape.

2.1   Kinematic Nomenclature

Fig. 2 shows the robot shape parameters and the corresponding frames. The deformation shape of ColoBot is characterized by three parameters as done in our previous prototype EDORA [3].

- L is the length of the virtual center line of the robotic tip
- α is the bending angle in the bending plane
- ϕ is the orientation of the bending plane

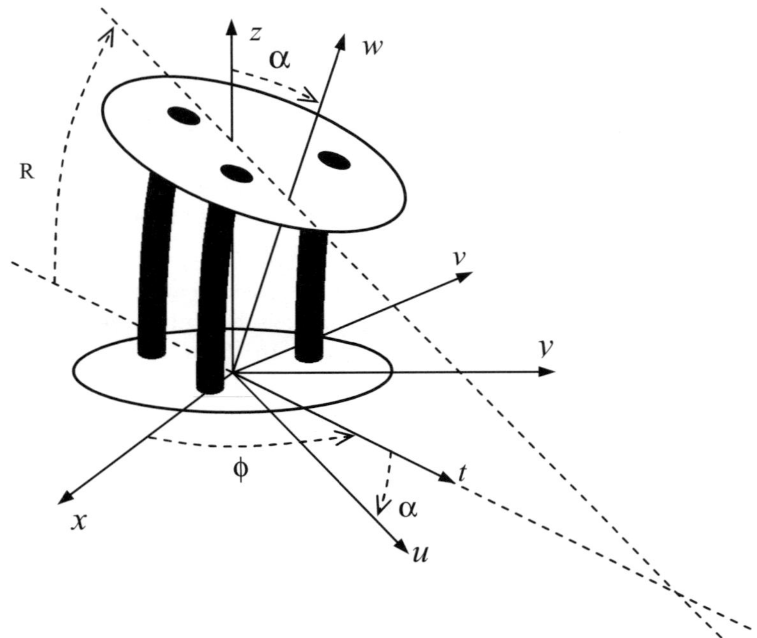

Fig. 2. Kinematic parameters of Colobot

The frame R_u (O-xyz) is fixed at the base of the actuator. The X-axis is the one that passed by the center of the bottom end and the center of the chamber 1. The XY-plane defines the plane of the bottom of the actuator, and the z-axis is orthogonal to this plane. The frame R_s (u, v, w) is attached to the top end of the manipulator. So the bending angle α is defined as the angle between the o-z axis and o-w axis. The orientation angle ϕ is defined as the angle between the o-x axis and o-t axis, where o-t axis is the project of o-w axis on the plan x-o-y. The notation is explained as follows:

i: chamber index, i = 1, 2, 3
R: radius of curvature of the centerline of the robotic tip
L_i : arc length of the ith chamber
L_0 : initial length of the chamber
P_i : pressure in the chamber i
S : effective surface of the chamber
R_i : radius of curvature of the ith chamber
ε: stretch length of the centerline
ΔL_i: the stretch length of the ith chamber

2.2   Characterization of the Pneumatic Actuator

Bellows-based flexible robots presented in [1] suppose that the length variation of each bellows is proportional to the applied pressure. In our case, the silicone-based

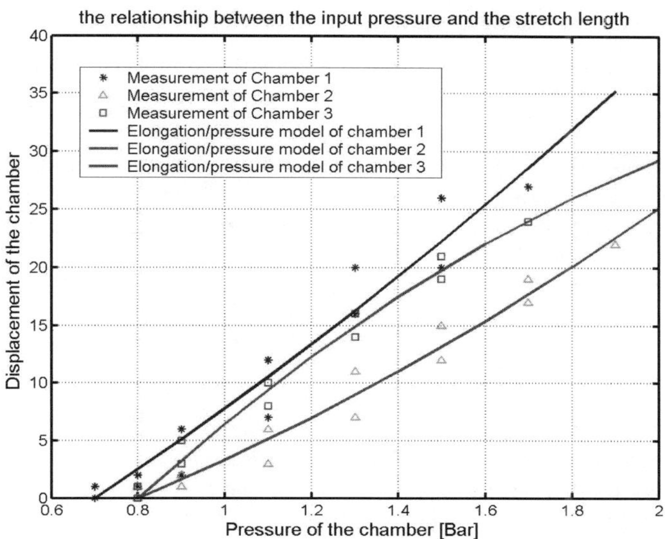

Fig. 3. Static characteristics of pneumatic chambers

actuator shows a strong nonlinearity relating the stretch length of the chamber to the pressure variation in preliminary experiments. This relationship is described as following:

$$\Delta L_i = f_i(P_i) \tag{1}$$

Where ΔL_i $(i = 1, 2, 3)$ is the stretch length of each chamber with corresponding pressure and f_i $(i = 1, 2, 3)$ is a nonlinear function of P_i. A third order polynomial approximation allows to fit significantly the actual data in the working zone of each chamber i.e. above the threshold pressure of each chamber (Fig. 3). The corresponding results can be written as:

$$\begin{cases} \text{if} \quad P_{1min} < P_1 < P_{1max} \\ \Delta L_1 = 37(P_1 - P_{1min})^3 - 54(P_1 - P_{1min})^2 \\ \qquad -9.5(P_1 - P_{1min}) \\ \text{if} \quad P_{2min} < P_2 < P_{2max} \\ \Delta L_2 = -9(P_2 - P_{2min})^3 - 18(P_2 - P_{2min})^2 \\ \qquad -11(P_2 - P_{2min}) \\ \text{if} \quad P_{3min} < P_3 < P_{3max} \\ \Delta L_3 = 0.8(P_3 - P_{3min})^3 - 8.9(P_3 - P_{3min})^2 \\ \qquad -34(P_3 - P_{3min}) \end{cases} \tag{2}$$

where P_{imin} $(i = 1, 2, 3)$ is the threshold of the working point of each chamber and their values equal: $P_{1min} = 0.7$ bar, $P_{2min} = 0.8$ bar, $P_{3min} = 0.8$ bar and P_{imax} $(i = 1, 2, 3)$ is the maximum pressure that can be applied into each chamber.

68   G. Chen, M.T. Pham, and T. Redarce

2.3 Kinematic Analysis

When the load effects are ignored and the deflected angle α is such as:

$$0 < \alpha < \pi \tag{3}$$

the deflected shape of the ColoBot at the bending moment is assumed to be an arc of a circle. Given this assumption, the geometry-based kinematic model [3] relating the robot shape parameters to the actuator inputs (chamber length) is expressed as following:

$$\begin{cases} L = \frac{1}{3}\sum_{i=1}^3 L_i \\ \phi = \text{atan2}\,\frac{\sqrt{3}(L_2-L_3)}{L_3+L_2-2L_1} \\ \alpha = \frac{2\sqrt{\lambda_L}}{3r} \end{cases} \tag{4}$$

where $\lambda_L = L_1{}^2 + L_2{}^2 + L_3{}^2 - L_1 L_2 - L_2 L_3 - L_3 L_1$ and r is the radius of the ColoBot. The detailed deduction of these equations can be found in [3]. By using the actuator models determined in section 2.2, the corresponding length of each chamber under the pressure variation can be expressed as following:

$$\begin{cases} L_1 = L_0 + \Delta L_1 = L_0 + f_1(P_1) \\ L_2 = L_0 + \Delta L_2 = L_0 + f_2(P_2) \\ L_3 = L_0 + \Delta L_3 = L_0 + f_3(P_3) \end{cases} \tag{5}$$

Thus the direct kinematic equations relating the robot shape parameters to the input pressures are then represented by:

$$\begin{cases} L = L_0 + \frac{1}{3}\sum_{I=1}^3 f_i(P_i) \\ \phi = \arctan 2\,\frac{\sqrt{3}(f_2(P_2)-f_3(P_3))}{f_3(P_3)+f_2(P_2)-2f_1(P_1)} \\ \alpha = \frac{2\sqrt{\lambda_p}}{3r} \end{cases} \tag{6}$$

where $\lambda_p = f_1(P_1)^2 + f_2(P_2)^2 + f_3(P_3)^2 - f_1(P_1)f_2(P_2) - f_2(P_2)f_3(P_3) - f_3(P_3)f_1(P_1)$

2.4 Task Space Representation of Kinematic Model

With the assumption that the bending shape of ColoBot is an arc of a circle, The cartesian coordinates (x, y, z) of the distal end of Colobot in the task space related to the robot bending parameters is obtained through a cylindrical coordinate transformation:

$$\begin{cases} x = \frac{L}{\alpha}(1 - \cos\alpha)\cos\phi \\ y = \frac{L}{\alpha}(1 - \cos\alpha)\sin\phi \\ z = \frac{L}{\alpha}\sin\alpha \end{cases} \tag{7}$$

And the state-space form of this model is given by:

$$X = f(Q_p) \tag{8}$$

where $X = (\alpha, \phi, L)^T, Q_p = (P_1, P_2, P_3)^T$.

3   Sensor-Based Motion Planning Algorithm for Autonomous Guidance

The objective of sensor-based motion planning is to calculate the safe position of the distal-end of Colobot compared to the colon wall in real-time based on the measurements of three distance sensors for guidance inside the colon. For more simplicity but without loss of generality, it is assumed that a colon is a cylindrical tube and its cross section is an ellipse in the sensor plane.

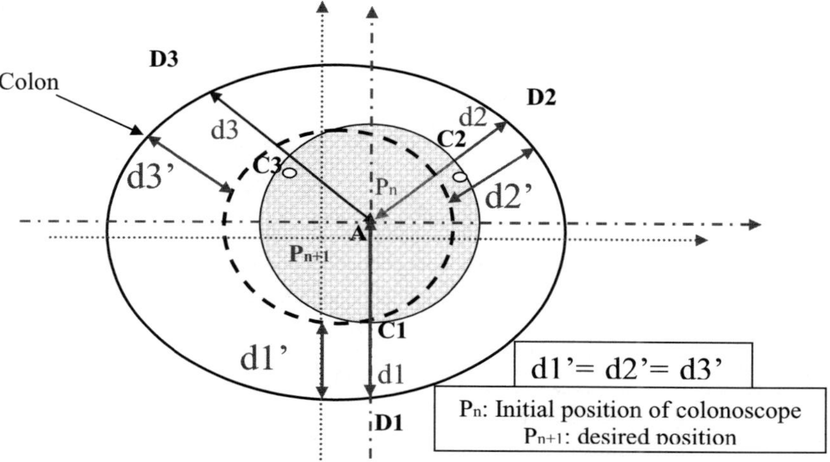

Fig. 4. Computation of the safe position

With these assumptions, the safe position will be the central axis of the colon. To approximate the colon axis, a method based on a circumscribed circle is proposed. Since three points D_1, D_2 and D_3 (Fig. 4) of sensor measurements form a triangle, the center of the circumscribed circle of this triangle is chosen as the safe position. This approach iterates as following:

1. Step 1: Three sensor measurements are collected.
2. Step 2: Position P_n in the frame R_s (Fig. 4) is evaluated with these three measurements.
3. Step 3: If P_n is a safe position, then go back step 1 for the next period; otherwise, next safe position P_{n+1} described in the Frame R_u (Fig. 4) is calculated through the circumscribed method and is fed to the kinematic control for execution.

Fig. 5 shows the algorithm of the planning and guidance control strategy of the distal end of the ColoBot inside the colon.

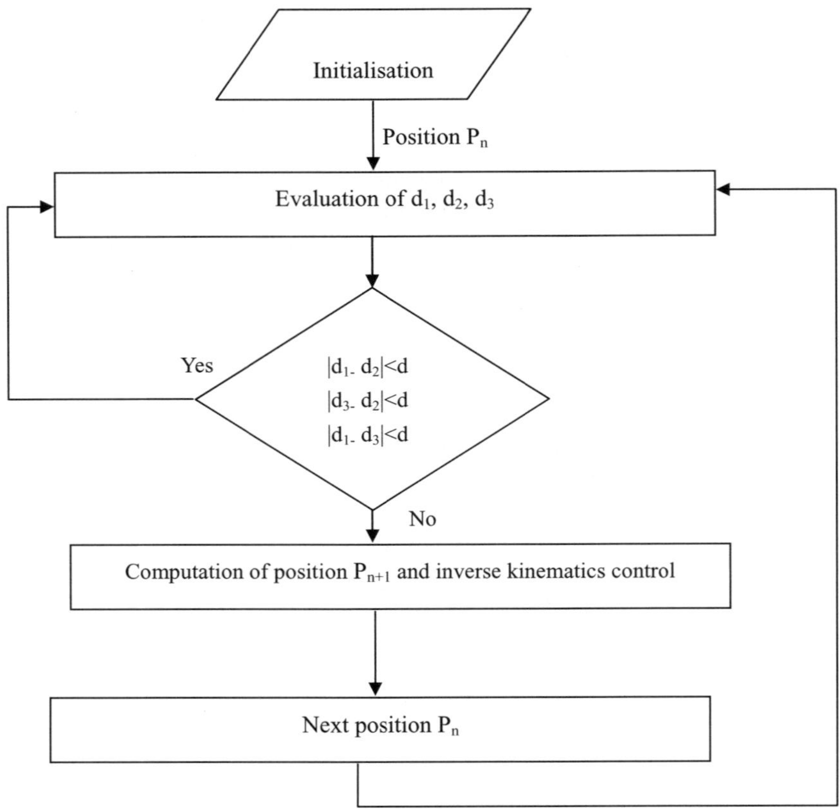

Fig. 5. Sensor-based planning and guidance control procedure

3.1   Positioning of the Distal End of Colobot Inside the Colon

Let A be the center of the ColoBot, C_i (i = 1,2,3) be three sensor points and D_i (i = 1,2,3) be the measurement point of each sensor on the colon wall. Fig. 4 shows the setting of these measurements. With these measurements, the distance d_i (i = 1,2,3) of the center of the distal end of Colobot with respect to the colon at each sensor direction can be calculated.

$$\begin{cases} d_1 = AD_1 = AC_1 + C_1D_1 = r + s1 \\ d_2 = AD_2 = AC_2 + C_2D_2 = r + s2 \\ d_3 = AD_3 = AC_3 + C_3D_3 = r + s3 \end{cases} \tag{9}$$

where R is the radius of the ColoBot and s_i (i= 1,2,3) is the ith sensor measurement.

The coordinates of points D_1, D_2 and D_3 can be obtained related to the local coordinate frame R_s (u,v,w) whose origin P_n as following (Fig. 2):

$$\begin{cases} D_1 : (s1 + r; 0; 0)_{R_s} \\ D_2 : (-(r + s2)\sin(\pi/6); -(r + s2)\cos(\pi/6); 0)_{R_s} \\ D_2 : (-(r + s3)\sin(\pi/6); (r + s3)\cos(\pi/6); 0)_{R_s} \end{cases} \tag{10}$$

The coordinates of P_n and the points D_1, D_2 and D_3 in the coordinate frame R_u (x, y, z) are expressed as following:

$$P_n = \left\{ \begin{array}{l} L\cos\phi\sin\alpha \\ -L\sin\phi\sin\alpha \\ L\cos\alpha \end{array} \right\}_{R_u}$$

$$OD_1 = \left\{ \begin{array}{l} \cos\theta[(s1+r)\cos\alpha + L\sin\alpha] \\ \sin\theta[(s1+r)\cos\alpha - L\sin\alpha] \\ (s1+r)\sin\alpha + L\cos\alpha \end{array} \right\}_{R_u}$$

$$OD_2 = \left\{ \begin{array}{l} -(s2+r)\sin(\pi/6)\cos\theta\cos\alpha+ \\ (s2+r)\cos(\pi/6)\sin\theta + L\cos\theta\sin\alpha \\ -(s2+r)\sin(\pi/6)\sin\theta\cos\alpha- \\ (s2+r)\cos\pi/6\cos\theta - L\sin\theta\sin\alpha \\ -(s2+r)\sin\pi/6\sin\alpha + L\cos\alpha \end{array} \right\}_{R_u} \quad (11)$$

$$OD_3 = \left\{ \begin{array}{l} -(s3+r)\sin(\pi/6)\cos\theta\cos\alpha- \\ (s3+r)\cos\pi/6\sin\theta + L\cos\theta\sin\alpha \\ -(s3+r)\sin(\pi/6)\sin\theta\cos\alpha+ \\ (s3+r)\cos\pi/6\cos\theta - L\sin\theta\sin\alpha \\ -(s3+r)\sin\pi/6\sin\alpha + L\cos\alpha \end{array} \right\}_{R_u}$$

3.2 Method Based on a Circumscribed Circle for Calculating the Safe Position

The center of the circumscribed circle formed by the triangle (D_1, D_2, D_3) has the equal distance to the three points. Thus, this center will be the intersection of the medians p_1 and p_2 of the line segment $[D_1 D_2]$ and $[D_2 D_3]$. In the coordinate frame $R_u(x, y, z)$, the equations of different straight lines are represented by:

$(D_1 D_2):$ $\quad y = a_1 x + b_1;$ $\qquad a_1 = \frac{Y_{D2} - Y_{D1}}{X_{D2} - X_{D1}}$

$(p_1):$ $\quad y = a_2 x + b_2$

is the perpendicular at $(D_1 D_2)$ and pass by $M_1(\frac{X_{D1} + X_{D2}}{2}, \frac{Y_{D1} + Y_{D2}}{2})$

thus, $a_2 = -\frac{1}{a_1}$ and $b_2 = Y_{M1} - a_2 X_{M1}$.

Similarily for $(D_2 D_3)$ and p_2, their equations are

$y = a_3 x + b_3$ and $y = a_4 x + b_4$.

Then it leads to:

$a_3 = \frac{Y_{D3} - Y_{D2}}{X_{D3} - X_{D2}}$; $M_2(\frac{X_{D3} + X_{D2}}{2}, \frac{Y_{D3} + Y_{D2}}{2})$

$b_4 = Y_{M2} - a_2 X_{M2}$; $a_4 = -\frac{1}{a_3}$

So P_{n+1} is the solution of these equations:

$$\begin{array}{l} y = a_2 x + b_2 \\ y = a_4 + b_4 \end{array} \quad (12)$$

Thus $P_{n+1} = \{\frac{b_4 - b_2}{a_2 - a_4}; a_2 \frac{b_4 - b_2}{a_2 - a_4} + b_2\}_{R_u}$

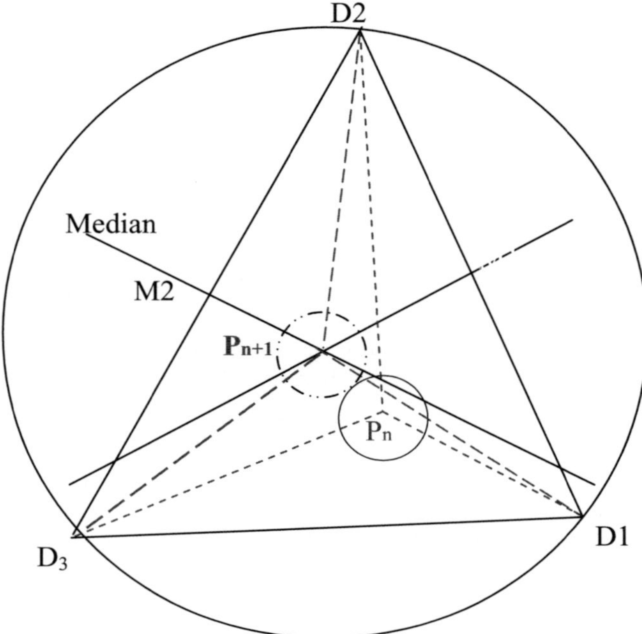

Fig. 6. Schema of the principle of the method

4 Formulation of the Control of ColoBot in the Task Space

4.1 Differential Kinematics

After determining the desired trajectory from sensor-based planning, the kinematic control of Colobot will be presented in this section. As described earlier, motion planning algorithm can only generate the trajectories represented by (x, y) coordinates. However, the Colobot has 3 degrees of freedom. So this manipulator becomes redundant with respect to the chosen task. The velocity kinematic equations are rewritten as following:

$$X = f(Q_p)$$
$$\dot{X} = \frac{\partial X}{\partial(\alpha,\phi,L)} \; \frac{\partial(\alpha,\phi,L)}{\partial Q_L} \frac{\partial Q_L}{\partial Q_P} \dot{Q}_p \tag{13}$$

Or:

$$\dot{X} = J_s J_l J_p \dot{Q}_p = J\dot{Q}_p \tag{14}$$

where $X = (x,y)^T$, $Q_L = (L_1, L_2, L_3)^T$, $Q_p = (P_1, P_2, P_3)^T$ and $J = J_s J_l J_p$ is the Jacobian matrix with relation to the three pressures in the chambers.

4.2 Resolution of the Differential Kinematics Inversion with Redundancy

In the case of a redundant manipulator with respect to a given task, the inverse kinematic problem admits infinite solutions. This suggests that redundancy can be conveniently exploited to meet additional constraints on the kinematic control problem in order to obtain greater manipulability in terms of manipulator configurations and interaction with the environment. A viable method is to formulate the problem as a constrained linear optimization problem. Work on resolved-rate control [19] proposed to use the Moore-Penrose pseudoinverse J^+ of the Jacobian matrix as:

$$\dot{Q}_p = J^+ \dot{X} = (J^T(JJ^T)^{-1})\dot{X} \tag{15}$$

In our case, however, there is a mechanical limit range for the elongation of each chamber and the corresponding pressure applied into the chamber of the ColoBot. The objective function is constructed to be included in the inverse Jacobian algorithm as the second criteria also called the null-space method [8, 14].

$$\dot{Q}_p = J^+ \dot{X} + \mu[I - J^+ J]\dot{g} \tag{16}$$

where μ is constant and g is the second criteria for optimization of the solution. This objective function evaluates the pressure applied in the chamber and the average pressure applied in the chamber. So the cost function $w(p)$ is expressed as following:

$$w(P) = \frac{1}{3} \sum_{i=1}^{3} (\frac{P_i - P_{iave}}{P_{imax} - P_{imin}})^2 \tag{17}$$

We can then minimize $w(p)$ by choosing:

$$\dot{g} = grad\,(w(P)) = \{\ \frac{\partial w}{\partial P_1}\ \frac{\partial w}{\partial P_2}\ \frac{\partial w}{\partial P_3}\ \} \tag{18}$$

5 Experimental Results

This section presents experimental results of the autonomous guidance control algorithm on an emulation platform of an operation of colonoscopy.

5.1 Implementation

Fig. 7 shows the low-level control system of ColoBot. The pressurized air comes through the compressor (1) and the general pressure is adjusted thanks to the device (2). The pressures in the chambers are controlled by three jet-pipe servovalves (3a, 3b and 3c). Three pressure sensors (4a, 4b and 4c) are connected between the servovalves and the ColoBot (5) for the pressure feedback control. Suitable drivers and amplifiers in the rack (6) were designed to amplify control signals applied to the actuator. A real-time controller is implemented through a Dspace card Real-Time Workshop with Simulink. The path planning and kinematics algorithms are expressed with a Simulink block diagram and compiled as a real-time executable under the DSP Processor of a Dspace card. The system runs at 500 Hz for a real-time control loop.

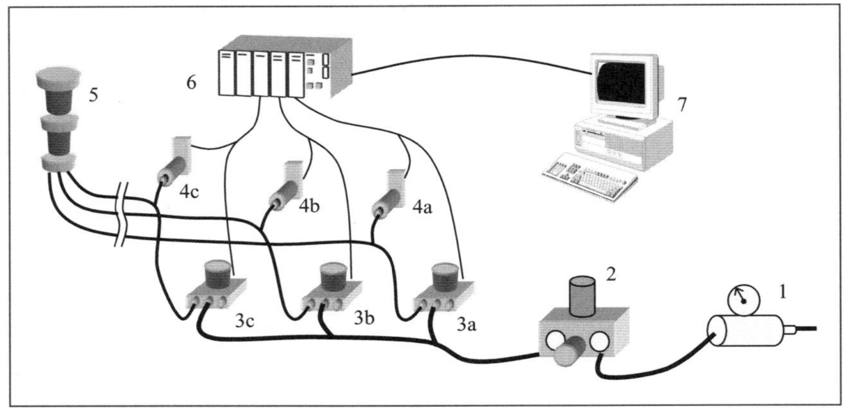

Fig. 7. Overall control system of ColoBot

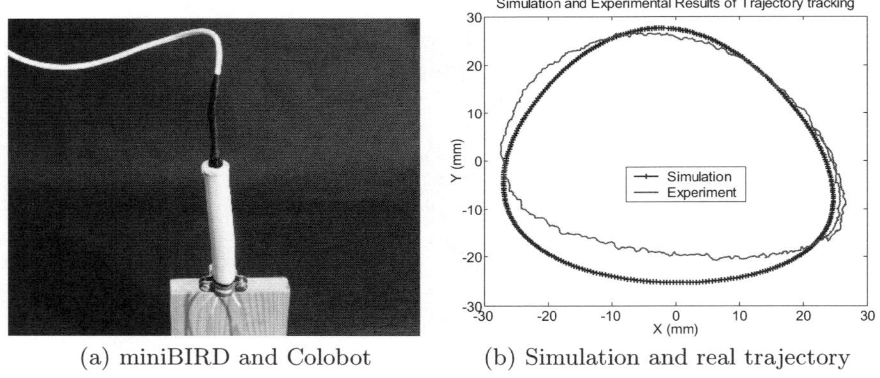

(a) miniBIRD and Colobot   (b) Simulation and real trajectory

Fig. 8. Kinematic control results

5.2   Kinematic Control

To validate the orientation control for ColoBot, the real experiment on the
Colobot as well as simulation have been carried out. The desired trajectory
is a circle with a diameter of 50 mm. The kinematic control alogithm is implemented under the assumption that displacement of ColoBot is sufficiently small
in a cycle. Thus the velocity kinematics relating state velocities and cartesian
velocities can be approximated by $\Delta X/\Delta T$ and $\Delta Q_p/\Delta T$. The figure on the
left of Fig. 9 shows the simulation control pressure of three chambers, while the
one on the right shows the actual control pressure to the three chambers. Fig.
8(b) shows the comparison of experimental results and simulation results of a
trajectory tracking. The position of the distal-end in the XOY plane for the
simulation is calculated using Eq. 7, while the position in the XOY plane for

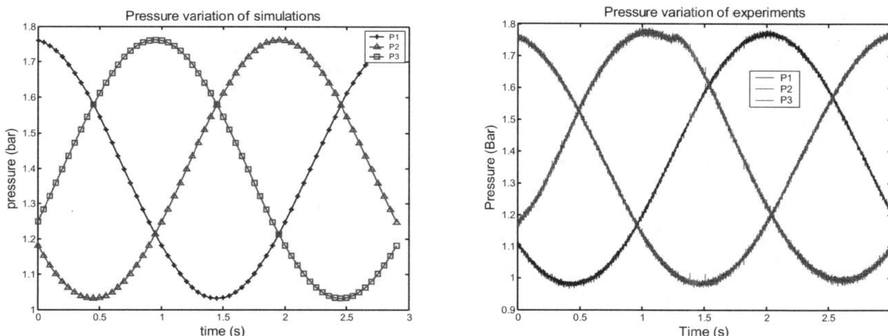

Fig. 9. Control pressures of the three chambers

real-time experiments is acquired using a 6 DOF magnetic tracker (miniBIRD) positioned at the top of the ColoBot. From this figure it shows that the two trajectories are very close.

5.3   Experimental Results on an Emulation Platform

For the purpose of testing the performance of sensor-based planning and guidance, an emulation platform (shown in Fig. 10) is designed and built to emulate the insertion process of a colonoscope in the colon. The base of this platform can manually move in two directions (X, Y) which simulates the movements of a colonoscope inside the colon. The bottom of ColoBot is fixed on that platform,

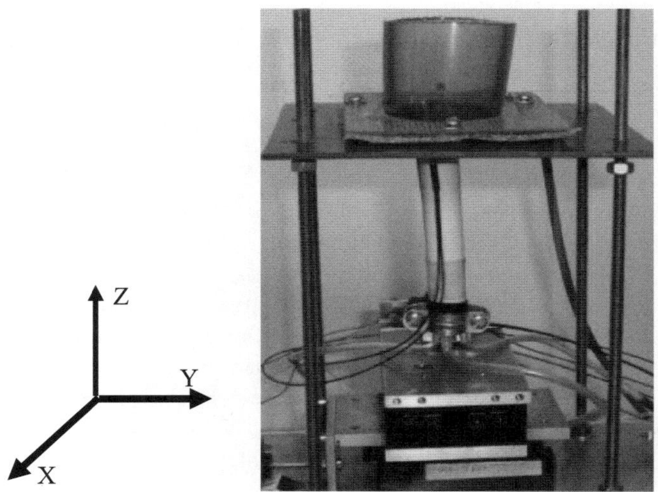

Fig. 10. Emulator for the operation of Colonoscopy

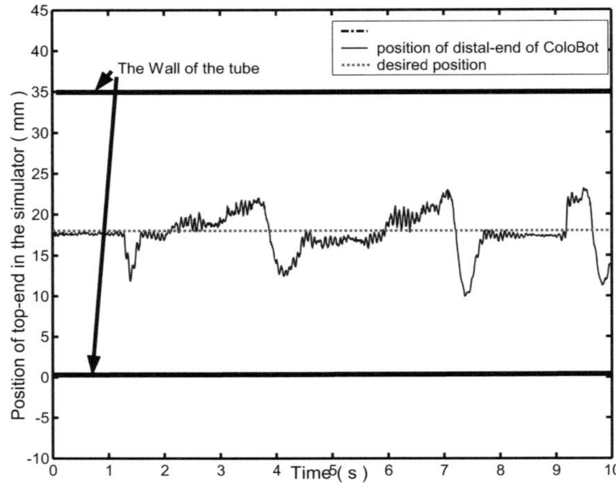

Fig. 11. Sensor-based planning and guidance control

ColoBot can thus move in the same way as does the platform. A tube with the diameter of 35mm is placed in a circle emulating the colon.

With this experimental setup, the test is to validate the sensor-based planning and guidance capability. The base of the platform moves stochastically in X and Y directions in order to emulate the movement of a colonoscope in the colon. Thus the best trajectory of distal end of ColoBot is the central axis of the tube. By calculating the trajectory as near as possilbe, ColoBot can avoid contacting with the tube wall. Fig. 11 plots the position of distal-end of ColoBot inside the tube with respect to the time. The pink line is the desired trajectory and the blue one is the calculated trajectory by using circumscribed method. It can be seen that the safe position is always around the desired position.

6   Conclusions and Future Works

This paper presented a sensor-based planning and guidance control strategy of a biologically-inspired continuum robot, called ColoBot, for semi-autonomous colonoscopy. The pneumatic driven Colobot is a silicon-based robotic manipulator which is driven through pneumatic actuators. Three optical fiber distance sensors are utilized for sensor-based safe planning under compliant and slippery environements of the colon. With the assumption that the measure plane is an ellipse, a method based on a circumscribed circle is utilized to calculate the reference position and orientation of ColoBot. While kinematic-based orientation control used this reference paths to adjust the position of ColoBot inside the colon to achieve guidance. Experiments on an emulation platform were presented to prove the feasibility of the guidance control strategy. In the near future, the proposed method will be tested in a tube-like environment which simulates the operation of colonoscopy.

Acknowledgments

The author acknowledges partial support by the European Commission through the Marie Curie Transfer of Knowledge project BRIDGET (MKTD-CD 2005029961).

References

1. Bailly, Y., Amirat, Y.: Modeling and control of a hybrid continuum active catheter for aortic aneurysm treatment. In: IEEE International Conference on Robotics and Automation, Barcelona, Spain, pp. 924–929 (2005)
2. Chen, G., Pham, M.T., Redarce, T.: Development and kinematic analysis of a silicone-rubber bending tip for colonoscopy. In: IEEE/RSJ International Conference on Intelligent Robots and Systems, Beijing, China, pp. 168–173 (2006)
3. Chen, G., Pham, M.T., Redarce, T., Prelle, C., Lamarque, F.: Design and control of an actuator for colonoscopy. In: 6th International Workshop on Research and Education in Mechatronic, Annecy, France, pp. 109–114 (2005)
4. Chirikjian, G.: Design and analysis of some nonanthropomorphic biologically inspired robots: An overview. Journal of Robotic Systems 18(12), 701–713 (2001)
5. Chirikjian, G., Burdick, J.: A modal approach to hyper-redundant manipulator kinematics. IEEE Transactions on Robotics and Automation 10(3), 343–354 (1994)
6. Dario, P., Paggetti, C., Troisfontaine, N., Papa, E., Ciucci, T., Carrozza, M., Marcacci, M.: A miniature steerable end-effector for application in an integrated system for computer-assisted arthroscopy. In: IEEE International Conference on Robotics and Automation, Albuquerque, USA, pp. 1573–1579 (1997)
7. Hirose, S.: Biologically inspired robots. Oxford University Press, Oxford (1993)
8. Hollerbach, J., Suh, K.: Redundancy resolution of manipulator through torque optimization. A.I.Memo 882, Massachussett Institute of Technology (1986)
9. Immega, G., Antonelli, K.: The KSI tentacle manipulator. In: IEEE International Conference on Robotics and Automation, Nagoya, Japan, pp. 3149–3154 (1995)
10. Jones, B., Walker, I.D.: Kinematics for multi-section continuum robots. IEEE Transactions on Robotics 22(1), 43–55 (2006)
11. Jones, B.A., Walker, I.D.: Practical kinematics for real-time implementation of continuum robots. IEEE Transactions on Robotics 22(6), 1087–1099 (2006)
12. Kapoor, A., Simaan, N., Taylor, R.: Suturing in confined spaces: Constrained motion control of a hybrid 8-dof robot. In: IEEE International Conference on Advanced Robotics, Washington, USA, pp. 452–459 (2005)
13. Lane, D., David, J., Robinson, G., O'Brien, D., Sneddon, J., Seaton, E., Elfstrom, A.: The amadeus dextrous subsea hand: Design, modeling, and sensor processing. IEEE Journal of Oceanic engineering 24(1), 96–111 (1999)
14. Nakamura, Y.: Advanced robotics, Redundancy and Optimization. Addison-Wesley, Reading (1991)
15. Piers, J., Reynaerts, D., Van Brussel, H., De Gersem, G., Tang, H.T.: Design of an advanced tool guiding system for robotic surgery. In: Proceedings of the International Conference on Robotics and Automation, Taipei, Taiwan, pp. 2651–2656 (2003)
16. Prelle, C., Lamarque, F., Mazeran, P.E.: Miniture high resolution position sensor for long stroke measurements. In: Proceedings of the 5th France-Japanese Congress and the 3rd European-Asian Congress on Mechatronics, Besançon, France, pp. 443–446 (2001)

17. Robinson, G., Davies, J.: Continuum robots - a state of the art. In: IEEE International Conference on Robotics and Automation, Detroit Michigan, USA, pp. 2849–2853 (1999)
18. Simaan, N., Taylor, R., Flint, P.: A dexterous system for laryngeal surgery- multi-backbone bending snake-like slaves for teleoperated dexterous surgical tool manipulation. In: IEEE International Conference on Robotics and Automation, New Orleans, USA, pp. 351–357 (2004)
19. Whitney, D.: Resolved motion rate control of manipulators and human prostheses. IEEE Transaction on Man-Machine systems 10(2), 47–53 (1969)

Fully-Isotropic T1R3-Type Redundantly-Actuated Parallel Manipulators

Grigore Gogu

Mechanical Engineering Research Group
Blaise Pascal University and French Institute of Advanced Mechanics
Clermont-Ferrand, France

Abstract. The paper presents T1R3-type fully-isotropic redundantly-actuated parallel manipulators (PMs) with four degrees of freedom. The mobile platform has one independent translation (T1) and three independent rotations (R3). A method is proposed for structural synthesis of fully-isotropic redundantly-actuated T1R3-type PMs based on the theory of linear transformations. A one-to-one correspondence exists between the actuated joint velocity space and the external velocity space of the moving platform. The Jacobian matrix mapping the two vector spaces of fully-isotropic T1R3-type PMs presented in this paper is 4×4 identity matrix throughout the entire workspace. The condition number and the determinant of the Jacobian matrix being equal to one, the manipulator performs very well with regard to force and motion transmission capabilities. Redundant actuation is used to obtain fully-isotropic solutions. As far as we are aware, this paper presents for the first time in the literature the use of redundancy to design fully-isotropic T1R3-type PMs.

1 Introduction

A parallel manipulator (PM) typically consists of a mobile platform connected to the base by at least two kinematic chains called legs. PMs can be considered a well-established option for many different applications of manipulation, machining, guiding, testing, control etc. With respect to serial manipulators, such mechanisms can offer advantages in terms of stiffness, accuracy, load-to-weight ratio, dynamic performances. Their disadvantages include a smaller workspace, complex command and a lower dexterity due to a high motion coupling and multiplicity of singularities inside their workspace. Uncoupled and fully-isotropic parallel manipulators can overcome these disadvantages. They have a very simple command.

Isotropy of a robotic manipulator is related to the condition number of its Jacobian matrix, which can be calculated as the ratio of the largest and the smallest singular values. A robotic manipulator is fully-isotropic if its Jacobian matrix is isotropic throughout the entire workspace, i.e., the condition number of the Jacobian matrix is equal to one. We know that the Jacobian matrix of a robotic manipulator is the matrix mapping (i) the actuated joint velocity space and the end-effector velocity space, and (ii) the static load on the end-effector and the actuated joint forces or torques. Thus, the condition number of the Jacobian matrix is a useful performance indicator characterizing the distortion of a unit sphere under this linear mapping [1], [2]. The isotropic design aims at ideal kinematic and dynamic performance of the manipulator [3]. In an

S. Lee, I.H. Suh, and M.S. Kim (Eds.): Recent Progress in Robotics, LNCIS 370, pp. 79–90, 2008.

isotropic configuration, the sensitivity of a manipulator is minimal with regard to both velocity and force errors and the manipulator can be controlled equally well in all directions. The concept of kinematic isotropy has been used as a criterion in the design of various parallel manipulators [4], [5].

Four types of PMs are identified in the literature [6]: (i) fully-isotropic PMs, if the Jacobian J is a diagonal matrix with identical diagonal elements throughout the entire workspace, (ii) PMs with uncoupled motions if J is a diagonal matrix with different diagonal elements, (iii) PMs with decoupled motions, if J is a triangular matrix and (iv) PMs with coupled motions if J is neither a triangular nor a diagonal matrix. Fully-isotropic PMs give a one-to-one mapping between the actuated joint velocity space and the external velocity space.

Redundantly-actuated PMs are mechanisms with more actuators than minimally required for doing the prescribed task. Redundant actuation in parallel manipulators can be achieved by: (i) actuating some of the passive joints within the existing legs, (ii) introducing additional actuated legs beyond the minimum necessary to actuate the manipulator and (iii) introducing some additional actuated joints within the legs beyond the minimum necessary to actuate the manipulator. We note that actuator redundancy does not affect the connectivity of the end-effector and only increase the number of actuators. In all cases M actuators have independent motion, M representing mechanism mobility. A selective choice of actuators with independent motion could have many advantages.

Redundancy in parallel manipulators is used to eliminate some singular configurations [7], to minimize the joint rates [8], to optimize the joint torques/forces [9], to increase dexterity workspace [10], stiffness [11], eigenfrequencies, kinematic and dynamic accuracy [12], to improve velocity performances [13] and both kinematic and dynamic control algorithms [14], to develop large forces in micro electro-mechanical systems [15], decoupling the orientations and the translations [16], to obtain reconfigurable platforms [17] and legs [18] or combined advantages [19]. Recently the author of this paper [20] proposed the use of redundancy for motion decoupling of parallel robots.

As far as we are aware, this paper focuses for the first time on the use of redundancy to design fully-isotropic T1R3-type PMs. Redundancy is achieved by introducing additional actuated joints within a leg beyond the minimum necessary to actuate the manipulator.

T1R3-type PMs are used in applications that require one translation (T1) and three independent rotations (3R) of the mobile platform. The major applications of this type of PMs are: flight and motion simulation [21], laparoscopic surgery [22], humanoid robotics [23].

We note that all solutions of T1R3-type PMs known in the literature [21]-[26] have coupled motions.

The general methods used for structural synthesis of parallel mechanisms can be divided into three approaches: the method based on displacement group theory [27] the methods based on screw algebra [28], [29] and the methods based on the theory of linear transformations [30], [31]. The approach proposed in this paper represents an extension of the methods founded on the theory of linear transformations to structural synthesis of redundantly actuated (Ra) fully-isotropic T1R3-type PMs. This approach integrates the new formulae of mobility, connectivity, redundancy and overconstraints of parallel manipulators proposed in [32], [33] and demonstrated via the theory of linear transformations.

The main aims of this paper are to present for the first time solutions of redundantly actuated T1R3-type PMs with decoupled and uncoupled motions along with fully-isotropic solutions and to show the structural synthesis approach that allowed us to obtain them.

2 Structural Synthesis of T1R3-Type RaPMs Via Theory of Linear Transformations

The basic kinematic structure of a T1R3-type RaPM is obtained by concatenating $k=4$ simple and complex legs G_i $(1_{Gi}\equiv0\text{-}...\text{-}n_{Gi}\equiv n)$, $i=1,...,4$. The first link (1_{Gi}) of each leg is the fixed platform (0) and the final link is the moving platform $(n_{Gi}\equiv n)$. The first $T_{Gi}+1$ joints of each leg G_i are actuated, where T_{Gi} represents the degree of redundancy of the leg G_i. In the RaPMs presented in this paper one leg has $T_{Gi}=2$. We denote by q_i and $\dot{q}_i$ the finite displacements and the velocities in the actuated joints, by $v_z=\dot{z}$, $\omega_\alpha=\dot{\alpha}$, $\omega_\beta=\dot{\beta}$ and $\omega_\delta=\dot{\delta}$ the linear and angular velocities of the mobile platform.

The structural synthesis approach proposed in this paper focuses on the synthesis of the legs G_i $(i=1,...,4)$ and is founded on the theory of linear transformations [6], [32], [33]. A mechanism L (that can be simple or complex, open or closed) achieves a linear transformation F from the finite dimensional vector space U (called joint velocity space) into the finite dimensional vector space W (called external velocity space). That is to each vector $\dot{q}$, in the joint velocity vector space U, the linear transformation F assigns a uniquely determined vector $F(\dot{q})$ in the vector space W:

$$F(\dot{q})=[J][\dot{q}] \tag{1}$$

Matrix J of this linear mapping is called Jacobian matrix. Vector $\dot{q}$ represents the joint velocity vector and it is composed of the relative velocities in the actuated joints. Vector $F(\dot{q})$ in W is the image of vector $\dot{q}$ in U under the transformation F. The image of U in W represents the range R_F of F and is called operational space. We can write

$$dim(R_F)=rank(F)=rank[J]. \tag{2}$$

Connectivity $S_{a/b}^{L}$ between two elements a and b in the mechanism L represents the number of independent relative velocities allowed by the mechanism between the two elements. It is given by the dimension of the vector space $R_{a/b}^{L}$ of relative velocities between elements a and b in the mechanism L

$$S_{a/b}^{L} = dim(R_{a/b}^{L}). \tag{3}$$

We note that the range $R_{a/b}^{L}$ is a sub-space of R^{L} and implicitly a sub-space of W. Connectivity is also called spatiality [6], [32].

The mechanism associated with a T1R3-type RaPM with uncoupled motions or fully-isotropic is denoted by Q. The existence of this mechanism involves the following conditions for the degree of mobility of the RaPM denoted by M_Q, leg mobility – M_{Gi}, the connectivity between the mobile and the fixed platforms in the RaPW - $S^Q_{n/1}$ and in each leg isolated from the RaPM - $S^{Gi}_{nGi/1}$, and for the base of the vector space of relative velocities of the mobile platform - ($R^Q_{n/1}$):

a) general conditions for any position of the mechanism (when $\dot{q}_1 \neq 0$, $\dot{q}_2 \neq 0$, $\dot{q}_3 \neq 0$, $\dot{q}_4 \neq 0$, $\dot{q}_5 = \dot{\alpha}$ and $\dot{q}_6 = \dot{\beta}$)

$$M_{Ai} \geq 4 \quad (i=1,\ldots,4), \tag{4}$$

$$4 \leq S^{Gi}_{nGi/1} \leq 6 \quad (i=1,\ldots,4), \tag{5}$$

$$M_Q > 4, \tag{6}$$

$$S^Q_{n/1} = 4 \tag{7}$$

$$(R^Q_{n/1}) = (v_z, \omega_\alpha, \omega_\beta, \omega_\delta), \tag{8}$$

b) particular condition when one actuator is locked $\dot{q}_i = 0$ $(i=1,\ldots,4)$

$$S^Q_{n/1} = 3, \tag{9}$$

$$(R^Q_{n/1}) = (\omega_\alpha, \omega_\beta, \omega_\delta), \text{ if } \dot{q}_1 = 0, \tag{10}$$

$$(R^Q_{n/1}) = (v_z, \omega_\beta, \omega_\delta), \text{ if } \dot{q}_2 = 0, \tag{11}$$

$$(R^Q_{n/1}) = (v_z, \omega_\alpha, \omega_\delta), \text{ if } \dot{q}_3 = 0, \tag{12}$$

$$(R^Q_{n/1}) = (v_z, \omega_\alpha, \omega_\beta), \text{ if } \dot{q}_4 = 0, \tag{13}$$

The characteristic point H of the moving platform is chosen in the intersecting rotation axes of the moving platform in such a way that bases ($R^Q_{n/1}$) defined by (8) and (10-13) represent the unique bases of the vector space of the relative velocities between the mobile and fixed platforms. This basis is given by:

$$(R^Q_{n/1}) = (R^{G1}_{nG1/1} \cap R^{G2}_{nG2/1} \cap R^{G3}_{nG3/1} \cap R^{G4}_{nG4/1}). \tag{14}$$

In the previous notations the kinematic chain L could be G_i, or Q and the last element n_L could be n_{Gi}, or n, with $i=1,\ldots,4$.

Connectivity (spatiality), mobility, redundancy and overconstraints of parallel manipulators are calculated by using the new formulas recently proposed in [32], [33] and demonstrated via theory of linear transformations.

The connectivity $S^Q_{n/1}$ between the mobile platform n and the reference link 1 in the parallel manipulator Q is given by the dimension of the vector space $R^Q_{n/1}$

$$S^Q_{n/1} = dim(R^Q_{n/1}). \tag{15}$$

Mobility M_{Gi} and connectivity $S_{nGi/1}^{Gi}$ of the kinematic chain associated with the leg G_i isolated from the parallel mechanism are given by:

$$M_{Gi} = \sum_{j=1}^{mGi} f_j - r_l^{Gi} \tag{16}$$

$$S_{nGi/1}^{Gi} = dim(\, R_{nGi/1}^{Gi}\,) \tag{17}$$

where m_{Gi} is the number of joints in leg G_i, f_j is the mobility of the jth joint, r_l^{Gi} the number of parameters that lost their independence in the closed loops integrated in complex leg G_i and $R_{nGi/1}^{Gi}$ is the vector space of the relative velocities between the mobile platform n_{Gi} and the reference link 1 in the kinematic chain associated with leg G_i isolated from the parallel manipulator. We note that for simple legs $r_l^{Gi} = 0$.

The mobility of the parallel manipulator Q is given by:

$$M_Q = \sum_{i=1}^{m} f_i - \sum_{i=1}^{k} S_{nGi/1}^{Gi} + S_{n/1}^{Q} - \sum_{i=1}^{k} r_l^{Gi}\,. \tag{18}$$

A leg with $M_{Gi} > S_{nGi/1}^{Gi}$ has the degree of redundancy:

$$T_{Gi} = M_{Gi} - S_{nGi/1}^{Gi} \tag{19}$$

and the parallel manipulator Q has the degree of redundancy

$$T_Q = M_Q - S_{n/1}^{Q}\,. \tag{20}$$

The degree of overconstraint of parallel manipulator Q is given by:

$$N_Q = 6q - \sum_{i=1}^{k} S_{nGi/1}^{Gi} + S_{n/1}^{Q} - \sum_{i=1}^{k} r_l^{Gi} \tag{21}$$

where q is the number of independent loops of the parallel mechanism.

If the Jacobian J is a diagonal matrix, the singular values are equal to the diagonal elements. The Jacobian of a fully isotropic mechanism has non zero identical singular values and a condition number equal to one. Consequently, a T1R3-type RaPM with uncoupled motions is fully isotropic if all diagonal elements of its Jacobian matrix are identical. In this case (1) becomes

$$\begin{bmatrix} \dot{z} & \dot{\alpha} & \dot{\beta} & \dot{\delta} \end{bmatrix}^T = \lambda [I] \begin{bmatrix} \dot{q}_1 & \dot{q}_2 & \dot{q}_3 & \dot{q}_4 \end{bmatrix}^T \tag{22}$$

where λ is the value of the diagonal elements, $[I]_{4\times4}$ is 4×4 identity matrix. Mechanism Q respecting (22) is fully-isotropic and implicitly has uncoupled motions. It achieves a homothetic transformation of coefficient λ between the velocity of the actuated joints and the velocity of the moving platform. When $\lambda = 1$ the Jacobian matrix $J = \lambda [I]_{4\times4}$ becomes 4×4 identity matrix. The condition number and the determinant of the Jacobian matrix being equal to one, the RaPM performs very well with regard to force and motion transmission.

3 T1R3-Type RaPMs with Decoupled Motions

We consider that the first $T_{Gi} +1$ joints of each leg G_i ($i=1,...,4$) are actuated (the underlined joints). The others joints of the leg are unactuated (passive joints). The RaPMs presented in this paper have G_4-leg with two redundant actuators ($T_{G4}=2$).

The simplest architecture of G_1–leg is a serial kinematic chain of type $\underline{P}S$ with a prismatic (P) and a spherical joint (S). This architecture has $m_{G1}=2$, $\sum_{j=1}^{2} f_j =4$, ($R_{n/1}^{Q}$)=(v_z, ω_α, ω_β, ω_δ), $M_{G1}= S_{nG1/1}^{G1}=4$, $r_{G1}=0$ and $T_{G1}=0$.

The simplest architecture of G_2–leg is a serial kinematic chain of type $\underline{P}'R'RR$ with a prismatic (P) and a three revolute joints (R) having orthogonal and concurrent axes. This architecture has $m_{G2}=4$, $\sum_{j=1}^{4} f_j =4$, $r_{G2}=0$, ($R_{n/1}^{Q}$)= (v_z, ω_α, ω_β, ω_δ), $M_{G2}= S_{nG2/1}^{G2}=4$ and $T_{G2}=0$.

The simplest architectures of serial kinematic chain G_3 have $m_{G3}=4$, $\sum_{j=1}^{4} f_j =6$, $\left(R_{nG3/1}^{G3}\right) = \left(v_x,v_y,v_z,\omega_\alpha,\omega_\beta,\omega_\delta\right)$, $M_{G3}= S_{nG3/1}^{G3}=6$, $r_{G3}=0$ and $T_{G3}=0$. These architectures may be of type $\underline{P}'R'R'S$ and $\underline{P}'R'PS$. Just revolute (R), prismatic (P) and spherical joint (S) are used in these solutions in which two consecutive revolute and prismatic joints have parallel or perpendicular axes.

The simplest architecture of kinematic chain G_4 is of type $\underline{RRR}UPU$ with $m_{G4}=8$, $\left(R_{nG4/1}^{G4}\right) = \left(v_x,v_y,v_z,\omega_\alpha,\omega_\beta,\omega_\delta\right)$, $\sum_{j=1}^{8} f_j =8$, $M_{G4}=8$, $S_{nG4/1}^{G4}=6$, $r_{G4}=0$ and $T_{G4}=2$. This leg integrates two redundant actuators with actuated velocities $\dot{q}_5$ and $\dot{q}_6$.

In the previous notations, the revolute and prismatic joints marked by apostrophe have parallel axes/directions and v_x,v_y,v_z represent the linear velocities of point $H \equiv O_0$ situated on the distal link ($n \equiv 9 \equiv 3_A \equiv 5_B \equiv 5_C \equiv 9_D$) of each leg isolated from the parallel mechanism.

We can see that legs G_2 and G_4 are actuated by rotating motors and the legs G_1 and G_3 by linear motors.

By connecting the legs G_i ($i=1,...,4$) to the output link $n \equiv n_{Gi}$ ($i=1,...,4$) we obtain the architectures of RaPWs with uncoupled motions presented in Fig. 1. To simplify the notations of links e_{Gi} ($i=1,...,4$ and $e=1,...,n$) by avoiding the double index in Fig. 1 and the following figures, we have denoted by e_A the elements belonging to leg G_1 ($e_A \equiv e_{G1}$), by e_B, e_C and e_D the elements of legs G_2 ($e_B \equiv e_{G2}$), G_3 ($e_C \equiv e_{G3}$) and G_4 ($e_D \equiv e_{G4}$). Link $n \equiv 9 \equiv 3_A \equiv 5_B \equiv 5_C \equiv 9_D$ is the moving platform.

The solution in Fig. 1 has: $m=18$, $k=4$, $q=3$, $\sum_{j=1}^{18} f_j =22$, $\sum_{i=1}^{4} S_{nGi/1}^{Gi} =20$, $\sum_{i=1}^{4} r_i^{Gi} =0$, $\left(R_{n/1}^{Q}\right)= \left(v_z,\omega_\alpha,\omega_\beta,\omega_\delta\right)$, $M_Q=6$, $S_{n/1}^{Q}=4$, $T_Q=2$ and $N_Q=0$. This solution two degrees of redundancy ($T_Q=2$) and is non overconstrained ($N_Q=0$).

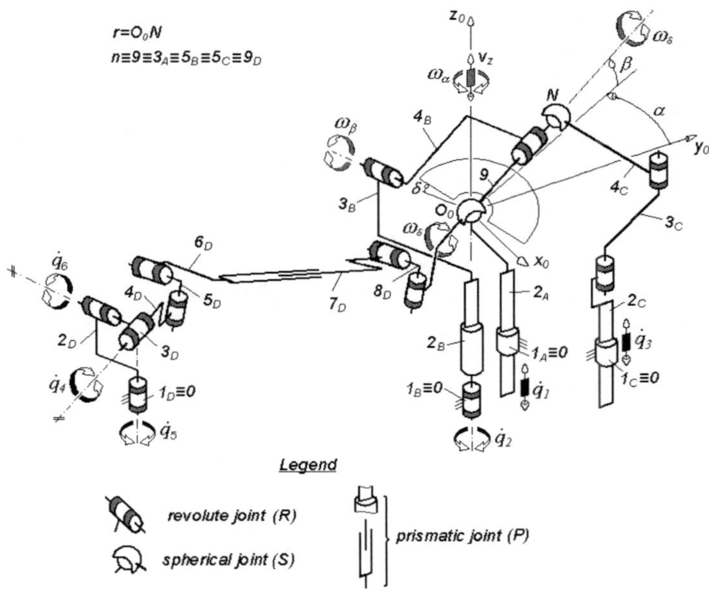

Fig. 1. Basic kinematic structure of non overconstrained T1R3-type RaPM with decoupled motions

For the solution in Fig. 1, the linear mapping (1) becomes:

$$
\begin{bmatrix} \dot{z} \\ \dot{\alpha} \\ \dot{\beta} \\ \dot{\delta} \end{bmatrix} = \begin{bmatrix} 1 & 0 & 0 & 0 \\ 0 & 1 & 0 & 0 \\ -a & 0 & a & 0 \\ 0 & 0 & 0 & 1 \end{bmatrix} \begin{bmatrix} \dot{q}_1 \\ \dot{q}_2 \\ \dot{q}_3 \\ \dot{q}_4 \end{bmatrix}, \quad a = \frac{1}{r \cos \beta}, \tag{23}
$$

where r is the length of the output link ($r=O_0N$). Point N represents the centre of the spherical joint of G_3-leg (Fig. 1). We can see that $\dot{z} = \dot{q}_1$, $\dot{\alpha} = \dot{q}_2$, $\dot{\beta} = a(\dot{q}_3 - \dot{q}_1)$, $\dot{\delta} = \dot{q}_4$ if

$$
q_5 = \alpha, \tag{24}
$$

$$
q_6 = \beta. \tag{25}
$$

Eqs. (24) and (25) represent the homokinetic condition for the double universal joint with telescopic shaft integrated in the leg G_4 giving $\dot{\delta} = \dot{q}_4$. To comply with this condition, the axis of the actuated rotation q_4 must be always parallel to the axis of the rotation δ of the moving platform.

4 T1R3-Type RaPMs with Uncoupled Motions

The leg G_3 in the RaPM with decoupled motions presented in the previous section connects the fixed and the mobile platform. If we connect leg G_3 between the first kinematic element of leg G_1 and the mobile platform ($1_C \equiv 2_A$) we can obtain PMs with the four motions of the mobile platform uncoupled (Fig. 2). For this solution the linear mapping (1) becomes:

$$\begin{bmatrix} \dot{z} \\ \dot{\alpha} \\ \dot{\beta} \\ \dot{\delta} \end{bmatrix} = \begin{bmatrix} 1 & 0 & 0 & 0 \\ 0 & 1 & 0 & 0 \\ 0 & 0 & a & 0 \\ 0 & 0 & 0 & 1 \end{bmatrix} \begin{bmatrix} \dot{q}_1 \\ \dot{q}_2 \\ \dot{q}_3 \\ \dot{q}_4 \end{bmatrix}, \quad a = \frac{1}{r\cos\beta}. \tag{26}$$

The third actuator of this solution with uncoupled motions is not attached to the fixed base. We can see that $\dot{z} = \dot{q}_1$, $\dot{\alpha} = \dot{q}_2$, $\dot{\beta} = a\dot{q}_3$ and $\dot{\delta} = \dot{q}_4$ if $q_5 = \alpha$ and $q_6 = \beta$.

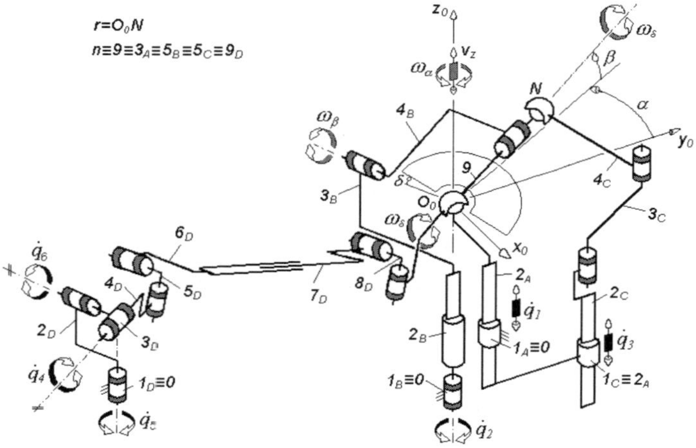

Fig. 2. Kinematic structure of non overconstrained T1R3-type RaPM with uncoupled motions

5 Fully-Isotropic T1R3-Type RaPMs

Fully-isotropic T1R3-type RaPMs can be obtained from the solution in Fig. 2 by using a leg G_3 actuated by a rotating motor. The fully-isotropic architecture presented in Fig. 3 uses the complex leg G_3 integrating planar extensible double parallelogram mechanism. The axes of the six revolute pairs of this mechanism are parallel to xy-plane. The leg G_3 is actuated by a rotating motor fixed on the base and integrated in a parallelogram loop.

The solution in Fig. 3a is derived directly from the solution with uncoupled motions in Fig. 2 by replacing the prismatic joint by the planar extensible double parallelogram mechanism. The solution in Fig 3b is derived from the solution in Fig. 3a by eliminating the spherical joint from G_1-leg and connecting it directly to G_2-leg by a revolute joint.

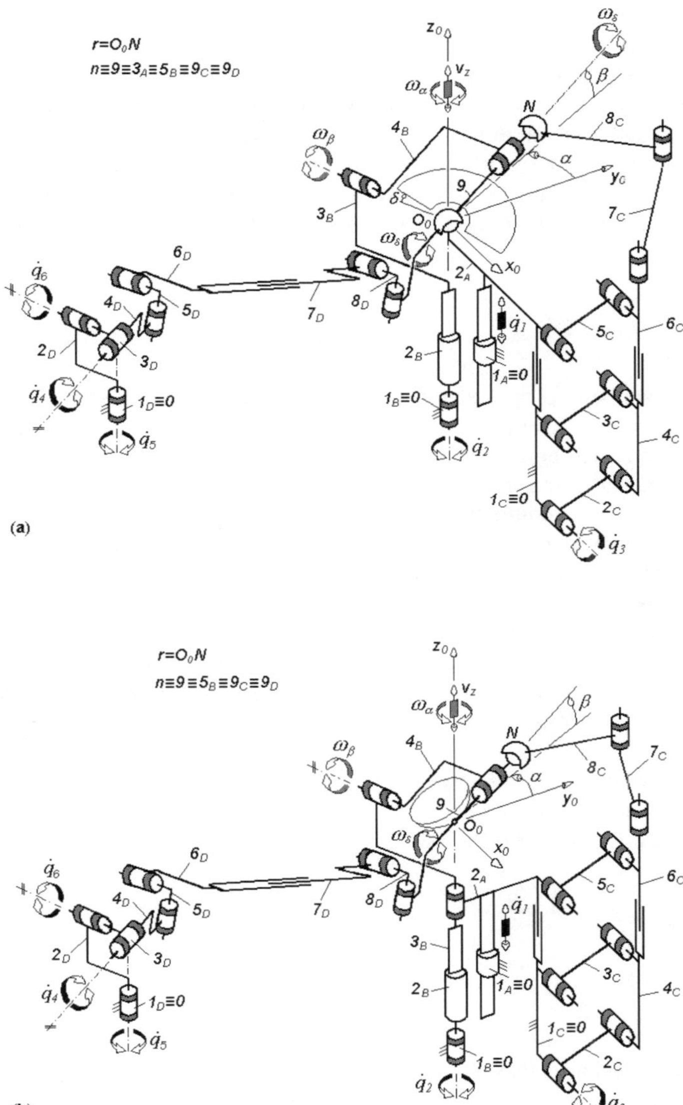

Fig. 3. Kinematic structures of overconstrained T1R3-type fully-isotropic RaPMs

Non overconstarined solutions may be derived from the solutions in Fig. 3 by introducing three idle mobilities (one translation and two rotations) in each parallelogram loop. Various solutions exist to introduce these idle mobilities, for example by replacing, in each parallelogram loop, two revolute joints by a cylindrical and a spherical joint.

The Jacobian matrix of the solutions presented in Fig. 3 is the *4×4* identity matrix throughout their entire workspace. Advantages of these fully-isotropic solutions include: (i) high stiffness enabling orientation of large loads with high angular velocities and accelerations, (ii) simplification of direct and inverse kinematic models, (iii) actuators situated on the fixed base. Moreover, solution in Fig. 3b gives the possibility to position the working device at the geometric centre of rotation thereby reducing inertia. The workspace of these solutions must be correlated with the angular capability of the homokinetic double universal joint and the translational capability of the telescopic shaft.

6 Conclusions

An approach has been proposed for structural synthesis of T1R3-type fully-isotropic redundantly actuated parallel manipulators. These solutions are derived from redundantly actuated non overconstrained solution with decoupled motions and simple legs. The Jacobian matrix mapping the joint and the operational vector spaces of the fully-isotropic redundantly actuated parallel manipulators presented in this paper is the identity matrix throughout the entire workspace. Examples of solutions with decoupled and uncoupled motions along with fully-isotropic solutions are also presented in this paper to illustrate the proposed approach. As far as we are aware, this paper presents for the first time in the literature the use of redundancy to design T1R3-type fully-isotropic parallel manipulators as well as fully-isotropic solutions. They are obtained by using an original method for structural synthesis founded on the theory of linear transformations.

Acknowledgments. This work was supported by the CNRS (The French National Council of Scientific Research) in the framework of the projects ROBEA-MAX (2002-2004) and ROBEA-MP2 (2004-2006).

References

[1] Salsbury, J.K., Craig, J.J.: Articulated hands: force and kinematic issues. Int. J. of Robotics Research 1(1), 1–17 (1982)

[2] Angeles, J.: Fundamentals of Robotic Mechanical Systems: Theory, Methods, and Algorithms. Springer, New York (1987)

[3] Fattah, A., Hasan Ghasemi, A.M.: Isotropic design od spatial parallel manipulators. Int. J. of Robotics Research 21(9), 811–824 (2002)

[4] Zanganeh, K.E., Angeles, J.: Kinematic isotropy and the optimum design of parallel manipulators. Int. J. of Robotics Research 16(2), 185–197 (1997)

[5] Tsai, K.Y., Huang, D.: The design of isotropic 6-DOF parallel manipulators using isotropy generators. Mech. Mach. Theory 38, 1199–1214 (2003)

[6] Gogu, G.: Structural synthesis of fully-isotropic translational parallel robots via theory of linear transformations. European Journal of Mechanics- A/Solids 23(6), 1021–1039 (2004)

[7] Wang, J., Gosselin, C.M.: Singularity analysis and design of kinematically redundant parallel mechanism. In: DETC 2002. Proceedings of ASME Design Engineering Technical Conferences (2002)

[8] Merlet, J.P.: Les robots parallèles, 2nd edn. Hermès, Paris (1997)

[9] Nokleby, N.S., Fischer, R., Podhorodeski, R.P., Firmani, F.: Force capabilities of redundantly-actuated parallel manipulators. Mech. Mach. Theory 40, 578–599 (2005)

[10] Marquet, F., Krut, S., Company, O., Pierrot, F.: ARCHI: a redundant mechanism for machining with unlimited rotation capacities. In: Proceedings of IEEE International Conference on Advanced Robotics, pp. 683–689 (2001)

[11] Chakarov, D.: Study of the antagonistic stiffness of parallel manipulators with actuation redundancy. Mech. Mach. Theory 39, 583–601 (2004)

[12] Valasek, M., Sika, Z., Bauma, V., Vampola, T.: The innovative potential of redundantly actuated PKM. In: Neugebauer, R. (ed.) Parallel Kinematic Machines in Research and Practice, The 4th Chemnitz Parallel Kinematics Seminar, pp. 365–383 (2004)

[13] Krut, S., Company, O., Pierrot, F.: Velocity performance indices for parallel mechanisms with actuation redundancy. Robotica 22, 129–139 (2004)

[14] Cheng, H., Yiu, Y.-K., Li, Z.: Dynamics and control of redundantly actuated parallel manipulators. IEEE Trans. Mechatronics 8(4), 483–491 (2003)

[15] Mukherjee, S., Murlidhar, S.: Massively parallel binary manipulators. Transactions of ASME, Journal of Mechanical Design 123, 68–73 (2001)

[16] Jin, Y., Chen, I.M., Yang, G.: Structure synthesis and singularity analysis of a parallel manipulator based on selective actuation. In: Proceedings of IEEE Conf. Robotics and Automation, pp. 4533–4538 (2004)

[17] Mohamed, M.G., Gosselin, C.M.: Design and analysis of kinematically redundant parallel manipulators with configurable platforms. IEEE Trans. Robot. Autom. 21(3), 277–287 (2005)

[18] Fischer, R., Podhorodeski, R.P., Nokleby, S.B.: Design of a reconfigurable planar parallel manipulator. Journal of Robotic Systems 21(12), 665–675 (2004)

[19] Nahon, M.A., Angeles, J.: Reducing the effect of shocks using redundant actuation. In: Proceedings of IEEE Conf. Robotics and Automation, pp. 238–243 (1991)

[20] Gogu, G.: Fully-isotropic redundantly-actuated parallel manipulators with five degrees of freedom. In: Husty, M., Schröcker, H.P. (eds.) Proceedings of the First European Conference on Mechanism Science (2006)

[21] Zlatanov, D., Gosselin, C.: A family of new parallel architectures with four degrees of freedom. In: Park, F.C., Iurascu, C. (eds.) Computational Kinematics, pp. 57–66 (2001)

[22] Zoppi, M., Zlatanov, D., Gosselin, C.M.: Analytical kinematics models and special geometries of a class of 4-DOF parallel mechanisms. IEEE Transactions on Robotics 21(6), 1046–1055 (2005)

[23] Lenarčič, J., Stanišić, M.M., Parenti-Castelli, V.: A 4-DOF parallel mechanism simulating the movement of the human sterum-clavicle-scapula complex. In: Lenarčič, J., Stanišić, M.M., (eds.) Advances in Robot Kinematics, pp. 325–332. Kluwer Academic Publishers, Dordrecht (2002)

[24] Huang, Z., Li, Q.C.: General methodology for type synthesis of symmetrical lower-mobility parallel manipulators and several novel manipulators. The International Journal of Robotics Research 21(2), 131–145 (2002)

[25] Kong, X., Gosselin, C.M.: Type synthesis of a 4-DOF SP-equivalent parallel manipulators: a virtual chain approach. In: CK 2005. Proceedings of International Workshop on Computational Kinematics (2005)

[26] Gallardo-Alvarado, J., Rico-Martinez, J.M., Alici, G.: Kinematics and singularity analysis of a 4-dof parallel manipulator using screw theory. Mech. Mach. Theory 41, 1048–1061 (2006)

[27] Hervé, J.M.: Design of parallel manipulators via the displacement group. In: Proceedings of 9th World Congress on the Theory of Machines and Mechanisms, pp. 2079–2082 (1995)

[28] Frisoli, A., Checcacci, D., Salsedo, F., Bergamasco, M.: Synthesis by screw algebra of translating in-parallel actuated mechanisms. In: Lenarčič, J., Stanišić, M.M., (eds.) Advances in Robot Kinematics, pp. 433–440. Kluwer Academic Publishers, Dordrecht (2000)

[29] Kong, X., Gosselin, C.M.: Type synthesis of 3-DOF spherical parallel manipulators based on screw theory. Transactions of the ASME, Journal of Mechanical Design 126, 101–108 (2004)

[30] Gogu, G.: Structural synthesis of fully-isotropic translational parallel robots via theory of linear transformations. European Journal of Mechanics- A/Solids 23(6), 1021–1039 (2004)

[31] Gogu, G.: Structural synthesis of fully-isotropic parallel robots with Schönflies motions via theory of linear transformations and evolutionary morphology. European Journal of Mechanics / A -Solids 26(2), 242–269 (2007)

[32] Gogu, G.: Mobility and spatiality of parallel robots revisited via theory of linear transformations. European Journal of Mechanics / A -Solids 24(5), 690–711 (2005)

[33] Gogu, G.: Mobility criterion and overconstraints of parallel manipulators. In: Proceedings of the International Workshop on Computational Kinematics, Paper 22-CK (2005)

Early Reactive Grasping with Second Order 3D Feature Relations

Daniel Aarno, Johan Sommerfeld, Danica Kragic[1], Nicolas Pugeault[2],
Sinan Kalkan, Florentin Wörgötter[3], Dirk Kraft, and Norbert Krüger[4]

[1] Centre for Autonomous Systems, Computational Vision and Active Perception, School of Computer Science and Communication, Royal Institute of Technology, 10044 Stockholm, Sweden
`{bishop, johansom, dani}@kth.se`
[2] School of Informatics, University of Edinburgh, 2 Buccleuch Place, Edinburgh EH8 9LW, United Kingdom
`npugeaul@inf.ed.ac.uk`
[3] Bernstein Center for Computational Neuroscience, University of Goettingen, Bunsenstr. 10, 37073 Goettingen, Germany
`{sinan, worgott}@bccn-goettingen.de`
[4] The Maersk Mc-Kinney Moller Institute, University of Southern Denmark, Campusvej 55, 5230 Odense M, Denmark
`{kraft, norbert}@mmmi.sdu.dk`

Summary. One of the main challenges in the field of robotics is to make robots ubiquitous. To intelligently interact with the world, such robots need to understand the environment and situations around them and react appropriately, they need context-awareness. But how to equip robots with capabilities of gathering and interpreting the necessary information for novel tasks through interaction with the environment and by providing some minimal knowledge in advance? This has been a longterm question and one of the main drives in the field of cognitive system development.

The main idea behind the work presented in this paper is that the robot should, like a human infant, learn about objects by interacting with them, forming representations of the objects and their categories that are grounded in its embodiment. For this purpose, we study an early learning of object grasping process where the agent, based on a set of innate reflexes and knowledge about its embodiment. We stress out that this is not the work on grasping, it is a system that interacts with the environment based on relations of 3D visual features generated trough a stereo vision system. We show how geometry, appearance and spatial relations between the features can guide early reactive grasping which can later on be used in a more purposive manner when interacting with the environment.

1 Introduction

For a robot that has to perform tasks in a human environment, it is necessary to be able to learn about objects and object categories. It has been recognized recently that grounding in the embodiment of a robot, as-well as continuous learning is required to facilitate learning of objects and object categories [4, 21]. The idea is that robots will not be able to form useful categories or object representations by only being a passive observer of its environment. Rather a robot should, like a human infant, learn

S. Lee, I.H. Suh, and M.S. Kim (Eds.): Recent Progress in Robotics, LNCIS 370, pp. 91–105, 2008.

about objects by interacting with them, forming representations of the objects and their categories that are grounded in its embodiment.

Central to the approach are three almost axiomatic assumptions, which are strongly correlated. These also represent the building blocks of our approach toward creating a cognitive artificial agent:

- Objects and Actions are inseparably intertwined; Entities ("things") in the world of a robot (or human) will only become semantically useful "objects" through the action that the agent can/will perform on them. This forms so-called Object-Action Complexes (named OACs) which are the building blocks of cognition.
- Cognition is based on recurrent processes involving nested feedback loops operating on, contextualizing and reinterpreting object-action complexes. This is done through actively closing the perception-action cycle.
- A unified measure of success and progress can be obtained through minimization of contingencies which an artificial cognitive system experiences while interacting with the environment or other agents, given the drives of the system.

To demonstrate the feasibility of our approach, we aim at building a robot system that step by step develop increasingly advanced cognitive capabilities. In this paper, we demonstrate our initial efforts towards this goal by designing a scenario for manipulation and grasping of objects.

One of the most basic interactions that can occur between a robot and an object is for the robot to push the object, i.e. to simply make a physical contact. Already at this stage, the robot should be able to form two categories: physical and non-physical objects, where a physical object is categorized by the fact that interaction forces occur. A higher level interaction between the robot and an object would exist if the robot was able to *grasp* the object. In this case, the robot would gain actual physical control over the object and having the possibility to perform controlled actions on it, such as examining it from other angles, weighing it, placing it etc. Information obtained during this interaction can then be used to update the robots representations about objects and the world. Furthermore, the successfully performed grasps can be used as ground truth for future grasp refinement, [4].

In this paper, we are interested in investigating an initial "reflex-like" grasping strategy that will form a basis for a cognitive robot system that, at the first stage, acquires knowledge of objects and object categories and is able to further refine its grasping behavior by incorporating the gained object knowledge, [1]. The grasping strategy does not require *a-priori* object knowledge, and it can be adopted for a large class of objects. The proposed reflex-like grasping strategy is based on second order relations of multi-modal visual features descriptors, called *spatial primitives*, that represent object's geometric information, e.g. 3D pose (position and orientation) as well as its appearance information, e.g. color and contrast transition etc. [9], see Fig. 1. Co–planar tuples of the spatial primitives allow for the definition of a plane that can be associated to a grasp hypothesis. In addition, these local descriptors are part of semi-global collinear groups [18]. Furthermore, the color information (by defining co–colority in addition to co–planarity of primitive pairs) can be used to further improve the definition of grasp hypotheses. In this paper, we employ the structural richness of the descriptors in terms

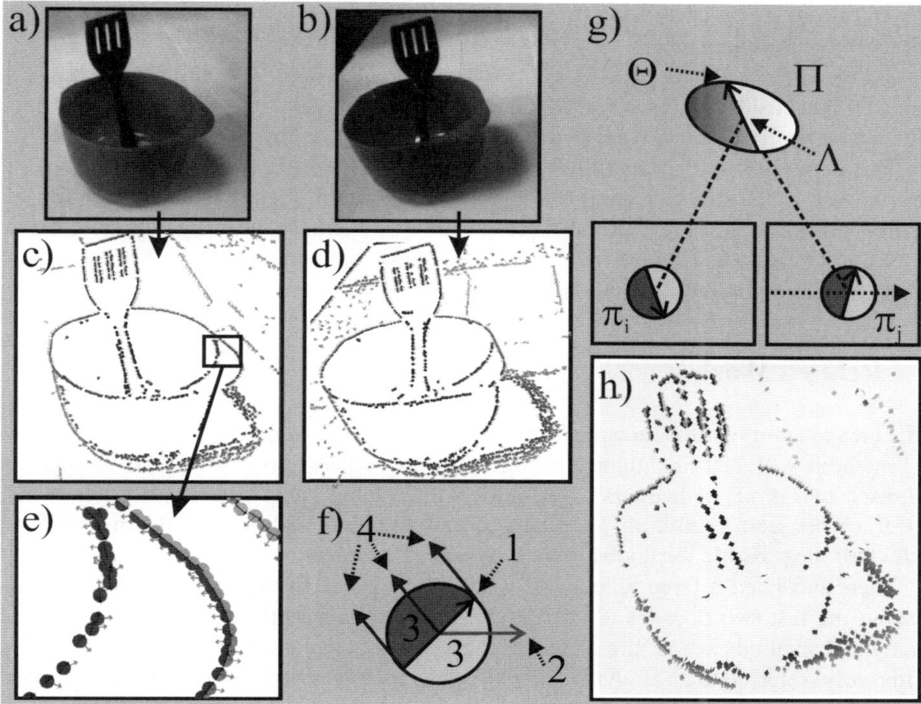

Fig. 1. Illustration of the vision module. a) and b) shows the images captured by the left and right cameras (respectively); c) and d) show the primitives extracted from these two images; in e) a detail of the primitive extraction is shown; f) illustrates the schematic representation of a primitive, where 1. represents the orientation, 2. the phase, 3. the color and 4. the optical flow. g) from a stereo–pair of primitives (π_i, π_j) we reconstruct a 3D primitive Π, with a position in space Λ and an orientation Θ; h) shows the resulting 3D primitives reconstructed for this scenario.

of their geometry and appearance as well as the structural relations co–linearity, co–planarity and co–colority to derive a set of grasping options from a stereo image.

We note that the purpose of this work is not to develop yet another grasping strategy for a specific setting, but rather to provide low-level grasping reflexes that can be used to generate successful grasps on arbitrary objects. These grasping reflexes are part of a larger framework on cognitive robotics where a robot is equipped only with a set of innate grasps which are used to develop more complex object manipulation abilities through interaction and reinforcement so that 1) more complex feature relations become associated to more precise and successful grasps, and 2) object knowledge becomes acquired and used to further refine the grasping process. We also have to stress out that no scene segmentation is performed, since the system does not even have a concept of an object to start with. In short, the contributions of our work are the generation of a set of grasp suggestions on unknown objects based on visual feedback, grouping of visual primitives for decreasing the size of the grasps and evaluation of grasps using the GraspIt! environment, [11].

In this work, "kitchen-type" objects such as cups, glasses, bowls and various kitchen utensils are considered. However, our algorithm is not designed for specific object classes but can be applied for any rigid object that can be described by edge–like structures.

This paper is organized as follows. In Section 2, we shortly review the related work and in Section 3 give a general overview of the system. Details about extraction of spatial primitives are presented in Section 4 and elementary grasping actions defined in Section 5. Results of the experimental evaluation are summarized in Section 9 and plans for future research outlined in Section 10.

2 Related Work

The idea to learn or refine grasping strategies is not new. Kamon *et al.* combined heuristic methods with learning algorithms to learn how to select good grasps [7]. Rössler *et al.* used two levels of learners to learn local and global grasp criteria [19], where the local learner learns about the local structure of an object and the global learner learns which of the possible local grasps are best given the object.

There has been a large amount of work presented in the area of robotic grasping during the last two decades [2]. However, much of this work has been dealing with analytical methods where the shape of the objects being grasped is known *a-priori*. This work, referred to as *analytical methods*, has focused primarily on computing grasp stability based on force and form-closure properties or contact-level grasps synthesis based on finding a fixed number of contact locations with no regard to hand geometry, [2],[3]. This problem is important and difficult mainly because of the high number of DOFs involved in grasping arbitrary objects with complex hands. Another important research area is grasp planning without detailed object models where sensor information such as computational vision is used to extract relevant features in order to compute suitable grasps, [5, 20, 14]. In this paper, we denote this approach as *sensor-driven*.

Related to our work, we have to mention systems that deal with automatic grasp synthesis and planning, [12],[17],[13],[15]. This work concentrates on automatic generation of stable grasps given assumptions about the shape of the object and robot hand kinematics. Example of assumptions may be that the full and exact pose of the object is known in combination with its (approximate) shape, [12]. Another common assumption is that the outer contour of the object can be extracted and a planar grasp applied, [13]. Taking into account both the hand kinematics as well as some *a-priori* knowledge about the feasible grasps has been acknowledged as a more flexible and natural approach towards automatic grasp planning [16],[12]. [16] studies methods for adapting a given prototype grasp of one object to another object. The method proposed in [12] presents a system for automatic grasp planning for a Barrett hand [6] by modeling an object as a set of shape primitives, such as spheres, cylinders, cones and boxes in a combination with a set of rules to generate a set of grasp starting positions and pregrasp shapes.

One difference between the analytical and sensor-driven approaches is that the former tend to use complex hands with many DOFs, while the latter use simple ones such as parallel yaw-grippers. One reason for this is that if the reconstruction of the object's shape is not very accurate, using a complex gripping device does not necessarily facilitate grasping performance. For sensor-driven approaches it is also very common to perform only planar grasps where all the contacts between the fingers and the object are confined to a plane. As an example, objects are placed on a table and grasped from above. This simplifies both the vision problem, since only the outer boundary of the object in the image plane has to be estimated, as well as the grasp planning by constraining the search space.

The main differences of our work compared to the abovementioned work are the following:

- We rely on 3D information based on three dimensional primitives extracted online. This allows us to compute arbitrary grasping directions compared to only planar grasps considered in, e.g. [13].
- The structural richness of the primitives (geometric and appearance based information, collinear grouping) allows for an efficient reduction of grasping hypotheses while keeping relevant ones.
- Our system focuses on generating a certain percentage of successful grasps on arbitrary objects rather than high quality grasps on a constrained set of objects. We will show that with our representations we are able to extract a sufficient number of successful grasping options to be used as initiator of learning schemes aiming at more sophisticated grasping strategies.

3   System Overview

The work presented in this paper serves as a building block for the development of a cognitive robot system. The robot platform considered is comprised of a set of sensors and actuators. The minimum requirements necessary to realize the work presented in this paper is that the sensors are able to deliver a set of visual primitives (section 4) and the configuration of the actuators. The required actuator is a manipulator, comprised of a robotic arm and a gripper device. In this context the term sensor is not necessarily related to a real physical sensing device, but rather an abstract measurement delivered to the system, possibly after performing computations on data sampled from a physical sensor.

The complete system is outlined in Fig. 2. In this paper we are interested in developing grasping reflexes. A grasping reflex is triggered by the vision system. The vision system continuously computes the spatial primitives described in section 4 which are feed as sensor input to the set of reflexes and to the cognitives system. If the grasping reflex has not been inhibited by the cognitive system and the sensor stimuli is strong enough, i.e. there are sufficiently many spatial primitives visible, the grasping reflex is performed. This reflex behavior computes a set of possible grasps and tries to perform them. Each grasp evaluated results in a reinforcement signal which can be used by the cognitive system to update its representation of the world. The following two sections describe the spatial primitives and the rules for generating the grasping actions.

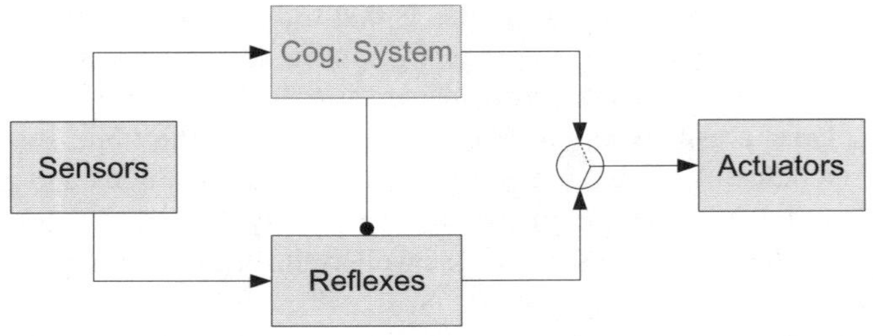

Fig. 2. System overview

4   Spatial Primitives

The image processing used in this paper is based on multi-modal visual primitives
[10, 9, 18]. First, 2D primitives are extracted sparsely at points of interest in the image (in this case contours) and encode the value of different visual operators (hereby
referred to as *visual modalities*) such as local orientation, phase, color (on each side
of the contour) and optical flow (see Fig. 1.d, 1.e and 1.f). In a second step, the 2D
primitives become extended to the spatial primitives used in this work. After finding
correspondences between primitives in the left and right image, we reconstruct a spatial
primitive, (see Fig. 1.g) that has the following components, (for details see [8, 18]):

$$\Pi = \{\Lambda, \Theta, \Omega, (\mathbf{c}_l, \mathbf{c}_m, \mathbf{c}_r)\},$$

where Λ is the 3D position; Θ is the 3D orientation; Ω is the phase (i.e., contrast transition); and, $(\mathbf{c}_l, \mathbf{c}_m, \mathbf{c}_r)$ is the representation of the color of the spatial primitive, corresponding to the left ($\mathbf{c}_l$), the middle ($\mathbf{c}_m$) and the right side ($\mathbf{c}_r$).

The sparseness of the primitives allows to formulate three *relations* between primitives that are crucial in our context:

- *Co–planarity:*
  Two spatial primitives Π_i and Π_j are co–planar iff their orientation vectors lie on
  the same plane, i.e.:

  $$cop(\Pi_i, \Pi_j) = 1 - |\mathbf{proj}_{\Theta_j \times \mathbf{v}_{ij}}(\Theta_i \times \mathbf{v}_{ij})|,$$

  where $\mathbf{v}_{ij}$ is defined as the vector $(\Lambda_i - \Lambda_j)$, and $\mathbf{proj}_{\mathbf{u}}(\mathbf{a})$ is defined as:

  $$\mathbf{proj}_{\mathbf{u}}(\mathbf{a}) = \frac{\mathbf{a} \cdot \mathbf{u}}{\parallel \mathbf{u} \parallel^2}\mathbf{u}. \tag{1}$$

  The co–planarity relation is illustrated in Fig. 3(a).
- *Collinear grouping (i.e., collinearity):*
  Two spatial primitives Π_i and Π_j are collinear (i.e., part of the same group) iff they
  are part of the same contour. Due to uncertainty in 3D reconstruction process, in
  this work, the collinearity of two spatial primitives Π_i and Π_j is computed using

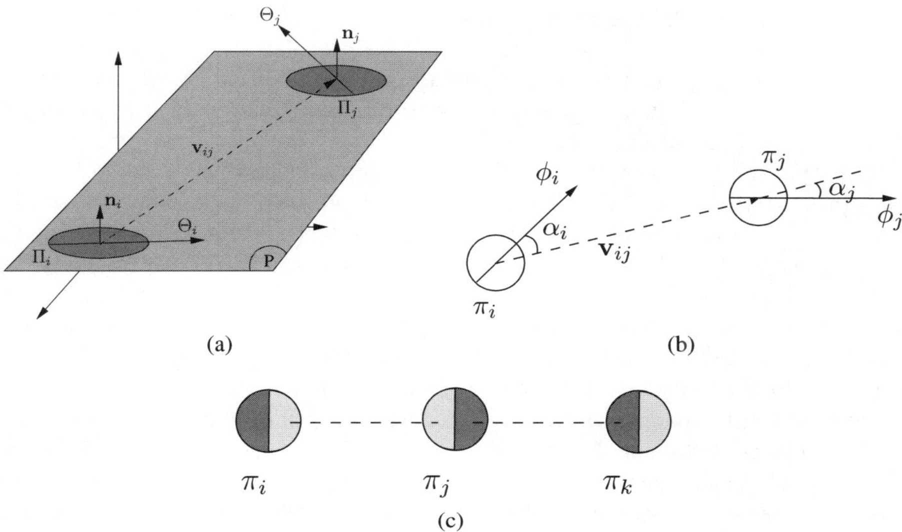

Fig. 3. Illustration of the relations between a pair of primitives. (a) Co–planarity of two 3D primitives Π_i and Π_j. (c) Co–colority of three 2D primitives π_i, π_j and π_k. In this case, π_i and π_j are cocolor, so are π_i and π_k; however, π_j and π_k are not cocolor. (b) Collinearity of two 2D primitives π_i and π_j.

their 2D projections π_i and π_j. We define the collinearity of two 2D primitives π_i and π_j as:

$$col(\pi_i, \pi_j) = 1 - \left| sin\left(\frac{|\alpha_i| + |\alpha_j|}{2} \right) \right|,$$

where α_i and α_j are as shown in Fig. 3(b), see [18] for more details on collinearity.

- *Co–colority:* Two spatial primitives Π_i and Π_j are co–color iff their parts that face each other have the same color. In the same way as collinearity, co–colority of two spatial primitives Π_i and Π_j is computed using their 2D projections π_i and π_j. We define the co–colority of two 2D primitives π_i and π_j as:

$$coc(\pi_i, \pi_j) = 1 - \mathbf{d}_c(\mathbf{c}_i, \mathbf{c}_j),$$

where $\mathbf{c}_i$ and $\mathbf{c}_j$ are the RGB representation of the colors of the parts of the primitives π_i and π_j that face each other; and, $\mathbf{d}_c(\mathbf{c}_i, \mathbf{c}_j)$ is Euclidean distance between RGB values of the colors $\mathbf{c}_i$ and $\mathbf{c}_j$. In Fig. 3(c), a pair of co–color and not co–color primitives are shown.

Co–planarity in combination with the 3D position allows for the definition of a grasping pose; Collinearity and co–colority allows for the reduction of grasping hypotheses. The use of the relations in the grasping context is shown in Fig. 4.

5 Elementary Grasping Actions

Coplanar relationships between visual primitives suggests different graspable planes. Fig. 4 shows a set of spatial primitives on two different contours l_i and l_j with co–planarity, co–colority and collinearity relations.

Five elementary grasping actions (EGA) will be considered as shown in Fig. 5. EGA1 is a "pinch" grasp on a thin edge like structure with approach direction along the surface normal of the plane spanned by the primitives. EGA2 is an "inverted" grasp using the inside of two edges with approach along the surface normal. EGA3 is a "pinch" grasp on a single edge with approach direction perpendicular to the surface normal. EGA4 is similar to EGA2 but its approach direction is perpendicular to the surface normal. Also it tries to go in "below" one of the primitives. EGA5 is wide grasp making contact on two separate edges with approach direction along the surface normal.

The EGAs will be parameterized by their final pose (position and orientation) and the initial gripper configuration. For the simple parallel jaw gripper, an EGA will thus be defined by seven parameters: $\mathrm{EGA}(x, y, z, \gamma, \beta, \alpha, \delta)$ where $\mathbf{p} = [x, y, z]$ is the position of the gripper "center" according to Fig. 6; γ, β, α are the roll, pitch and yaw angles of the vector $\mathbf{n}$; and δ is the gripper configuration, see Fig. 6. Note that the gripper "center" is placed in the "middle" of the gripper.

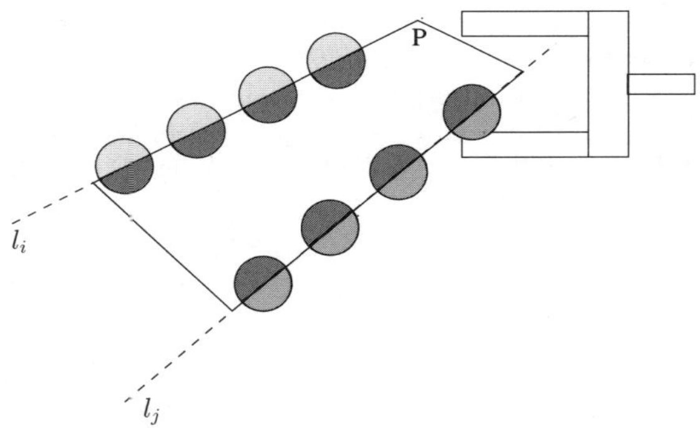

Fig. 4. A set of spatial primitives on two different contours l_i and l_j that have co–planarity, co–colority and collinearity relations; a plane P defined by the co–planarity of the spatial primitives and and an example grasp suggested by the plane

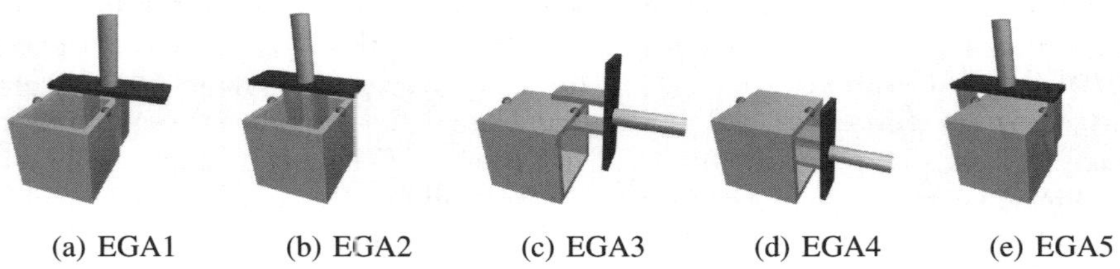

(a) EGA1   (b) EGA2   (c) EGA3   (d) EGA4   (e) EGA5

Fig. 5. Elementary grasping actions, EGAs

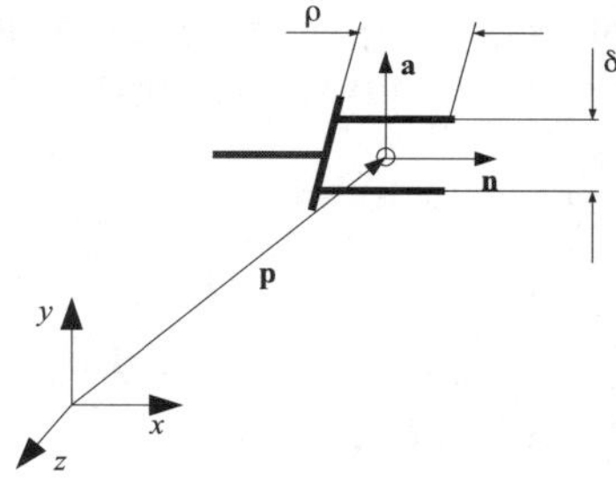

Fig. 6. Parameterization of EGAs

The main motivation for choosing these grasps is that they represent the simplest possible two fingered grasps humans commonly use. The result of applying the EGAs can be evaluated to provide a reinforcement signal to the system. The number of possible outcomes of each of the EGAs are different and will be explained below.

For all of the EGAs the possibility of an *early failure* exists. That is, the EGA fails before reaching the target configuration. This will result in a reinforcement R_{fe}. Furthermore, it is possible for all EGAs to fail a grasping procedure.

For EGA1, EGA3 and EGA5, a failed grasp can be detected by the fact that the gripper is completely closed. This situation will result in a reinforcement R_{fl}.

For EGA1 and EGA3, the expected grasp is a pinch type grasp, i.e. narrow. Therefore, they can also "fail" if the gripper comes to a halt too early, that is $\delta > \Delta_{min}$. This will result in a reinforcement R_{ft}.

EGA2 fails if the gripper is fully opened, meaning that no contact was made with the object. This gives a reinforcement R_{fh}.

To detect failure of EGA4, a tactile sensor is required on the side of the "fingers". If, after positioning and opening the gripper, there is no contact between the object and the tactile sensor, the EGA has failed. This results in a reinforcement R_{fc}.

If none of the above situations is encountered, a positive reinforcement R_g is given, and the EGA is considered successful.

6　Computing Action Parameters

Let $\Gamma = \{\Pi_1, \Pi_2\}$ be a primitive pair for which the coplanar relationship is fulfilled. Let Γ_i be the i:th pair and $\mathbf{p}$ the plane defined by the coplanar relationship of the primitives of Γ_i. Let $\Lambda(\Pi)$ be the position of Π and $\Theta(\Pi)$ be the orientation of Π. The parameterization of the EGAs is given with the gripper normal $\mathbf{n}$ and the normal of the surface between the two fingers $\mathbf{a}$ as illustrated in Fig. 6. From this, the yaw, pitch and roll angles can be easily computed.

For EGA1, there will be two possible parameter sets given the primitive pair $\Gamma = \{\Pi_1, \Pi_2\}$. The parameterization is as follows:

$$\mathbf{P}_{\text{gripper}} = \Lambda(\Pi_i)$$

$$\mathbf{n} = \nabla(\mathbf{p})$$

$$\mathbf{a} = \mathbf{perp_n}(\Theta(\Pi_i))/ \parallel \mathbf{perp_n}(\Theta(\Pi_i)) \parallel \quad \text{for } i = 1, 2$$

where $\nabla(\mathbf{p})$ is the normal of the plane $\mathbf{p}$ and $\mathbf{perp}_\mathbf{u}(\mathbf{a})$ is the projection of $\mathbf{a}$ perpendicular to $\mathbf{u}$. That is $\mathbf{perp}_\mathbf{u}(\mathbf{a}) = \mathbf{a} - \mathbf{proj}_\mathbf{u}(\mathbf{a})$, where $\mathbf{proj}_\mathbf{u}(\mathbf{a})$ is defined according to (1).

For EGA2, there is only one parameter set.

$$\mathbf{d} = \Lambda(\Pi_2) - \Lambda(\Pi_1)$$
$$\mathbf{p}_{\text{gripper}} = \Lambda(\Pi_1) + \mathbf{d}/2$$
$$\mathbf{n} = \nabla(\mathbf{p})$$
$$\mathbf{a} = \mathbf{n} \times \mathbf{d}/ \parallel \mathbf{n} \times \mathbf{d} \parallel$$

For EGA3, there will be two possible parameter sets for $i=1, j=2$ and $i=2, j=1$.

$$\mathbf{d} = \Lambda(\Pi_j) - \Lambda(\Pi_i)$$
$$\mathbf{n} = \mathbf{d}/ \parallel \mathbf{d} \parallel$$
$$\mathbf{p}_{\text{gripper}} = \Lambda(\Pi_i)$$
$$\mathbf{a} = \mathbf{n} \times \nabla(\mathbf{p})$$

For EGA4, there will be two possible parameter sets for $i = 1, j = 2$ and $i = 2, j = 1$. Where ϵ is a step size parameter that will depend on the gripper used.

$$\mathbf{d} = \Lambda(\Pi_j) - \Lambda(\Pi_i)$$
$$\mathbf{n} = \mathbf{d}/ \parallel \mathbf{d} \parallel$$
$$\mathbf{p}_{\text{gripper}} = \Lambda(\Pi_i) - \nabla(\mathbf{p}) \cdot \epsilon$$
$$\mathbf{a} = \mathbf{n} \times \nabla(\mathbf{p})$$

EGA5 will have the same parameters as EGA2 except that the gripper opening will be $\delta = \parallel \Lambda(\Pi_2) - \Lambda(\Pi_1) \parallel + \Delta$.

7 Limiting the Number of Actions

For a typical scene, the number of coplanar pairs of primitives is in the order of $10^3 - 10^4$. Given that each coplanar relationship gives rise to 8 different grasps from the five different categories, it is obvious that the number of suggested actions must be further constrained. Another problem is that coplanar structures occur frequently in natural scenes and only a small set of them suggest feasible actions, e.g. objects placed on a table create a lot of 3D line structures coplanar to the table but can not grasped directly by a grasping direction normal to the table. In addition, there exist many coplanar pairs of primitives affording similar grasps.

To overcome some of the above problems, we make use of the structural richness of the primitives. First, their embedding into collinear groups naturally clusters the grasping hypotheses into sets of redundant grasps from which only one needs to be tested. Furthermore, co–colority, gives an additional hypothesis for a potential grasp.

8   Using Grouping Information

From the 2D primitives (before stereo reconstruction) collinear neighbors can be found. The collinear neighbors can be mapped to corresponding 3D primitives. These small neighborhoods form the set of *small groups*, $\{g_1, g_2, ..., g_N\}$. The *large groups*, $\{G_1, G_2, ..., G_M\}$, are formed by the grouping of the small groups such that if Π_i and Π_j are part of group g_x and Π_j and Π_k is part of group g_y then g_y and g_x is part of the same large group G_z. Using this grouping information it is possible to add additional constraints on the generation of EGA s.

First, all primitives that are not part of a sufficiently large group G_i are discarded. Secondly, the relations co–planarity and co–colority between small groups of primitives are computed such that primitive $\Pi_i \in g_x$ and $\Pi_j \in g_y$ are only considered to have a co–planarity or co–colority relation if all primitives in g_x are coplanar or cocolor w.r.t all primitives in g_y. Finally, it is possible to constrain the generation of EGAs to only one EGA of each type for each large group.

9   Experimental Evaluation

Fig. 8, Fig. 9 and Fig. 10 show some of the grasps generated for the scenes evaluated here. Fig. 7 shows visual features generated by the stereo system and a selection of generated actions. Fig. 8 shows a simple plate structure for which the outer contour is generated since the object is homogeneous in texture. Fig. 9 shows a scene with a single, but a more complex object than the previous one. Fig. 10 shows two scenes with two (cup and knife) and three objects (box, cup and bottle9.

On each of the scene, after the spatial primitives have been extracted, elementary actions shown in Fig. 5 are tested. There are few reasons for which a certain grasp may fail:

- The system does not have the knowledge of whether the object is hollow or not, so testing EGA2 will results with a collision and thus failure.
- Since no surface is reconstructed, EGA1 will fail for hollow objects which are grasped from "below".

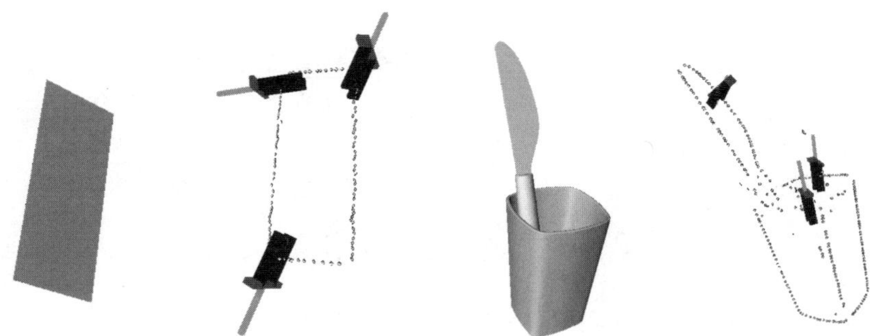

Fig. 7. Two example scenes designed for testing and a selection of the generated actions

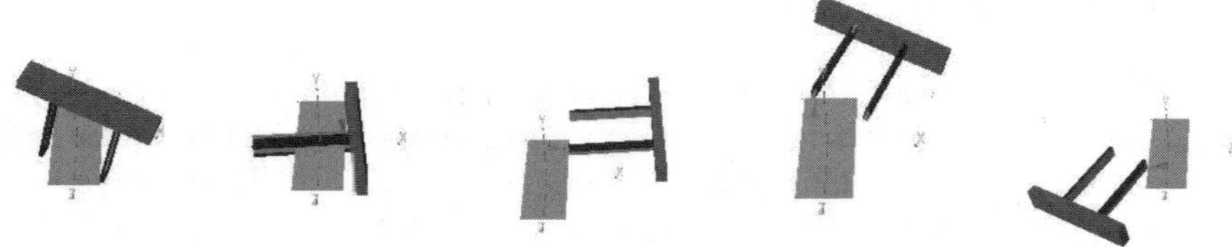

Fig. 8. Examples of tested grasps on a plate (from left): successful grasp using EGA5, and a few early failures using EGA1, EGA3 and EGA5, res5respectively

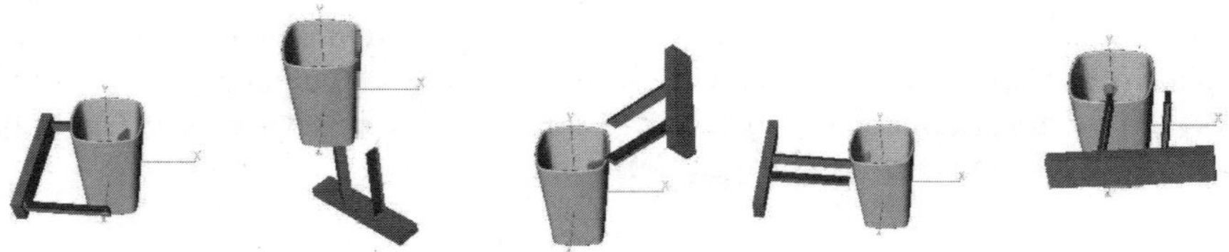

Fig. 9. Examples of tested grasps on a cup (from left): a successful grasp using EGA1, and a few early failures using EGA1, EGA1, EGA2 and EGA3, respectively

Table 1. Experimental evaluation of the grasp success rate where the following notation is used: pl (co-planarity), gr (grouping), cl (co-colority) and (succesfull/tested) grasps

Scene	gr	pl+gr	col+gr	gr+pl+col
Plane	90% (9/10)	67% (4/6)	86% (6/7)	80% (4/5)
Cup	21% (14/66)	27% (10/37)	27% (13/49)	29% (7/24)
Cup/Kn	20% (9/45)	9% (3/32)	26% (9/35)	15% (3/20)
3 objects	6%(27/434)	7% (7/98)	9% (12/139)	9% (9/53)

Table 2. The number of generated action hypotheses where the following notation is used: no gr (no grouping), pl (co-planarity), gr (grouping), cl (co-colority)

Scene	(no gr)	(no gr)+pl	(no gr)+col	(no gr)+pl+co	gr	gr+pl	gr+col	gr+pl+coll
Plane	46 224	35 608	38 512	30 224	80	48	56	40
Cup	172 224	96 112	89 392	56 120	528	296	392	192
Cup/knife	269 360	140 920	139 136	79 104	360	256	280	160
3 objects	927 368	303 960	315 336	166 008	3472	784	1112	424

- If the hand, during the approach, detects a collision on one of the fingers, the grasping process is stopped. In reality, this grasp may happen to be successful anyway if the object is moved so that it is centered between the fingers.

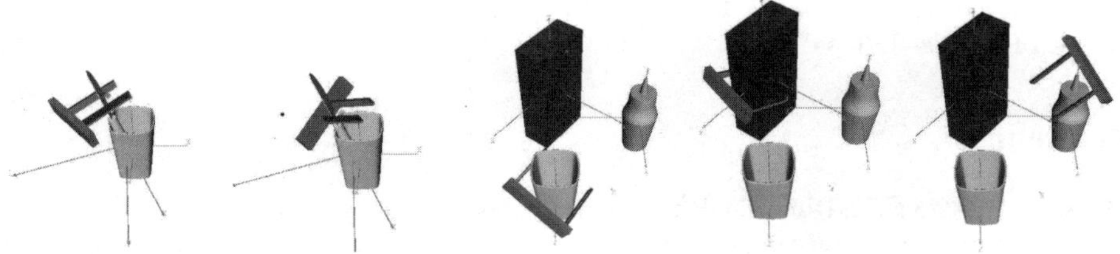

Fig. 10. Examples of successful grasps with two and three objects

Table 1 summarizes the results for the generated success rate regarding a number of successful grasps given no knowledge of the object shape. We note that the results are a summary of an extensive experimental evaluation since, given different types and combinations of spatial primitives all generated actions had to be evaluated. It can be seen that for a scene of low complexity (plate) the average number of successful grasps is close to 80%. For more complex scenes this number is dependant on the number and type of objects. It is also important to note not only the percentage but the number of evaluated grasps. Although, in some cases, the success rate is lower when primitives are integrated, there are much fewer hypotheses tested. These results should also be considered together with the results presented in Table 2 where we show how the integration of grouping, co-colority and co-planarity affects the number of generated hypotheses (affordances). Another thing to point out related to Table 1 is that most of the unsuccessful grasps happened due to an "early failure" such as that a contact was detected before the grasp was executed. Again, this failure may in some cases result with a successful grasp anyway. Another big source of failure was that there was nothing to lift, i.e. EGA3 could not have been applied.

10   Conclusions

Robots should be able to extract more knowledge through their interaction with the environment. The basis for this interaction should not be a detailed model of the environment and lots of *a-priori* knowledge but the robot should be engaged in an exploration process through which it can generate more knowledge and more complex representations. In this paper, we have presented one of the building blocks necessary in such a system.

In particular, we have designed an early grasping system, based on a set of innate reflexes and knowledge about its embodiment. We relied on 3D information based on primitives extracted online and showed how the structural richness of primitives can be used for an efficient reduction of grasping hypotheses while keeping relevant ones. Rather than dealing with high quality grasps on a constrained set of known objects, we have demonstrated that the system is able of generating a certain percentage of successful grasps on arbitrary objects. This is important for our future research that will develop complex learning schemes aiming at more sophisticated grasping strategies and knowledge representation.

Acknowledgement. This work has been supported by EU through the project PACO-PLUS, FP6-2004-IST-4-27657.

References

1. Azad, P., Asfour, T., Dillmann, R.: Combining appearance-based and model-based methods for real-time object recognition and 6d localization. In: IEEE International Conference on Intelligent Robots and Systems (2006)
2. Bicchi, A., Kumar, V.: Robotic grasping and contact: A review. In: IEEE International Conference on Robotics and Automation, pp. 348–353 (2000)
3. Ding, D., Liu, Y.H., Wang, S.: Computing 3-d optimal formclosure grasps. In: IEEE International Conference on Robotics and Automation, pp. 3573–3578 (2000)
4. Fitzpatrick, P., Metta, G., Natale, L., Rao, S., Sandini, G.: Learning About Objects Through Action - Initial Steps Towards Artificial Cognition. In: IEEE International Conference on Robotics and Automation, pp. 3140–3145 (2003)
5. Hauck, A., Rüttinger, J., Sorg, M., Färber, G.: Visual Determination of 3D Grasping Points on Unknown Objects with a Binocular Camera System. In: IEEE/RSJ International Conference on Intelligent Robots and Systems, pp. 272–278 (1999)
6. http://www.barrett.com/robot/products/hand/handfram.htm
7. Kamon, I., Flash, T., Edelman, S.: Learning Visually Guided Grasping: A Test Case in Sensorimotor Learning. IEEE Transactions on Systems, Man and Cybernetics 28(3), 266–276 (1998)
8. Krüger, N., Felsberg, M.: An explicit and compact coding of geometric and structural information applied to stereo matching. Pattern Recognition Letters 25(8), 849–863 (2004)
9. Krüger, N., Lappe, M., Wörgötter, F.: Biologically motivated multi-modal processing of visual primitives. The Interdisciplinary Journal of Artificial Intelligence and the Simulation of Behaviour 1(5), 417–428 (2004)
10. Krüger, N., Wörgötter, F.: Multi-modal primitives as functional models of hyper-columns and their use for contextual integration. In: De Gregorio, M., Di Maio, V., Frucci, M., Musio, C. (eds.) BVAI 2005. LNCS, vol. 3704, pp. 157–166. Springer, Heidelberg (2005)
11. Miller, A.T., Allen, P.: Graspit!: A versatile simulator for grasping analysis. In: ASME International Mechanical Engineering Congress and Exposition (2000)
12. Miller, A.T., Knoop, S., Allen, P.K., Christensen, H.I.: Automatic grasp planning using shape primitives. In: IEEE International Conference on Robotics and Automation, pp. 1824–1829 (2003)
13. Morales, A., Chinellato, E., Fagg, A.H., del Pobil, A.: Using experience for assessing grasp reliability. International Journal of Humanoid Robotics 1(4), 671–691 (2004)
14. Morales, A., Recatalá, G., Sanz, P.J., del Pobil, Á.P.: Heuristic Vision-Based Computation of Planar Antipodal Grasps on Unknown Objects. In: IEEE International Conference on Robotics and Automation, pp. 583– 588 (2001)
15. Platt Jr., R., Fagg, A.H., Grupen, R.A.: Extending fingertip grasping to whole body grasping. In: International Conference on Robotics and Automation, pp. 2677–2682 (2003)
16. Pollard, N.S.: Parallel methods for synthesizing whole-hand grasps from generalized prototypes. PhD thesis, Dept. of Electrical Engineering and Computer Science, Massachusetts Institute of Technology (1994)
17. Pollard, N.S.: Closure and quality equivalence for efficient synthesis of grasps from examples. International Journal of Robotic Research 23(6), 595–613 (2004)
18. Pugeault, N., Wörgötter, F., Krüger, N.: Multi-modal scene reconstruction using perceptual grouping constraints. In: Proceedings of the 5th IEEE Computer Society Workshop on Perceptual Organization in Computer Vision (in conjunction with IEEE CVPR 2006) (2006)

19. Rössler, B., Zhang, J., Knoll, A.: Visual Guided Grasping of Aggregates using Self-Valuing Learning. In: IEEE International Conference on Robotics and Automation, pp. 3912–3917 (2002)
20. Rutishauser, M., Stricker, M.: Searching for Grasping Opportunities on Unmodeled 3D Objects. In: British Machine Vision Conference, pp. 277–286 (1995)
21. Stoytchev, A.: Behavior-Grounded Representation of Tool Affordances. In: IEEE International Conference on Robotics and Automation, pp. 3060–3065 (2005)

Chapter II

Perception Guided Navigation and Manipulation

Summary of Perception Guided Navigation and Manipulation

Sukhan Lee

The main performance feature of the next generation of robots may be the capability of perception guided navigation and manipulation. This is especially so when we consider professional and personal/domestic service robots that should carry out tasks in an unstructured environment. The advancement of perception guided navigation has been well demonstrated to date by SLAM (Simultaneous Localization and Map Building) capability of various field and service robots and, in particular, of those sensor-integrated autonomous vehicles contested successfully in the DARPA Grand Challenge. In spite of such notable achievements, service robots today are yet to show what is comparable to human capability of perception guided navigation and manipulation. Human seems heavily favored in the ability of recognizing, modeling and understanding its environment, in the dependability of navigation and manipulation performance under a wide range of environmental variations, as well as in the efficiency of processing perceptual, in particular, visual information. It is noted that the advancement of perception guided manipulation seem more or less lagging behind that of perception guided navigation. This is because perception guided manipulation relies more heavily on a higher level of intelligence, including more precise and comprehensive modeling and understanding of, often heavily cluttered, 3D workspace, more sophisticated manipulation and grasp planning in connection with modeling and understanding of workspace, more direct interactions with environment that require the coordination of multiple sensor modalities such as vision, force and tactile, and more adaptation to unexpected local variations.

This chapter reports on the recent advancement of perception guided navigation and manipulation with 13 contributions. The 13 contributions are organized into two parts: the first 6 articles on navigation and the latter 7 on manipulation.

Part 1: Perception Guided Navigation

In their article entitled, "Link Graph and Feature Chain Based Robust Online SLAM for fully Autonomous Mobile Robot Navigation System using Sonar Sensors," Amit Kummar Pandey and K. Madhava Krishna propose a method of robust SLAM based only on sonar data. To address the robustness of sonar based SLAM despite the issues related to data association, they have developed the notion of feature chain with link graph and of combining features onto occupancy grid in such a way as to remove false feature associations as well as overcoming the limitation from independence assumption between cells. In their article entitled, "Service Robots Navigating on Smart Floors," Thomas Kampke, Boris Kluge, Erwin Prassler, and Matthias Strobel show in demonstration that a smart floor of an area-wide dense network of thousand

S. Lee, I.H. Suh, and M.S. Kim (Eds.): Recent Progress in Robotics, LNCIS 370, pp. 109–111, 2008.
springerlink.com

of RFID transponder can support accurate and reliable navigation of service robots. An interesting theoretical work on controlling multiple agents by moving their targets is presented by Timothy Bretl. Timothy Bretl shows that, assuming each agent moves at constant speed toward the closest target, only two targets are necessary to maintain separation between four agents, or, even between any number of agents, given suitable initial conditions. Martin Persson, Tom Duckett, and Achim Lilienthal present, in their article entitled "Improved Mapping and Image Segmentation by Using Semantic Information to Link Aerial Images and Ground-Level Information," a method of linking aerial images and ground-level information that facilitates both a faster detection of building outlines for a mobile robot and, at the same time, a provision of elevation data in aerial image segmentation. The mobile robot estimates walls from a ground perspective, which are then matched with edges from an aerial image for building detection. Francois Saidi, Olivier Stasse and Kazuhito Yokoi present in their paper entitled "Active Visual Search by a Humanoid Robot" a framework of visual search behavior for an object of interest in 3D environment for their HRP-2 humanoid robot. They propose to construct a visibility map representing feasible robot configurations and poses for exploring the space of interest under visual and mechanical constraints. Planning of a search behavior is based on maximizing target detection probability while minimizing energy, distance and time. Yuta Yoshihata, Kei Watanabe, Yasushi Iwatani and Koichi Hashimoto investigate in their article entitled, "Visual Control of a Micro Helicopter under Dynamic Occlusions," how to use multiple redundant cameras stationed on ground to track and control a hovering helicopter dependably despite the markers on the helicopter are subject to dynamic occlusion.

Part 2: Perception Guided Manipulation
Romeo Fomena Tatsambon and Francois Chaumette propose in their article entitled, "Visual Servoing from Spheres with Paracatadioptric Cameras," a new optimal combination of visual features for visual servoing based on paracatadioptric cameras. They show that the new combination of visual features not only allow feature motions to avoid the dead zone, but also make a classical control law to be globally stable even in the presence of modeling error. In their article entitled, "Dynamic Target Detection for Robotic Applications Using Panoramic Vision System," Abedallatif Baba and Raja Chatila use panoramic images for the detection of dynamic targets such as people, taking advantage of panoramic images offering global scenes around a robot. Moving targets are detected by a modified optical flow as binary images of foreground pixels. Two approaches for determining the direction of detected targets: one for an unfolded image, another for a raw panoramic image are provided. In their article entitled, "Vision-Based Control of the RoboTenis System," Luis Ángel, A. Traslosheros, J.M. Sebastian, L. Parí, R. Carelli and F. Roberti present visual servoing of a parallel robot for the tracking of faster moving objects with unknown trajectories. A predictive control law is formulated based on the prediction of the future position and velocity of the moving object and the compensation of the delay introduced by the vision system. In their article entitled, "A Particle Filter Based Probabilistic Fusion Framework for Simultaneous Recognition and Pose Estimation of 3D Objects in a Sequence of Images," Jei-hun Lee, Seung-Min Baek, Changhyun Choi and Sukhan Lee propose a particle filter framework for robust recognition and pose

estimation of 3D objects based on fusing multiple evidences from a sequence of images. In order to ensure robustness against a wide range of environmental variations, they also devise a method for selecting and collecting an optimal set of features through the in-situ monitoring of the environment. In their article entitled, "Preliminary Development of a Line Feature-based Object Recognition System for Textureless Indoor Objects," Gunhee Kim, Martial Herbert and Sung-Kee Park address the issue of recognizing texture-less objects in an indoor environment based on line features. Segmented line-typed features and pair-wise geometric relationships between these features are used for recognition, observing that most of indoor objects of the same category are of consistent relations among line features. Peter Slaets, Wim Meeussen, Herman Bruyninckx, and Joris De Schutter present in their article entitled, " Modelling of Second Order Polynomial Surface Contacts for Programming by Human Demonstration," automatic model building of quadratic objects in the context of programming by human demonstration. The motion of human operator using a demonstration tool is captured by a 3D camera and a force/torque sensor. Particle filter is used to recognize discrete objects and estimate their geometric parameters simultaneously. Joon-Young Park, Kyeong-Keun Baek, Yeon-Chool Park and Sukhan Lee present in their article, "Robot Self-Modeling of Rotational Symmetric 3D Objects Based on Generic Description of Object Categories," a method for a robot to self-model the objects that are referred to or pointed out by human. The approach starts with the generic description of a category of objects as a geometric as well as functional integration of parts the 3D shape. Given the 3D point clouds from the scene, 3D edge detection segments out the target object referred by human by removing out the background and/or neighboring objects. It is demonstrated that a rotational symmetric part or object can be modeled precisely based on an iterative estimation of rotational axis.

Link Graph and Feature Chain Based Robust Online SLAM for Fully Autonomous Mobile Robot Navigation System Using Sonar Sensors

Amit Kumar Pandey[1] and K. Madhava Krishna[2]

[1] Robotics Research Centre, International Institute of Information
Technology-Hyderabad (IIIT-H), Hyderabad, India
`akpandey@research.iiit.ac.in`
[2] Robotics Research Centre, International Institute of Information
Technology-Hyderabad (IIIT-H), Hyderabad, India
`mkrishna@iiit.ac.in`

Summary. Local localization of a fully autonomous mobile robot in a partial map is an important aspect from the view point of accurate map building and safe path planning. The problem of correcting the location of a robot in a partial map worsens when sonar sensors are used. When a mobile robot is exploring the environment autonomously, it is rare to get the consistent pair of features or readings from two different positions using sonar sensors. So the approaches, which rely on readings or features matching, are prone to fail without exhaustive mathematical calculations of sonar modeling and environment modeling. This paper introduces link graph based robust two step feature chain based localization for achieving online SLAM (Simultaneous Localization And Mapping) using sonar data only. Instead of relying completely on matching of feature to feature or point to point, our approach finds possible associations between features to localize. The link graph based approach removes many false associations enhancing the SLAM process. We also map features onto Occupancy Grid (OG) framework taking advantage of its dense representation of the world. Combining features onto OG overcomes many of its limitations such as the independence assumption between cells and provides for better modeling of the sonar providing more accurate maps.

Keywords: SLAM, Sonar, Autonomous Mobile Robot, Feature Chain, Link Graph.

1 Introduction

IN this paper we will focus on issue of continuous localization of a mobile robot, which is inevitable requirement for any fully autonomous mobile system to navigate safely in a partially known environment and to explore and build accurate and safe map of the environment. During the process of exploration, robot takes decisions to move to different locations to get a new 360 *degree* scan of the unknown portion of the environment to increment the map and plans a safe path within the known region to reach there. During the process of scanning and moving, the robot looses track of its 3D position (x, y, θ) due to unavoidable odometric error, which needs to be corrected before any further exploration action has to be taken. This Simultaneous Localization and Mapping problem

S. Lee, I.H. Suh, and M.S. Kim (Eds.): Recent Progress in Robotics, LNCIS 370, pp. 113–131, 2008.
springerlink.com

(SLAM) is considered as chicken and egg problem in mobile robotics; because for building an accurate map and planning a safe and correct path robot need to know its exact position, which requires localization, but to localize the robot need to have the map. Figure 1 shows a map built by an autonomous robot, which is actually the map of a straight corridor, but due to the localization error the straight parallel walls are appearing as curved walls. Offline SLAM processes tend to delay the localization process as they await the detection of features such as corners and edges or more data. But in an exploration process such delays can become unacceptable and requires for correction of robot pose at frequent intervals. In sonar based exploration, due to very limited number of accurately detected features in one scan, frequent pose correction is difficult. This is because edges and corners are highly unreliable (sometimes may not be detected also) and wall segments are the only reliable features. Our approach creates feature chain having a root feature and other link features using wall segments, which is used for localization. And the link graph approach removes the false associations between wall segments, and corner and edge readings and gives a reliable set of associations suitable for SLAM. This it does by doing angle fit and distance fit analysis over several possible association patterns and chooses the best set of associations.

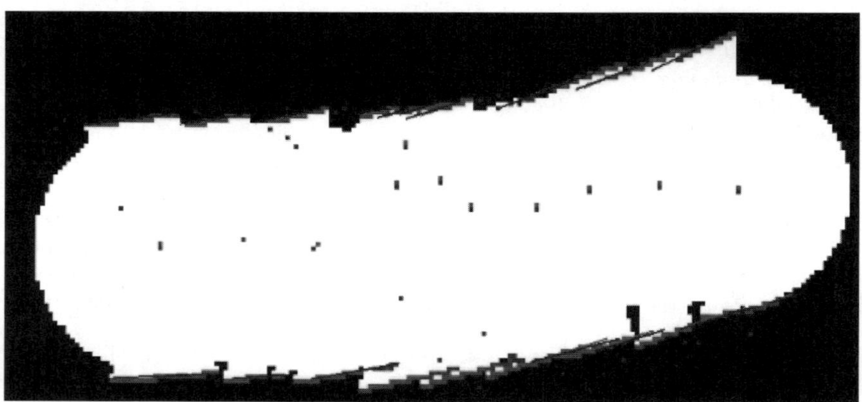

Fig. 1. Map of a straight corridor without correcting localization error during map building process. Straights walls are appearing as a curved.

There are basically two broad approaches for SLAM. One is based on feature matching and another is based on raw data matching. But these approaches are basically dependent on the assumption that the inputs from the sensors are reliable and consistent. Generally a combination of sensors like vision and sonar [1] are used or a laser sensor [2, 3] is used or camera [4] is used. But very few authors [5, 6] have addressed the SLAM problem in indoor environment with only sonar sensors. The reason is that, by using sonar sensors, the situation rapidly deteriorates since sonar readings are susceptible to high degree of uncertainty especially due to angular and radial errors along with specular reflection problems. In fact one cannot get a pair of consistent readings for a particular world location from two different positions.

As shown in the fig.2, the world point P_w has been hit by two sonar beams, from positions P_{t1} and P_{t2} of the robot at successive time instants. One may expect to obtain readings r_1 & r_2 as shown by solid lines. But practically one gets instead some unpredicted readings like r_{1extd} & r_{2extd}. The algorithms which assume that these inconsistencies in the reading are due to robots odometric error will not work, since for a given environment it is very difficult to predict whether the inconsistency in the reading is due to sonar error or due to error in robot pose. This problem is not seen however with the laser sensor.

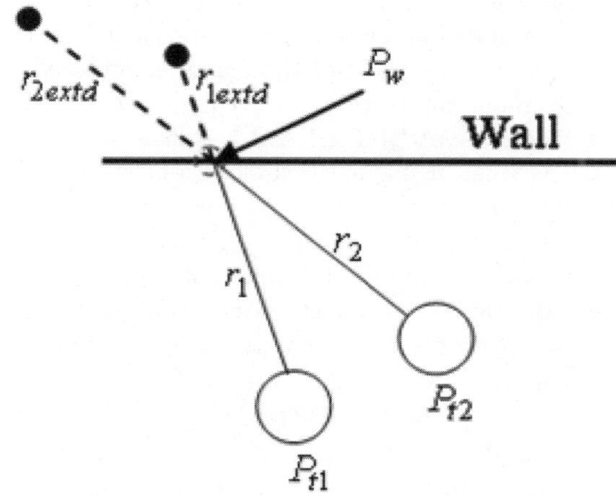

Fig. 2. From two position for sonar sensor instead of getting readings shown by solid line, we will get some unpredictable readings shown by extended dotted line, for a same world point P_w

Fig. 3. By sonar sensor for two instants even from nearly same positions we can get two different wall segment features, which can be overlapping but need not to be the same (a) or non-overlapping (b)

In exact feature matching based SLAM; once again there are problems with sonar sensors. The features which we can extract from sonar sensor are wall segments, corners and edges [7, 8]. Each of such features has a sparse representation and a robot's scan from two proximal places does not guarantee matching between features. If we assume for a corridor like simple environment, with only array of sonar sensors, wall segments will be the feature, which can be detected. And as shown in fig.3(a), in the best case it can detect two different wall segments with some overlapping parts and in worst case we will get non overlapping segments as shown in fig.3(b). Hence not only matching of features between successive scans is difficult, the usual image registration algorithms that try to solve for the rotational and translational displacement between successive scans do not work since overlap between features is minimal and it is also difficult to predict how much the actual overlapping is.

While these problems do not exist with laser range finders and vision based sensors, sonar is still an attractive proposition due to its low cost and small size. Also Extended Kalman Filter (EKF) based online SLAM approaches [9, 10] which use edges, corners and other point like features as landmarks to localize will not work for the corridor like environment (fig:1) or the environment where edges and corners are sparse and most of the features detected are line segments. Further the EKF approach tends to work well the less ambiguous the point landmarks are. For this reason, EKF SLAM requires significant engineering of feature detectors, sometimes using artificial beacons as features [9]. So, either the robot has to visit the previously seen edge or corners to localize or it has to use some manually put landmarks to localize or it has to delay localization until it detects some edges, corners or other landmarks. And the situation becomes worse with sonar sensors because assuring edges and corners like point features from an individual 360 *degree* scan by sonar are very difficult whereas getting line segment like feature is easy.

In order to overcome the above problems and to achieve SLAM with sonar sensor only, we have proposed a novel Feature Chain based two-step localization. In fact with sonar, even edges, corners or circular objects can get classified as line segment features [8]. This obviously results into false association with an actual wall and hence deteriorates the localization accuracy. So to remove these false associations and to achieve robustness a Link Graph based global association analysis approach has been introduced. The resulting maps show that with the presented approach the robot is accurately able to complete the loop autonomously without any external path guiding.

2   Background Work

In [7], we have presented a novel approach to get a safe and more accurate map using sonar sensors only, based on feature detection and mapping onto occupancy grid framework. We will use that approach for updating the map. For the continuity, the basics of the approach will be briefly discussed here.

In general the reading from sonar is reliable only when at least one ray of the sonar beam hits normally to the surface. As shown in fig.4, three different sonar beams are hitting the walls (shown as different colors) with different axis angles. But the distance returned by all the three beams will be the perpendicular distance shown by the middle solid line, because only that ray will return back to the sonar sensor. Hence we will get a set of approximately same reading (within a threshold) for all the sonar beams which have been fired within the angular range $\Delta\omega$, where $\Delta\omega$ is beam width of one sonar beam. This region is called the *Region of Constant Depth (RCD)*, readings of which are the most reliable readings in the case of sonar sensors.

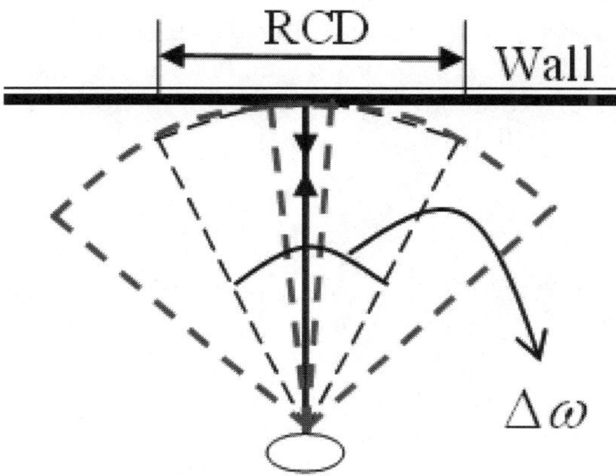

Fig. 4. Concept of RCD. For all the three sonar beams shown as different colors, the perpendicular reading shown as middle solid arrowed line will return to the sonar sensor

There are special patterns of RCDs for walls and corners for a 360 *degree* scan of the environment by sonar sensor. In [7] we have utilized these properties to extract the features like walls and corners and able to build the accurate and safe map of the environment. In [7], a detailed description of a Bayesian framework based approach for mapping is presented. Here we will assume that features like wall segments and corners have been extracted in each 360 *degree* scan by sonar sensors, and we will directly use this information for our present work.

3   Methodology

As explained in section 1, it is rare to get consistent pair of features or readings by sonar from two different positions. So, instead of relying on feature matching, present method relies on feature association and creates chains of associated features.

Definition 1 (Feature Chain). *A feature chain is a set of features obtained in different scans, where each feature is associated with at least one feature of the chain by fulfilling the criteria of association.*

Each feature chain will be having a *root* and many *links*. *Root* will be the most reliable feature of the chain.

3.1 Seed Chain Creation

Robot makes a first 360 *degree* scan of the environment from its current position. Instead of updating the map instantly with each reading, features are extracted from the raw range data [7]. Each wall segment feature will serve as a root of a new feature chain and will be assigned a probability $P(root\_of\_first\_scan) = 1$. Initially all chains are having only one feature which is the root itself. Any feature which will be detected in future scans will be either associated with any chain or will become a new root for a new feature chain. If associated, then that feature will be a part of the feature chain and called as a *link* of that feature chain. In that fashion a feature chain will grow. Also a chain can grow in both the directions.

3.2 Feature Association

During the process of exploration robot moves to a new location and takes a new 360 *degree* scan. Most of the time each such scan will result into some new wall segment features. These new features will be first tried to associate with any link or root of any existing feature chain. As shown in fig.5, for associating overlapping wall segments, a mean distance $D1$ is calculated, which is the average distances from both end points and middle point of a wall segment extracted in the current scan onto other existing wall segments which are link or root of feature chains. Also the angle α between the two wall segments is calculated. If they are within some threshold, then both are associated. For non overlapping walls one more criterion is tested, the distance $D2$ between the nearest end points of both the walls. If this is also within some threshold only then both features are being associated. In our case we have experimentally fixed $D1$ as 20 *cm*, $D2$ as 50 *cm* and α as 20 *degrees*. Note that it is not necessary that features will be associated with the root of a chain; they can be associated with any other link feature of an existing chain. And it may also be the case that a feature will

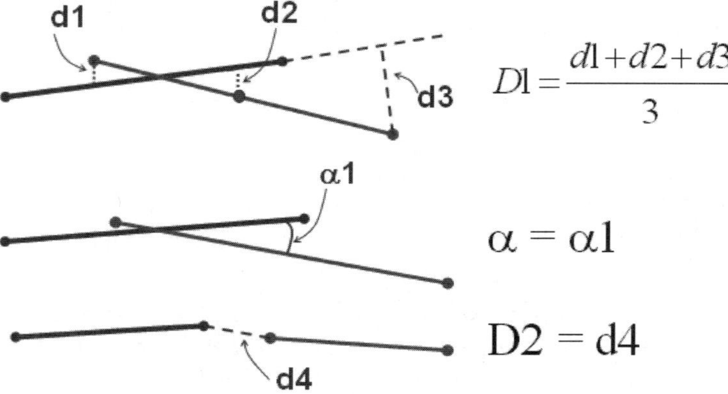

Fig. 5. Association criteria for overlapping and non-overlapping features

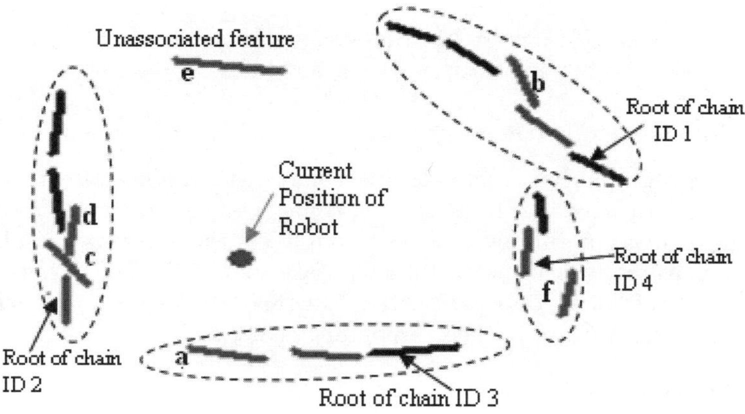

Fig. 6. 4 different feature chains (encircled), Features extracted in current scan (Red Segments), Links of existing chains to which current features has been associated (Green Segments)

not be associated at all, which has been addressed later in this paper. For better understanding we will use fig.6 which is a typical scenario from sonar scan. Red segments in fig.6 show new features extracted in a particular scan and green segments show the links of an existing feature chain to which the current feature has been associated based on above mentioned criteria. Other links of the chains have been shown in black. The corresponding root of each chain has been marked separately. Note that in some cases current feature has been associated with the root of the chain and in others with link of the chain.

Using the initial set of associated features obtained as above; we create a two-layered graph, which we are calling as *link graph*. Vertices at top layer of the link graph show the IDs of the feature chains to which at least one feature

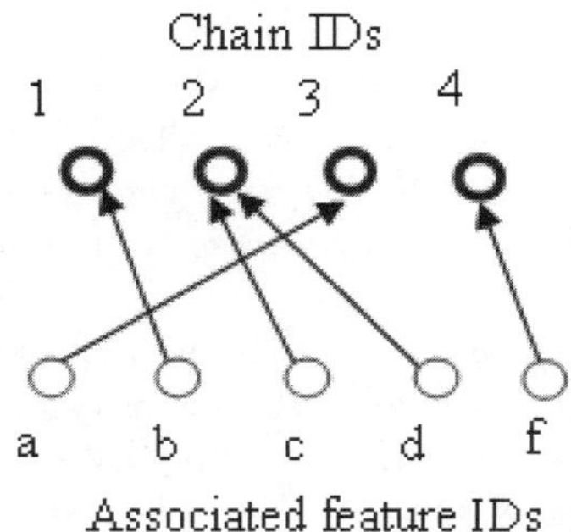

Fig. 7. Initial link graph of associations of fig6

extracted in current scan has been associated. Vertices at the bottom layer show the current associated features ID. Edges show the association. Figure 7 shows the initial link graph of associations shown in fig.6.

3.3   Achieving Robustness

The initial link graph, constructed as above may contain some false associations, which if used for correcting the pose (x, y, θ) may result in incorrect localization. In fact the association till now is loosely coupled in the sense they are based on individual feature association. So to detect false association we have to analyze the associations from a global perspective. For this, a two-step individual feature fit based analysis, with root of corresponding chain, is done.

Angle Fit Analysis

First *angle fit analysis* is done. For each associated feature of initial link graph, robot is virtually oriented in such a way that the angle difference between that particular feature and its associated root becomes zero. And then with this new orientation of robot, the orientation and position of all other features are being calculated. Then for this particular fit, the angle difference of each associated feature present in initial link graph with its root is calculated. Note that this angle difference is calculated with the root of the chain whereas for association, the angle difference was checked with each link of the different chains. Then another link graph is created with those associations whose angle differences with root are within a window of $+th$ to $-th$. In our case we have chosen th as $(Avg\_ang + \epsilon)$, where Avg_ang is average angle difference between all the initially associated features and the root of the associated chains; ϵ is error limit calculated experimentally, in our case ϵ is 2 degree.

Figure 8 shows link graphs by fitting angle of each associated feature of fig.6 one at a time. The link graph having maximum number of links will be the winner link graph from angle fit analysis. In our example the link graphs 8(B), 8(D) and 8(E) are having maximum number of links, so they will be the winner from angle fit analysis and also will be a candidate for final winner link graph.

The rotational error between two wall RCDs of the same wall are generally due to the rotational error of the robot between the two locations. However the same can not be said of corner and edge RCDs [8]. One may think it is easy to discard such corner and edge RCDs from the association process but many a time such corner and edge RCDs are wrongly classified as walls. This in turn leads to wrongly classified walls getting associated with previous walls. But the errors of such false associations will not be globally consistent with the errors of other true associations. If we will take these types of false associations for correcting location and orientation we end up with incorrect localization. There exist several other sources of false associations whose errors are not globally consistent with errors due to true associations. The angle fit analysis explained above takes advantage of this inconsistency to remove these false associations. The other sources of such errors are pillar coming out of the wall, circular objects close to wall and so on.

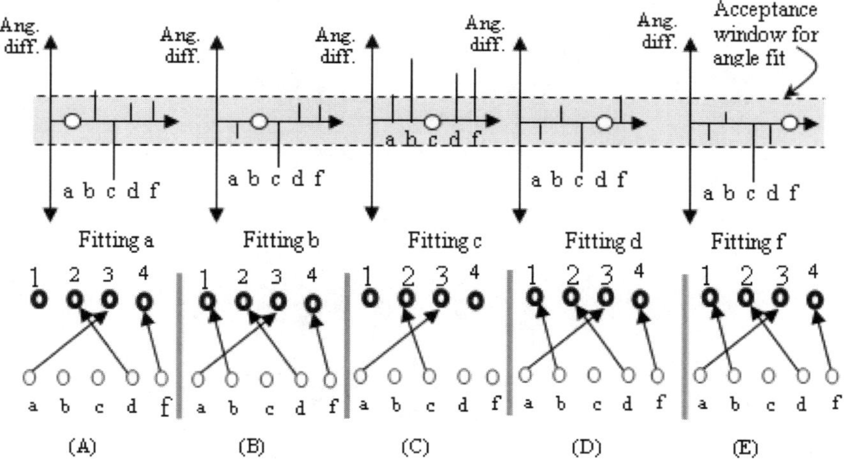

Fig. 8. Top row : Angle difference between feature and the root after exactly fitting one feature at a time by making angle difference 0. Bottom row : Corresponding link graphs with the links which are within the green window of the accepted angle difference.

Distance Fit Analysis

The winner link graphs from angle fit obtained as above may contain some associations, which are not agreeing globally from the point of view of perpendicular distance between feature and the corresponding root. As shown in the fig.6, the current feature f has been associated with feature chain 4 and it is also present in the winner link graph of angle fit. But it may be the case that the associated chain is of an open door, which is parallel to a wall and feature f is a new feature of that wall but falsely associated with feature chain of door. These types of false association can occur in other cases where there is some step like structure in the environment. To remove those associations we will generate a set of a different kind of link graph based on the distance fit as explained below.

In *distance fit analysis*, firstly, all the associated features will be made parallel to their root features by virtually rotating them individually with respect to robot. Then the average normal distance from middle point of each feature to the line formed by the associated roots will be found. Let us say it as actual average normal dist, *Act_Avg_Norl_Dist*. Now each associated feature will be made parallel to the root feature one at a time and the normal distance from middle point of the feature to root line segment will be found. This normal distance will be broken into Δx and Δy components. All features will be shifted by this $(\Delta x, \Delta y)$. It is basically fitting one feature exactly based on the normal distance. Then other feature will be again made parallel to their corresponding roots with respect to robot from their new shifted position and again the individual normal distance from the middle point of each feature to the corresponding root line will be found. All those association for which this new normal distance is within

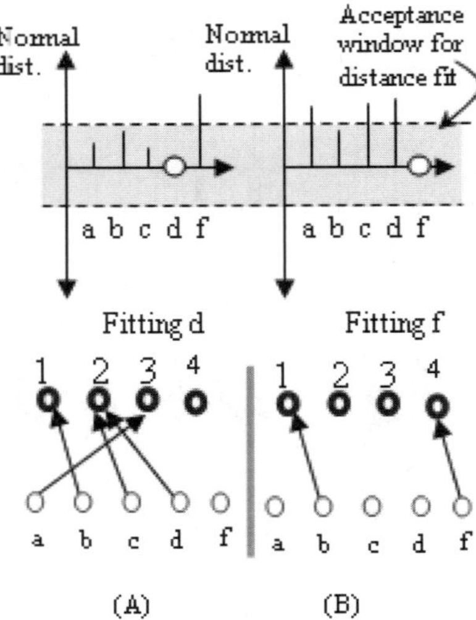

Fig. 9. Top row: Normal distance from middle of feature and the root after exactly fitting one feature at a time by making normal distance 0. Bottom row: Corresponding link graphs with the links which are within the green window of the accepted normal distance.

some threshold $-th2$ to $+th2$ will form an edge in the link graph corresponding to that particular fit. We have taken th2 as $(Act\_Avg\_Norl\_Dist + \epsilon2)$, $\epsilon2$ is calculated practically, in our case it is 5 cm.

Figure 9 shows link graph made by distance fit. Figure 9 (A) shows the link graph when feature d has been exactly fit, we will get similar link graphs for fitting features a, b and c. Figure 9 (B) shows the link graph by fitting feature f. So, obviously the winner link graph for distance fit is 9(A).

Winner Link Graph

Now to decide the final winner link graph, various pairs of winner graph of angle fit and distance fit will be made by taking one from each set at a time. For each pair, another link graph is created by taking the common edges from both the graphs. The final winner will be the link graph from the pair, which will be having maximum number of common edges. And the common edges will be the final edges of the winner link graph. In our example we will get various pairs having same number of common edges, the pair 8(B), 9(A) is one of them. So we will choose the common part of this pair as the final winner link graph as shown in fig.10. It has removed the false associations of features c and f obtained in current scan.

3.4 Calculating Probability of Association

After having a reliable set of association which will be used for correcting pose (x, y, θ) of the robot, we will assign some measure of certainty of association

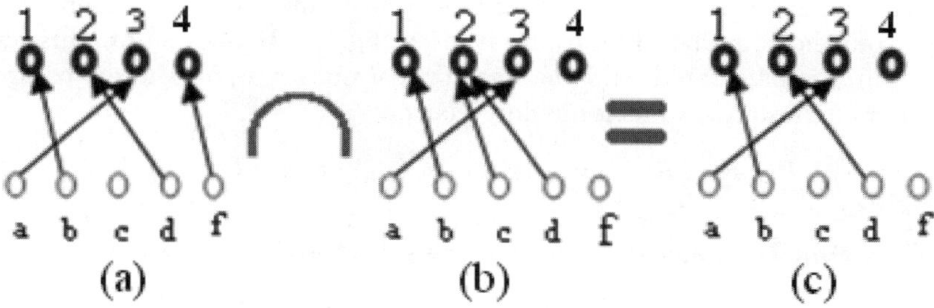

Fig. 10. (a) : one of the winner graph by angle fit, (b) : one of the winner graph by distance fit, (c) : Final winner link graph by taking common edges which has removed false association of features c and f

to a particular feature with the corresponding associated chain. We will assign higher weight to those associations, which are more agreeing in a global sense. Because it may be the case that angle difference of a particular feature with its root is less but for 3 other associated features it is falling in a higher range within some limits. So those 3 associations are more agreeing about the amount of orientation error even when their angle differences are more. Hence we find clusters of features based on the angle difference with corresponding roots. The cluster having maximum number of features will be given highest weight. A probability is assigned to each cluster by equation 1:

$$P(cluster) = \frac{Number\_of\_elements\_in\_the\_cluster}{Total\_number\_of\_elments} \tag{1}$$

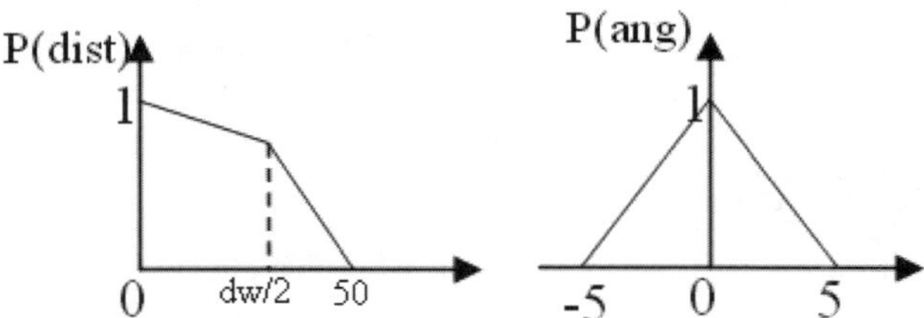

Fig. 11. (a) graph for calculating probability with respect to nearest end point distance; (b) with respect to angle difference from mean of the cluster

A second type of probability is calculated based on the distance of the associated feature to the nearest end of the associated link (not necessarily the root) in the chain. This probability $P(dist)$ is calculated by the graph shown in fig. 11 (a). Here dw is the average width of the commonly existing doors. If the difference is less than half of the door width then probability will be higher to ascertain more certainly that the current feature as the part of the same wall. A third type of probability is calculated based on angle difference from the mean of the cluster. Note that cluster itself has been found by angle difference of a feature from the corresponding root. Graph in fig. 11 (b) shows how this probability $P(ang)$ is calculated.And the final probability of association is computed as a union of these probabilities as in equation 2:

$$P(association) = P(cluster) \cup P(dist) \cup P(ang) \tag{2}$$

3.5 Creating New Chains from Unassociated Features

It is common and also encouraging to have some unassociated features in a particular scan. It is an indication of some new unseen portion or structure of the environment. Each unassociated feature will create a new chain by making itself as the root of that chain. But this time the probability of that root will not be 1 as in the case of first scan. $P(root)$ is basically the certainty of location of that root feature itself and since robot has moved from its initial position, some error must have been introduced. There are two sub-cases: (a) some other features of the current scan have been associated with some existing feature chain. (b) None of the features of the current scan could be associated with any of the existing chains. For the first case the probability to root will be assigned by equation 3 :

$$P(new\_root) = \frac{\sum_{i=1}^{n} P(root_{i^{th}\_associated\_chain})}{n} \tag{3}$$

Where n is the total number of features of the current scan which has been associated to any of the existing chain. For the case (b), the $P(new\_root)$ will be half of the probability which has been calculated for the most recent case (a) by equation 3. And it will be assigned a tag of being a dangling chain. This will be the region where uncertainty in localization will be the most. Our exploration strategy should be such that it should encourage getting some unassociated features but of type (a), discourage getting dangling chains and also it should try to associate dangling chains with some normal chains by finding the missing link if possible. A special case has been discussed at subsection 3.7.

3.6 Local Localization

After having a reliable feature association with association probabilities, a two step localization process has been introduced to correct the position (x, y) and orientation θ of the robot. First the orientation of the robot is corrected followed by (x, y) correction.

Correcting Orientation

For correcting the angle assumption of the robot we take a weighted sum of angle difference from the roots (not with the associated links), and find the robots orientation, which will minimize equation 4 :

$$\theta = arg\_min(\frac{\sum_{i=1}^{n}((w_1 + w_2 + w_3) * ADR(asso_i))}{n}) \tag{4}$$

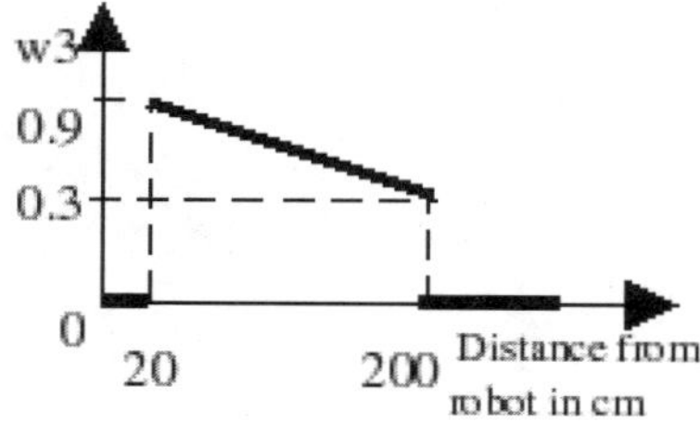

Fig. 12. Graph for calculating weight w_3 based on distance to feature

Where arg_min returns the value of argument θ which will minimize the quantity inside the bracket, n is the number of associated features instances. $ADR(asso_i)$ is the angle difference of the feature i with the root of the associated feature chain.[1] Weights w_1 is simply the probability of root, $P(root)$, of the associated chain, calculated by equation 3; w_2 is half of the probability of association $P(association)$ calculated by equation 2. If a feature has been detected far from the robots current position the accuracy of orientation and position of feature compared to the nearby-detected features will be less. So another weight w_3 takes care of this fact. Figure 12 shows the graph to calculate w_3 with respect to distance from the robot. 20 cm and 200 cm are the minimum and maximum range of getting reliable reading by our sonar sensor.

Correcting Position (x, y)

For correcting the (x, y) co-ordinate of the robot, we have to align each individual associated feature completely with its associated link in the chain. Then the normal distance from the middle point of the current feature onto the associated link feature will be calculated as shown in figure 13. For each valid association this distance will be decomposed into $(\Delta x, \Delta y)$. This decomposition is required, because if we directly use the normal distance for searching the correct position and if all the associated features are parallel to either x or y axis then we will end up with a line segment as the possible correct position instead of getting a point. But by using maximum values of Δx and Δy from this decomposition as a window for our search in x and y direction, if features are parallel to any of the axis, as may be the case for corridor, one of the components will nearly equal to 0 for all the associated features. Hence even if robot will not be able to correct the position along one axis, it will also not increase the error along that axis by avoiding the ambiguity.

Now to get the correct (x, y), we search in a window of $-$ $maximum$ of Δx and $-$ $maximum$ of Δy to $+maximum$ of Δx and $+maximum$ Δy, from the current assumed position of the robot. For each possible position of the robot in

---

[1] It is important to note that the association has been done with link of the chain but the correction will be done with the root of the chain, because the roots are more reliable than any other link of the same chain.

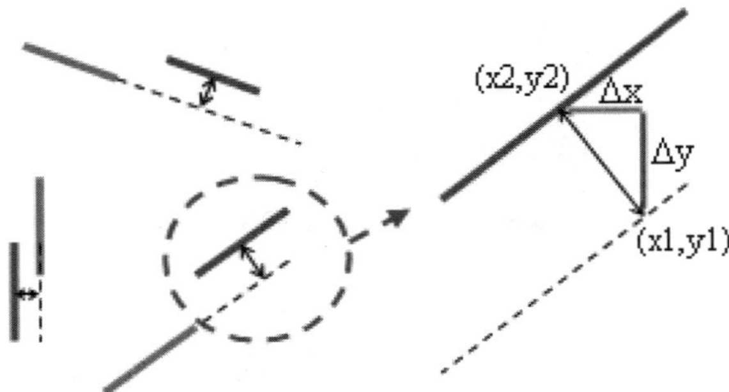

Fig. 13. Finding normal distance and decomposing the shift into (x,y) component

that window, new position of the middle points (which has been found earlier by making all the associated features virtually parallel to root with respect to robot) is calculated. The shift in x and y, which will minimize the weighted sum of the normal distances given by equation 5, will be the final shift to be added in the current position to get the correct position of robot.

$$(\Delta x, \Delta y) = arg\_min(\frac{\sum_{i=1}^{n}((w_1 + w_3) * NDR(asso_i))}{n})$$ (5)

Where w_1 and w_3 are the weights calculated in equation 5 and $NDR(asso_i)$ is the normal distance from the middle of the feature (made virtually parallel) to the associated root.

After getting corrected (x, y, θ), all the readings and the features are processed to make them consistent with the corrected location and orientation of the robot. And then map is incremented using our map building approach presented in [7].

3.7   Handling Duel Association

A special case can arise when a same current feature has been initially associated with two different feature chains. Then each association will be treated as separate instance and all the analysis for removing false associations will be done as usual. If even after false association removal, that feature is still being associated with both the chains; it indicates either the loop completion or the missing link of a feature due to which same wall is being treated as two different features so far. Also one chain may be the dangling chain mentioned in the section 3.5.

In all the cases the two chains will be merged and the new root will be the root which is having higher probability. Also if one chain is dangling chain then that tag will be removed from that chain, because it has found its base chain. So the exploration strategy will not try to find its base chain any more for reducing the region of uncertainty.

4   Experimental Results and Analysis

We have tested our algorithm in different real environments. We are using Amigobot equipped with 8 sonar sensors fixed at certain angular intervals. For taking a 360 *degree* scan we are rotating the robot with 4 *degree* interval and storing the readings from one sensor only. Remaining sonar sensors are used during exploration and path planning.

Figure 14 shows the corridor environment. Figure 15(a) shows the map build by using present approach of SLAM in a fully autonomous fashion. Figure 15(b) shows the map built without correcting the localization error. For comparing both the maps, a dotted rectangle has been overlaid on them. The shift of the walls from the sides of the rectangle is much in 15(b), which is without SLAM and the shift in 15(a) is negligible showing the accuracy of map closer to the ground truth. The gray region at the bottom right in fig.15(a) is showing the safe region where robot can move and the green cells are showing the shortest path planned by robot to move to next location for further exploration.

Figure 16(a) and fig. 16(b) show two portions of a corridor as it bends around a corner, which is not at right angles. Figure 17(a) shows the map built using the present method of SLAM. Figure 17(b) shows the map obtained by robot

Fig. 14. A corridor environment

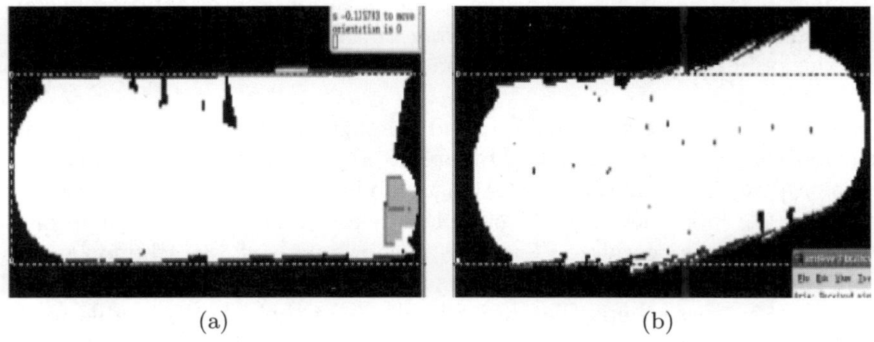

(a)   (b)

Fig. 15. Comparison of the maps obtained with SLAM fig.(a) and without SLAM fig.(b), with an overlaid dotted rectangle resembling the actual environment as shown in fig.14

(a)   (b)

Fig. 16. Two portion of a corridor

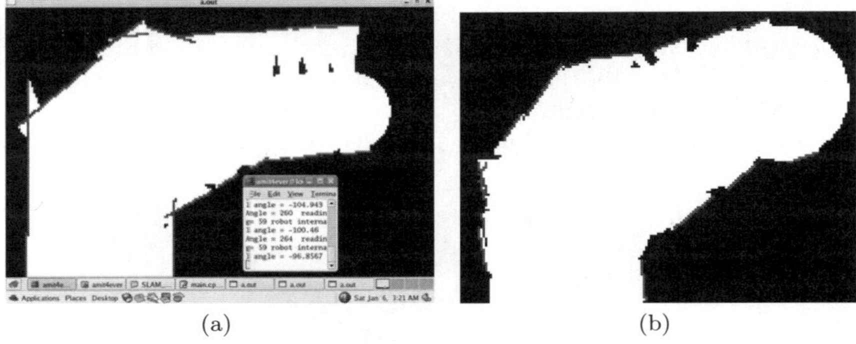

(a)   (b)

Fig. 17. (a): map of environent shown in fig.16, by using present approach of SLAM, (b) : map of the same environment built without correcting localization error

of the same environment but without robots position being corrected. The enhancement is visible. In fig. 17(b) the robots position and orientation becomes more inaccurate after some scans resulting in a map that is shifted and tilted. Using the present approach however the map in fig. 17(a) is closer to ground truth results.

As a proof of concept we have tested our algorithm in two different environments, which are having loops. Figure 18(a) and 18(b) shows two such environmental setups. Environment 18(b) is having one side open, so that after completing the loop robot should be able to explore and build the map of the open portion of the corridor. Figure 19(a) shows map of 18(a) build by using present approach and fig. 19(b) shows map of 18(b) using present approach. It is evident from these maps that in both the cases the robot was able to successfully complete the loop by exploring the environment autonomously without any path guiding, and the maps are also near to ground truth. Without correcting localization error robot was even not able to explore the loop environment completely without external path guidance.

Fig. 18. Two environments which requires the robot to complete the loop, 18(b) is having one end open

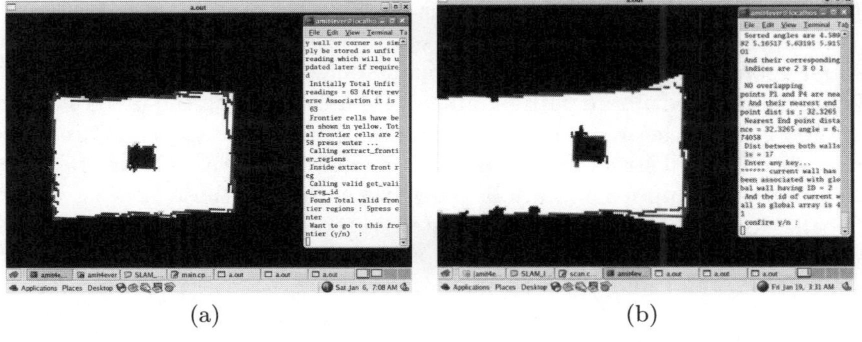

Fig. 19. Figures (a) is Map built of fig.18(a); figure (b) is map of fig.18(b), in both case loop has been completed successfully

Fig. 20. Another test-environment having rooms and corridor

Fig. 21. Map built by using present approach of SLAM. Without correcting the localization error robot was even not able to come out of the room autonomously.

Figure 20 shows another environment having corridor and rooms. Figure 21 shows the map built by present approach of SLAM. Robot had started from the first room and then autonomously came out of the room by exploring and building the map with safe path planning. Then by building the map of corridor it entered into another room to explore and build map. During the entire process due to various small structures in the environment robot was getting lots of false associations. Our approach was able to detect and remove those false associations to localize properly. Also when we tested without correcting the localization error, robot was even not able to come out from the first room autonomously to explore the corridor and other rooms.

5   Conclusion

We have presented a robust feature chain based approach for online SLAM by using data from sonar sensors alone. The readings from sonar sensors are unpredictable beyond some minimum angle of incidence of the sonar beam onto the obstacle and hence getting consistent pair of features or readings from two different positions is rare. The proposed feature chain based approach is able to handle various issues with sonar data and successfully localizes and builds a more accurate map. One of the benefits of maintaining the feature chains having roots and other links is that the robot does not need to re-visit the previously detected features each time it has to localize. It can use the information of earlier features to localize from the current position itself even if those features have not been detected in the present scan. Furthermore for associating a current feature the closest link of the chain is used but for correcting the localization error we traverse back to the root of the chain and then the position and orientation of the root is used for minimizing the cost functions, thus enhancing the accuracy. Also the developed two-step global consistency association analysis based on link graphs of angle fit and distance fit is able to remove false associations; these false associations are practically unavoidable when using sonar data. The present approach enables the robot to explore the environment autonomously without any external aid. The robot is able to build maps involving loops, demonstrating its efficacy.

References

1. Choi, J., Ahn, S., Choi, M., Chung, W.K.: Metric SLAM in home environment with visual objects and sonar features. In: IEEE International Conference on Intelligent Robots and Systems (2006)
2. Gutmann, J.-S., Schlegel, C.: AMOS: Comparision of scan matching approaches for self localization in indoor environment (1996)
3. Stachniss, C., Hahnel, D.: Exploration with active loop closing for FastSLAM. In: IEEE International Conference on Intelligent Robots and Systems (2004)
4. Maohai, L., Bingrong, H.: Novel mobile robot simultaneous localization and mapping using rao-blackwellised particle filter. International Journal of Advance Robotics System (2006)
5. Tardos, J.D., Neira, J., Newman, P.M., Leonard, J.J: Robust mapping and localization in indoor environment using sonar data. International Journal of Robotics Research (2002)
6. Wijk, O., Christensen, H.I.: Triangulation based fusion of sonar data with application in robot pose tracking. IEEE Transactions on Robotics and Automation (2000)
7. Pandey, A.K., Madhava Krishna, K., Nath, M.: Feature based occupancy grid maps for sonar based safe mapping. In: IJCAI. International Joint Conference on Artificial Intelligence, pp. 2172–2177 (2007), Available:
   `http://www.ijcai.org/papers07/contents.php`
8. Lacroix, S., Dudek, G.: On the identification of sonar features. In: IEEE International Conference on Intelligent Robots and Systems (1997)
9. Thrun, S., Burgard, W., Fox, D.: Probabilistic Robotics, pp. 312–316, 324–332. The MIT Press, Cambridge (2005)
10. Douillard, B.: Design and implementation of an SLAM algorithm on an Amigobot. M.S. Thesis, Dept. of Mechanical and Aerospace engineering, State university of New York at Buffelo (2005)

Services Robots Navigating on Smart Floors

Erwin Prassler[1,2], Thomas Kämpke[1], Boris Kluge[1], and Matthias Strobel[1]

[1] InMach Intelligente Maschinen GmbH, 89081 Ulm, Germany
   {kaempke, kluge, prassler, strobel}@inmach.de
[2] Univ. of Applied Science Bonn-Rhein-Sieg, 53757 Sankt Augustin, Germany
   erwin.prassler@fh-brs.de

Summary. We describe an approach to position estimation for mobile service robots and automatically guided vehicles which is based on RFID technology. The core idea of the approach is to structure the work space by means of a *Smart Floor* in a manner which enables and supports reliable navigation and positioning with absolute accuracy over large distances. The "intelligence" of the floor is in a dense, area-wide network of thousands of RFID transponder, which are mounted underneath the regular floor covering and quasi serve as radio beacons. The navigation system described in this paper, has been presented at CeBIT 2006 in Hannover by invitation of the German Ministry of Education and Research. InMach Intelligente Maschinen has further received the Walter Reis Innovation Award for Service Robotics for this innovative solution.

1 Introduction

In this paper we describe an approach to position estimation for mobile service robots and automatically guided vehicles which is based on RFID technology. The core idea of this approach is similar to a well-established solution for vehicles which are guided by wires or magnetic beacons. These "navigation aids" are buried in the floor to structure the work space in a manner which enables and supports reliable navigation and positioning with absolute accuracy over large distances.

Structuring of the work space in our case does not involve the application of any visible markers or wires. Instead the workspace is equipped with a so-called smart floor – a key component of our approach – which neither in its basic physical consistency nor in its optical appearance differs much from a regular floor covering. The "intelligence" of the floor is in a dense, area-wide network of thousands of RFID transponder, which are mounted underneath the regular floor covering and quasi serve as radio beacons. A commercial version of such a smart floor has been developed by the German company Vorwerk Teppichwerke, which is a partner of InMach Intelligent Maschinen GmbH.

The second key component in our approach is the actual RFID based vehicle navigation system. While the vehicle is in motion, this navigation system senses the identifiers of RFID transponders in its proximity and computes an estimate of the absolute position and orientation of the vehicle and uses these data to

S. Lee, I.H. Suh, and M.S. Kim (Eds.): Recent Progress in Robotics, LNCIS 370, pp. 133–144, 2008.

guide the vehicles locomotion along a planned path. As a side effect of this absolute position accuracy the data which are perceived, for example, by the vehicles low-cost range sensors, can be communicated by wireless LAN to a central server and there be integrated in to a central, distortion-free, global map of the environment.

Through their synergy these two key components *smart floor* and *RFID based navigation* provide a low-cost, flexible guidance system with absolute position accuracy for mobile service robots and AGVs. This solution has the potential not only to re-animate but also to boost the market for mobile robots for professional services which has ben almost disappearing in the past years.

The navigation system described in this paper, has been presented at CeBIT 2006 in Hannover by invitation of the German Ministry of Education and Research. InMach Intelligente Maschinen has further received the Walter Reis Innovation Award for Service Robotics for this innovative solution.

2   Related Work

The RFID based navigation system described here is not so much the result of a basic research activity or a prove of concept. It is an industrial prototype which is currently under field testing and which is supposed to reach series-production readiness by the end of 2006. It will be used in two professional applications first: in an *automatically guided transportation vehicle* (AGV) for hospital logistics and in a robot for professional cleaning. This short review of related work will therefore focus on professional guidance and navigation systems for mobile (service) robots and only briefly discuss basic research on RFID based navigation and smart floors. The state of the art in professional navigation systems will be shown in a few selected, highly representative Systems.

2.1   Mobile Service Robot Guidance and Navigation

Using mobile robots for professional services for example in the area of facility management is not a very original idea. Since the beginning of the nineties several attempts have been undertaken, to automatize burdensome, dirty, low-qualified tasks by means of mobile service robots. The following systems are representative for the state of the art in this field.

Helpmate™ - Logistics in Hospitals

Helpmate was the first mobile service robot, which was specifically designed for logistic and transportation tasks in hospitals (see Fig. 1). HelpMate had a rather comprehensive sensor configuration, which included amongst others a laser range finder, a set of sonar sensors, and a gyroscope. It used a floor plan and permanently installed markers for navigation. Helpmate was preferably used for transportation of food, medication, and laundry. In spite of the public and media interest, the commercial use of HelpMate has not become a success story. In total only 150 units were manufactured and installed [1]. Helpmate cost approximately

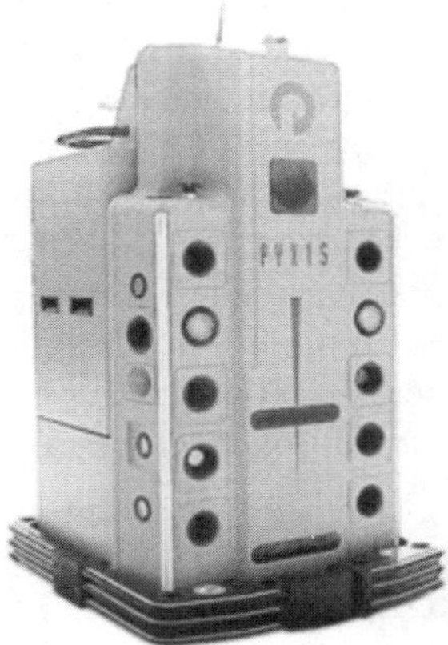 

Fig. 1. Logistics robot HelpMate and robotic floor scrubber Hefter Cleantech ST82R

US\$ 110.000 per unit. Typically the system were not sold but leased for an hourly rate of US\$ 4 to 6. The yearly lease was US\$ 110.000. It is not completely unlikely that the low number of installed systems is related to the rather high initial cost and the insufficient price-performance ratio.

Robots for Professional Cleaning

A representative of the more recent developments in the area of service robots for professional cleaning is the robotic floor scrubber ST82 R Variotech of Hefter Cleantech, Germany. Its navigation system SINAS (Siemens Navigation System) included two laser range finder, an array of sonar sensors, a gyroscope and an odometry system. Using its sensors the system was capable of generating a map of its work space. ST82 R Variotech was distributed for approximately 50.000 Euro. Amongst others it was installed in a Dutch supermarket chain. One unit which was installed at the airport of Manchester, UK, was give to the Queen Elizabeth II Hospital in Welwyn Garden City, Hertfordshire in 2001. The manufacturing and sale of ST82 R Variotech has been discontinued [2].

2.2   Recent Developments in the Area of AGVs

For many years automatic guidance of vehicles by induction wires or magnetic beacons or optical markers were state of the art ([3, 5, 4]). Not until recently a sensor, which has significantly influenced and fostered basic research in mobile robotics, namely the SICK PLS and LMS sensor, has also expanded into the development of automatically guided vehicles (AGVs). A representative development is the system Transcar LTC-2 by Swisslog Telelift GmbH ([4]). Transcar LTC-2 uses two laser range finder (one in the front and one in the rear area) to sense and measure the environment. To estimate the position of the vehicle the perceived range images are compared and aligned with a give floor plan. One would expect that the system by using laser range finders has become significantly more flexible than traditional, wire-guided systems. This flexibility has its limits, however. For reliable position estimation and navigation a complete metric model (map) of the environment is required. In a second topological map way points for navigation are defined. The approach quasi implements a virtual automatic guidance. Cost for pre-structuring the work space can be avoided this way. However every change of the environment for consistency reasons requires an update of the metric map of the work space. Such an update may require a significant effort and cause significant cost. Automatically mapping the workspace, which has its own deficiencies, is not foreseen in this approach. Navigation in an unknown environment seems to be impossible.

2.3   RFID Navigation and Smart Floors in Basic Research

There is not much basic research in RFID navigation and smart floor. The literature is accordingly sparse. Based on the patent situation it seems that industry has discovered this topic before the academic community. There is actually an significant number of patents or patent applications out there which address similar or almost identical problems: given we can distribute a sufficient number of RFID transponders over the workspace, how can a robot use such an infrastructure and the information encoded in the transponders to estimate its position and execute its mission. The approaches described in [10, 9, 6, 7, 8] address the same problem and provide very similar solutions. Typically the position estimation is solved by aligning and/or fitting the observations of one or more tags and their approximate position sensed by odometry to an established frame with known tag positions.

A frequently cited academic publication is the work by Hähnel et al. [11]. They generate maps of RFID tags with mobile robots and compute a posterior about the position of a tag after the trajectory and the map has been generated with a highly accurate FastSLAM algorithm for laser range scans. A concept of a Smart Floor equipped with RFID transponders was presented in [12]. This smart floor is primarily used for identifying and tracking people.

3   Technical Approach

The core idea of this approach is to pre-structure the work space like in earlier approaches using induction loops and magnetic beacons in order to enable reliable navigation and absolute positioning with high accuracy over long distances. Our navigation system operates on a so called smart floor, a dense, area-wide network of thousands of RFID transponder, which are mounted underneath the regular floor covering and quasi serve as radio beacons.

The guidance system gets by without expensive range sensors like laser range finder and uses regular sonar sensors as they are used in parking assistance systems in the automotive industry. An extra "sensor" which is used for position estimation is an RFID reader. These devices are commercially available for prices starting at 100 Euros.

The core idea of this approach is to pre-structure the work space like in earlier approaches using induction loops and magnetic beacons in order to enable reliable navigation and absolute positioning with high accuracy over long distances. Our navigation system operates on a so called smart floor, a dense, area-wide network of thousands of RFID transponder, which are mounted underneath the regular floor covering and quasi serve as radio beacons.

The guidance system gets by without expensive range sensors like laser range finder and uses regular sonar sensors as they are used in parking assistance systems in the automotive industry. An extra "sensor" which is used for position estimation is an RFID reader. These devices are commercially available for prices starting at 100 Euros.

3.1   RFID Makes the Floor Smart

The core idea of the Smart Floor is to distribute and integrate RFID transponders in the floor in an area-wide manner and use the transponders for various purposes. Since the integration in the floor covering itself would complicate the manufacturing and increase the price of the floor covering and would have to be implemented separately and specifically for every type of floor covering, our partner Vorwerk Teppichwerke has developed a so-called *Smart Underlay* (see Fig. 2). The smart underlay is separate meadow-like textile, in which the RFID transponders are integrated typically in a regular grid. This underlay can be placed nearly below every non-metallic floor covering. The material has furthermore an acoustic insulation effect. At the IT fair CeBIT 2006 in Hannover, for example, smart underlay was placed below wood (laminate), plastic, tiles, and carpet. Together with the floor covering on top of it the smart underlay forms the so-called *Smart Floor*.

The operation principle of RFID transponder is as follows. A reading device (reader) creates an electro-magnetic field over an antenna. RFID transponders located in this field will be charged by the emitted energy and enables to respond with an electro-magnetic echo. This echo typically carries an identifier plus various information such as co-ordinates, traffic information, environmental information, service information and alike, which is received by the reader. Transponders cannot only be read but also be written with information.

The piece costs for transponders basically depend on the storage capacity. The price per tag is currently around 0,30 to 0,50 Euro. Due to very tight competition a decay of this price to approx. 0,10 Euro per transponder is not unlikely.

The RFID tags integrated in a smart underlay or a smart floor can serve various purposes:

- They can be used for position estimation e.g. in large buildings for robots as much as for humans.
- They can store location specific information regarding the execution and performance of a service. For example, one could store the date of the last cleaning service and the code of a specific cleaning liquid or procedure.
- They could store information concerning the guidance of robot vehicles such as speed limits, temporary or permanent restricted areas.

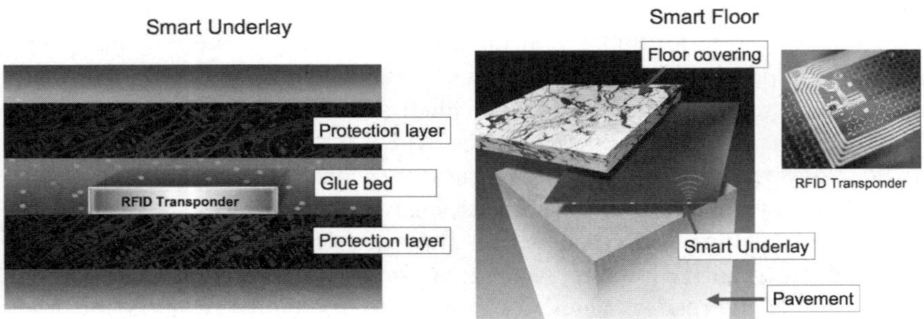

Fig. 2. Design Smart Underlay and Smart Floor

The smart floor quasi can be impregnated with information which is of relevance for robot navigation and the execution of the mission and the service.

It is worth noting, that location-based information regarding the environment and the objects therein comes with absolute position accuracy. This property allows one to fuse and integrate the perceptions of many robots with many different sensor suits over time in a central map. Through this integration of time and the perceptions of many robots an increasingly accurate and complete image of the environment arises. All robots can access this central map "for free" and at any time and use it for safe locomotion and execution of their mission. The RFID readers, which are used to read the information stored in the tags, quasi serve as additional, low-cost but highly accurate sensor.

In the extreme case, a robot could navigation almost blindly only by using its RFID readers, the RFID transponders, and the central map. This, however, requires the information in the central map to be up to date and to represent a correct and complete image of the environment.

3.2   Self-localization on a Smart Floor

A robot that uses RFID transponders for planar self-localization may be equipped with one or several RFID readers each having one or several antennas. Operating several antennas from one reading device can be facilitated by time-multiplexing. Any antenna is directed towards the floor. Obviously, a large antenna lobe allows for reading all transponders from a large area but each at a small spatial resolution and a small antenna lobe allows for the opposite.

Single Point Observations

Let the spatial resolution of transponder observation be high enough to allow for estimates as single points in robot coordinates. Also, it is assumed that at most one transponder can be seen at any moment and transponder observation does not provide any directional information. These assumptions correspond to a "point-shaped" antenna.

When the robot moves for a certain time or distance, it records all transponder readings in one local map. Whenever self-localization is invoked, the local map is matched to a global map of the smart floor. The global map contains the positions of all transponders in world coordinates.

Self-localization under the present conditions admits the computation of the position only. The orientation is inferred from the robot odometry.

The present map matching is nothing but point cloud matching with the correspondence problem being trivially solved since transponder identifiers are unique. Translation and rotation of the local map to the global map adhere to the least squares approach

$$\min_{\varphi \in [0,360^o),\, \vartheta \in I\!\!R^2} \sum_{i=1}^{n} ||T_i^{world} - (Rot_\varphi T_i^{robot} + \vartheta)||^2 .$$

The summation index n equals the number of transponder observations, with multiple observations of the same transponder being handled as observations of different transponders with possibly the same coordinates. T_i^{robot} and T_i^{world} are the coordinates of transponder i in the robot and the world frame respectively. The translation vector ϑ may vary unboundedly in the plane and the rotation is denoted by a standard rotation matrix

$$Rot_\varphi = \begin{pmatrix} \cos\varphi & -\sin\varphi \\ \sin\varphi & \cos\varphi \end{pmatrix} .$$

The optimal rotation angle is computed from two candidates values. The first value is $\varphi_1 \in [0, 180^o)$ with

$$\tan\varphi_1 = \frac{\sum_{i=1}^{n} det(T_i^{robot} - \bar{T}^{robot} \quad T_i^{world} - \bar{T}^{world})}{\sum_{i=1}^{n} (T_i^{robot} - \bar{T}^{robot})^t (T_i^{world} - \bar{T}^{world})}$$

if the denominator is different from zero, else $\varphi_1 = 90^o$ and the second candidate angle is $\varphi_2 = \varphi_1 + 180^o$. The superscript t denotes the transpose of vectors and the superscript $^-$ denotes average vectors.

The optimal rotation angle is obtained from value insertion as

$$\varphi_0 = \underset{\varphi_{j=1,2}}{\mathrm{argmin}} \sum_{i=1}^{n} ||T_i^{world} - \bar{T}^{world} - (Rot_{\varphi_j} T_i^{robot} - \bar{T}^{robot})||^2$$

and the optimal translatoin vector is $\vartheta_0 = \bar{T}^{world} - Rot_{\varphi_0} \bar{T}^{robot}$.

Position and orientation of the robot frame are eventually transformed into the world frame by

$$P^{world} = Rot_{\varphi_0} P^{robot} + \vartheta_0$$
$$\varphi^{world} = \varphi^{robot} + \varphi_0 \bmod 360^o .$$

Single Line Observations

A suitable antenna shape for mobile robots is that of a rod antenna or a slim rectangle that comes close to a rod. The antenna is mounted crossways to the machine direction and it should extend over the whole robot width. The latter results in a high likelihood to observe all transponders that are overpassed.

A straight robot motion is considered first. Whenever a transponder is traversed, it can be assumed to lie on a line segment that indicates the rod position at the observation moment. All the segments are parallel since the motion is straight. The distances between successive observations of transponders $i - 1$ and i are recorded as $D_{i-1,i}$, see figure 3.

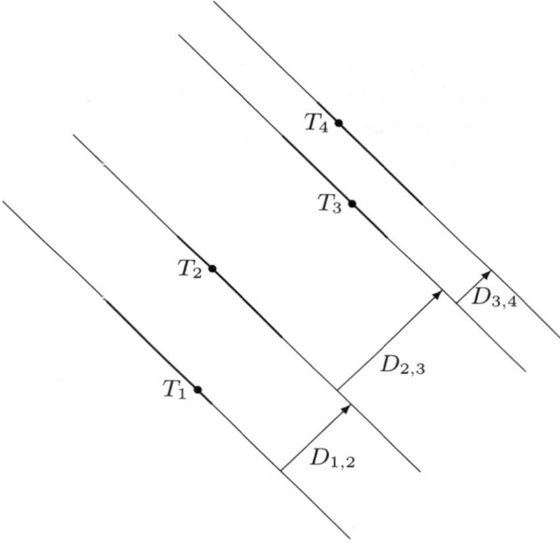

Fig. 3. Straight robot motion with transponder overpasses in the order $1, 2, 3, 4$ with successive distances $D_{1,2}, D_{2,3}, D_{3,4}$. The rod positions (bold segments) on the parallels are unknown and the exact positions of the transponders relative to the rod positions are unknown.

The observed distance between transponder traversals being directional rather than Euclidean distances between transponder positions allows to compute the robot orientation in world coordinates. Let the robot make n transponder traversals along one straight line. Its direction in world coordinates is a solution of the minimization problem

$$\min_{\varphi \in [0,360^{o})} \sum_{i=2}^{n} \left((\cos \varphi, \sin \varphi) \left(T_i^{world} - T_{i-1}^{world} \right) - D_{i-1,i} \right)^2.$$

This problem can be solved numerically by iterated bisection. The robot position is not unique with respect to the observations but it can be narrowed down

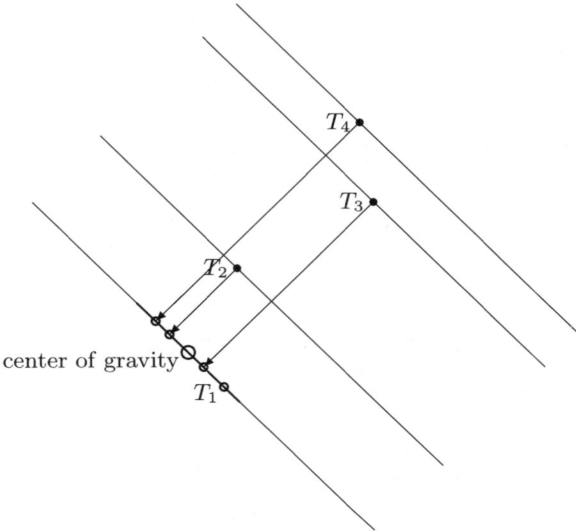

Fig. 4. Projections of transponder positions onto the line through the first transponder position. The center of gravity of these projections is a reasonable approximation for the robot position in world coordinates at the moment of observation of transponder 1.

to a segment such that all projected transponder positions fall in that segment, see figure 4.

From this segment, the center of gravity of all projections can be chosen as position estimate of the robot while it observed the first transponder. The current position in world coordinates is then inferred as in the foregoing section.

The approach can be extended from motion along a single line to other paths, in particular to parallel lines as they are typically used by cleaning robots.

4   Field Evaluation

The results of a first quantitative evaluation of the position estimation performance of our approach is shown in Fig. 5. For this evaluation a test arena of 4 m × 3 m was set up, with RFID tags laid out in intervals of 50 cm. Within this test arena the robot was sent on a rectangular course for about 10 minutes at a speed of approximately 30 cm/sec. During the experiment the robot traveled approximately 170 meters. The position was estimated approximately every 200 millisecond or every 5 to 6 cm. As can be seen in Fig. 5, the trajectory fitted to the estimated position forms a smooth path.

In Fig. 6, the error between ground truth position and estimated position is shown. Ground truth was obtained by means of a specifically designed measuring system involving a laser range finder, which of course also contributes some error to the measurement. In the diagram in Fig. 6 we can see that there is some significant position errors in the beginning of the experiment. Such significant errors occur only during the initialization phase of the algorithm. Once the system has the first fix, the errors drop drastically. It is also worth noting that the error in the described experiment stayed within bounds defined by the tag distance and the reading distance of the RFID reader. This, however, is not a guaranteed bound. If the reader does not discover a tag for a longer period of time, the position error might as well grow beyond this bound.

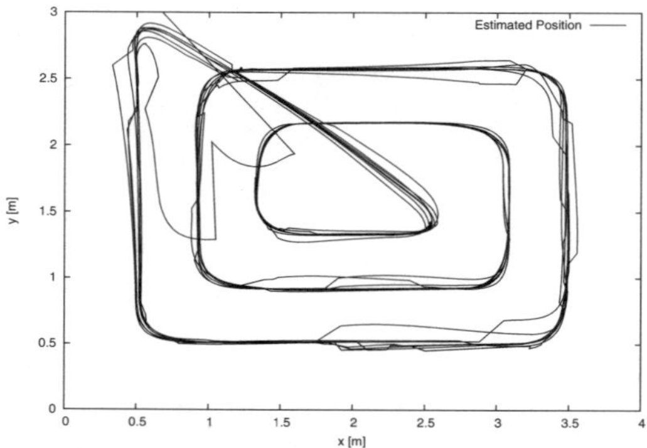

Fig. 5. Positions estimated along a rectangular course over a traveled distance of approx. 170 meters

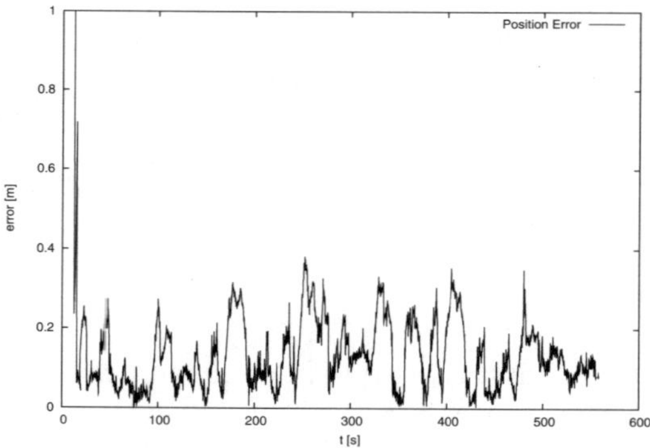

Fig. 6. Position error between ground truth position and estimated position

The RFID based navigation system described in this paper has matured to a level which allows and suggests excessive field testing. Three cleaning robots equipped with prototypes of the navigation system have been demonstrated during the international IT fair CeBIT 2006 in Hannover (see Fig. 7). During this exhibit the robots were in operation for approximately 6 to 8 hours per day, seven days long. Occasional mulfunctions of the robots were due to mechanical problems and problem with the collision detection sensors. The RFID navigation

Fig. 7. CeBIT 2006: first field tests for RFID navigation of robo40 cleaning robots

component itself worked without any problems. Currently an extended field test and a permanent installation of a smart floor in a gym and a cleaning robot navigating with RFID is in preparation.

5   Conclusion

We have presented a novel position estimation system for mobile service robots using RFID transponders in a so-called smart floor. The approach has several advantages over existing solutions. It provides absolute position accuracy over arbitrary distances at low-cost. The sensor(s) needed for absolute position estimation are one or more RFID reader devices at price of 50 to 200 Euro each (depending on range and frequency). Additional features and advantages are easy installability and moderate installation cost, flexibility, and a spectrum of functions provided by the programmable smart floor.

Altogether the described RFID based robot navigation system has the potential not only to revitalize but to boost the technology and the stagnating market in the field of automatically guided transportation vehicles for industrial applications as much as the development of mobile service robots for professional applications.

Acknowledgment

This work was partially supported by the German Ministry of Education under grant no. 01IME01J in the context of the German Service Robotics Initiative DESIRE.

References

1. Robot navigation technology,
   http://www.atp.nist.gov/eao/sp950-1/helpmate.htm
2. BBC News: Hospital recruits robot cleaner,
   http://news.bbc.co.uk/1/hi/health/1396433.stm

3. Fraunhofer IPA Technologieforum. Fahrerlose Transportsysteme (FTS) (2005)
4. Fottner, J.: FTS-Einsatz in der Krankenhauslogistik. In: Fraunhofer IPA Technologieforum. Fahrerlose Transportsysteme (2005)
5. Osterhoff, W.: Technologische Innovationen und Trendentwicklungen bei FTS. In: Fraunhofer IPA Technologieforum. Fahrerlose Transportsysteme (2005)
6. Song, J.-G., Lee, S.-Y.: Mobile Robot Using RF Module. US 6.597,143 B2. Samsung Kwangju Electronics Co. (2001)
7. Sung-Il, J.: Automatic Guided System and Control Method Thereof. US 6.493.614 B1. Samsung Electronics Co. (2002)
8. Lee, Y.-K., Choe, W.: Apparatus and Method for Measuring Position of Mobile Robot. US 2003/0236590 A1 Samsung Electronics Co. (2003)
9. Piotrowski, T.: RFID Navigation System. US 2003/0080901 A1. Koninkljike Philips Electronics N.V. (2003)
10. Douglas, B., Bencel, J.: System for Automatic Guided Vehicle Navigation. US 6.049.745. FMC Corporation (1997)
11. Hähnel, D., Burgard, W., Fox, D., Fishkin, K., Philipose, M.: Mapping and Localization with RFID Technology. In: ICRA. Proc. IEEE Int. Conf. on Robotics and Automation, New Orleans (2004)
12. Orr, R.J., Abowd, G.D.: The Smart Floor: A Mechanism for Natural User Identification and Tracking. In: Proc. Conf. on Human Factors in Computing Systems (2000)

Control of Many Agents by Moving Their Targets: Maintaining Separation

Timothy Bretl

Department of Aerospace Engineering, University of Illinois at Urbana-Champaign,
Urbana, Illinois, 61801
tbretl@uiuc.edu

Summary. Consider a large group of agents chasing a small group of moving targets.
Assume each agent moves at constant speed toward the closest target. This paper
studies the problem of controlling the agents indirectly by specifying the motion of the
targets. In particular, it considers the problem of maintaining a minimum separation
distance between each pair of agents, something that is impossible to do with only one
target. This paper shows that only two targets are necessary to maintain separation
between four agents. It also shows results in simulation to support the conjecture that
only two targets are necessary to maintain separation between any number of agents,
given suitable initial conditions.

1 Introduction

This paper is motivated by recent work in which groups of microorganisms are
used to manipulate small objects [22]. The microorganisms, of the genus *Paramecium*, are controlled by cyclic application of electric fields. A single paramecium
responds to an applied electric field by swimming roughly in the direction from
anode to cathode. This response is known as *galvanotaxis* [21]. Given a sufficiently accurate model of its dynamics [18], galvanotaxis can be used as a
basis for regulating position or tracking a trajectory with a paramecium [17].
Of course, since all paremecia respond in approximately the same way to an
applied electric field, members of a large group can not be controlled individually. Instead, the micro-manipulation presented in [22] relies on the tendency of
paramecia to cluster during galvanotaxis. Electric fields are applied to move the
center of this cluster in a heuristic circulation pattern ("collide" and "return")
in order to push a small object.

A number of questions are raised by this work. For example, is it possible to
stretch or shrink the cluster of paramecia along one dimension? Is it possible to
maintain two or more stable clusters at the same time? More generally, how can
we decide whether an arbitrary configuration of paramecia is reachable, and how
can we construct a control policy to achieve any reachable configuration? Is it
always necessary to use closed-loop control, or is an open-loop (even sensorless)
policy sometimes sufficient? These questions are relevant to a variety of other
biological multi-agent systems in addition to paramecia: for example, guiding

S. Lee, I.H. Suh, and M.S. Kim (Eds.): Recent Progress in Robotics, LNCIS 370, pp. 145–156, 2008.
springerlink.com

crowds during emergency egress [15, 19], herding cattle with a robotic sheep-dog [24, 25] and invisible fences [7], or interacting productively with cockroaches using a mobile robot [8].

In this paper we consider a simple multi-agent system and focus on the task of maintaining two or more stable clusters of agents. In our model, there are a small number of targets whose trajectory we can specify, and a much larger number of agents that each move at constant speed toward the closest target. We will prove the following two statements: first, that at least two targets are necessary to maintain more than one stable cluster of agents; and second, that only two targets are necessary to maintain four stable clusters of agents. We will also show results in simulation to support the conjecture that only two targets are necessary to maintain any number of stable clusters, given suitable initial conditions.

2   Related Work

2.1   Control of Multi-agent Systems

Distributed control architectures have been applied successfully to a variety of multi-agent systems, including mobile sensor networks, automated air traffic control systems, and graphical simulations of flocking birds and schooling fish. In each case, an implicit assumption is that we can choose how individual agents respond to each other and to the environment. For example, we might specify that each agent regulates distance from nearest neighbors (to achieve a cohesive swarm) [20, 12], updates its heading based on the average heading of its neighbors (to achieve movement snychronization) [13], or moves toward the circumcenter of its neighbors (to achieve rendezvous) [9].

In this paper, we are interested in multi-agent systems where the dynamics of each agent are fixed. As a result, we can only control agents indirectly by changing external stimuli. This problem has received less attention, mostly in the context of formation control using virtual leaders [14] (in particular when the response of each agent is a linear function of its relative state [23]). Other strategies include commanding the location of a formation's centroid or the variance of its distribution [1, 11]. It is not well understood how to extend these heuristics to more general types of group motion.

2.2   Pursuit

The questions raised in the introduction are more general than any particular multi-agent system. But because of the system model we consider in this paper, we will be interested here in proving things about the motion of agents that chase moving targets at constant speed. In fact, the paths followed by such agents have been studied for over a century in the mathematics literature, where they are known as *curves of pursuit*. A classic series of papers by Arthur Bernhart provides analytic descriptions of these curves in certain cases, such as when the target is moving along a line or around a circle [2, 3, 5, 4]. Consolidations of this work appear in many textbooks on differential equations, such as [10]. In addition to having obvious military application, pursuit curves have been used for anything from explaining why ant trails are straight [6] to generating formations of mobile robots [16]. Of course, most of this past work assumes that agents are chasing the same targets for all time. In this paper we consider cases in which agents switch from one target to another. Indeed, we will show that this switching is necessary in order to maintain more than two stable clusters of agents.

3   Multi-agent System Model

Consider a collection of n agents and m targets. In general, we will assume n is much bigger than m. Let $x_1, \ldots, x_n \in \mathbb{R}^2$ be the position of each agent, and let $z_1, \ldots, z_m \in \mathbb{R}^2$ be the position of each target. All of these variables are functions of time. In our model, each agent moves at constant velocity toward the closest target. For simplicity, we will assume this velocity has unit magnitude. So we can describe the dynamics of each agent i as follows:

$$\frac{dx_i}{dt} = \frac{z_j - x_i}{\|z_j - x_i\|} \quad \text{where} \quad j = \arg\min_k \|z_k - x_i\|.$$

Our goal might be to specify a trajectory $z_j(t)$ for each target $j = 1, \ldots, m$ that achieves a desired trajectory $x_i(t)$ for each agent $i = 1, \ldots, n$. However, it is clear that if $n > m$ then the multi-agent system is not controllable. As a result, there are some agent trajectories than cannot be achieved (for example, try maintaining each agent at a fixed location). Moreover, if $n \gg m$, then the multi-agent system is highly underactuated, so even if there exist target trajectories that result in desired agent trajectories, it is not at all clear how to find them. So instead, we restrict our focus to finding a class of target trajectories that maintains separation between agents.

4   Separation Cannot Be Maintained Using Only One Target

Assume there is only one target (so $m = 1$). We can show that for any trajectory $z(t)$ of this target, it is in general impossible to maintain separation between pairs of agents. Define the squared distance between two agents by

$$f(x_i, x_j) = \|x_i - x_j\|^2 = (x_i - x_j)^T (x_i - x_j)$$

First we show that f never increases with time:

$$\begin{aligned}
\frac{1}{2}\frac{df}{dt} &= (x_i - x_j)^T \left(\frac{dx_i}{dt} - \frac{dx_j}{dt}\right) \\
&= (x_i - x_j)^T \left(\frac{z - x_i}{\|z - x_i\|} - \frac{z - x_j}{\|z - x_j\|}\right) \\
&= -((z - x_i) - (z - x_j))^T \left(\frac{z - x_i}{\|z - x_i\|} - \frac{z - x_j}{\|z - x_j\|}\right).
\end{aligned}$$

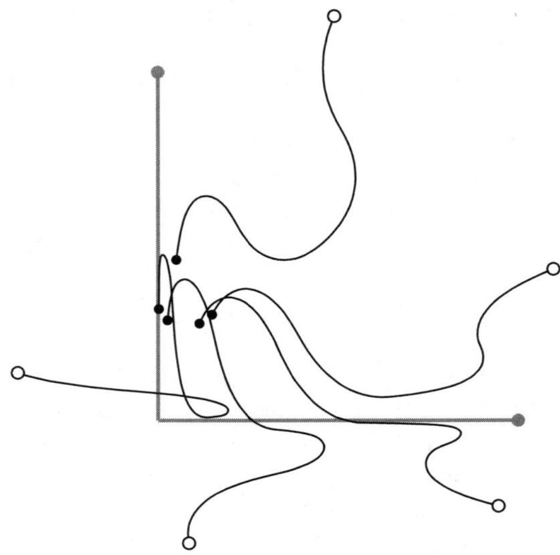

Fig. 1. Five agents following one target that moves out and back along the x-axis and then out and back along the y-axis

Let $a = z - x_i$ and $b = z - x_j$. Then it remains to show that

$$(a - b)^T \left(\frac{a}{\|a\|} - \frac{b}{\|b\|} \right) \geq 0.$$

But notice that

$$(a - b)^T \left(\frac{a}{\|a\|} - \frac{b}{\|b\|} \right) = \frac{\|a\| + \|b\|}{\|a\| \|b\|} \left(\|a\| \|b\| - a^T b \right).$$

Then since the Cauchy-Schwarz inequality tells us that $|a^T b| \leq \|a\| \|b\|$, we have

$$(a - b)^T \left(\frac{a}{\|a\|} - \frac{b}{\|b\|} \right) \geq 0$$

as desired. And in fact, we know equality only holds when $a = cb$, or equivalently $z - x_i = c(z - x_j)$, for some scalar $c \in \mathbb{R}$. So we can say that all pairs of agents get closer together (regardless of the target trajectory) unless they are collinear with the target. Under the mild assumption that there exists $\Delta t > 0$ such that for any interval of time $[t, t + \Delta t]$ we can find $\delta > 0$ and $\epsilon < 1$ satisfying

$$|(z - x_i)^T (z - x_j)| \leq \epsilon \|z - x_i\| \|z - x_j\|$$

on a subinterval

$$[t', t' + \delta \cdot \Delta t] \subset [t, t + \Delta t]$$

then $\|x_i - x_j\| \to 0$ as $t \to \infty$ for all i, j. In other words, if the target stays some minimum distance away from the line containing each pair of agents for a nonzero fraction of time, then all agents approach the same location. As an example, consider the target trajectory shown in Fig. 1, in which a group of agents quickly begins to converge.

5 Separation Can Be Maintained between Four Agents Using Two Targets

Now assume there are two targets (so $m = 2$). We can show that for certain trajectories $z_1(t)$ and $z_2(t)$ of these targets, it is possible to maintain separation between four agents. In particular, let $v > 1$ be the constant speed of each target. Then for any time $t > 0$ we take $s = t \mod (4/v)$ and define

$$z_1(s) = \begin{cases} (vs, 0) & \text{if } 0 \leq s \leq 1/v \\ (2 - vs, 0) & \text{if } 1/v < s \leq 2/v \\ (0, vs - 2) & \text{if } 2/v < s \leq 3/v \\ (0, 4 - vs) & \text{if } 3/v < s \leq 4/v \end{cases}$$

$$z_2(s) = -z_1(s).$$

So the first target moves from the origin to $(1, 0)$ and back, then from the origin to $(0, 1)$ and back, all at constant speed v. Similarly, the second target moves from the origin to $(-1, 0)$ and back, then from the origin to $(0, -1)$ and back, also at speed v. Both targets repeat these motions for all time.

When this system is simulated, the resulting agent trajectories fall into one of four limit cycles (see Fig. 2). Each limit cycle has a beautiful hourglass shape, and lies entirely in one quadrant of the plane. The limit cycles are also symmetric about the origin, about each axis, and about each 45° diagonal. Moreover, these limit cycles are passively stable—target trajectories need not be modified in response to perturbations in the trajectory of each agent.

We would like to characterize these limit cycles analytically. Let $\Delta t = 4/v$. For any agent i, we denote the map from $x_i(k\Delta t)$ to $x_i((k + 1)\Delta t)$ by $\phi(x)$. We want to show that there are exactly four fixed points of ϕ, one in each quadrant.

5.1 Interpretation as Linear Pursuit Curves

A *curve of pursuit* is the path taken by an agent that chases a moving target by traveling directly toward it at constant speed. A pursuit curve is called *linear* if the target is moving along a straight line. Notice that the target trajectories we defined above each consist of four straight line segments. As a result, the trajectory of each agent is a sequence of linear pursuit curves.

The shape of a single linear pursuit curve, as shown in Fig. 3, can be described analytically [10]. For the purposes of this derivation, let the target's position be $(0, \eta)$ and the agent's position be (x, y). Assume the target travels at speed v,

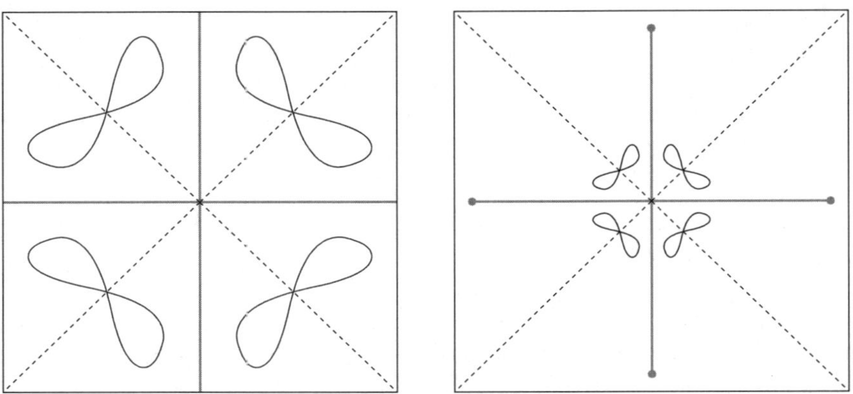

Fig. 2. The agent trajectories resulting from cyclic motion of two targets. These targets repeatedly move out and back, in opposite directions, along first the x-axis and then the y-axis. The top image is a close-up of the bottom image.

so $\eta = vt$, and that the agent travels at unit speed, so the arc-length $s = t$. By definition of arc length we have

$$\left(\frac{ds}{dt}\right)^2 = \left(\frac{dx}{dt}\right)^2 + \left(\frac{dy}{dt}\right)^2.$$

Since we also have

$$\frac{d\eta}{dt} = v\frac{ds}{dt}$$

then

$$1 + \left(\frac{dy}{dx}\right)^2 = \frac{1}{v^2}\left(\frac{d\eta}{dx}\right)^2. \tag{1}$$

The agent moves directly toward the target, so

$$(y - \eta) = \frac{dy}{dx}(x - 0) = x\frac{dy}{dx}. \tag{2}$$

Deriving this expression with respect to x we have

$$\frac{d\eta}{dx} = -x\frac{d^2y}{dx^2}.$$

Plug this into (1) and we find

$$1 + \left(\frac{dy}{dx}\right)^2 = \frac{1}{v^2}\left(-x\frac{d^2y}{dx^2}\right)^2.$$

Let $p = dy/dx$. Then we have

$$1 + p^2 = \frac{1}{v^2}\left(-x\frac{dp}{dx}\right)^2$$

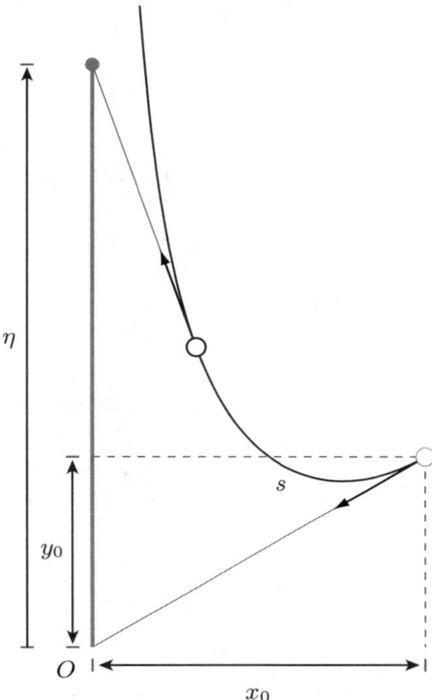

Fig. 3. A curve of linear pursuit. The target starts from the origin and moves along the y-axis at constant speed v. The agent starts from (x_0, y_0) and moves directly toward the target at unit speed.

which we can write as

$$\sqrt{1 + p^2} = -\frac{x}{v}\frac{dp}{dx}.$$

This expression is integrable:

$$-v\int \frac{dx}{x} = \int \frac{dp}{\sqrt{1 + p^2}}$$

$$\Rightarrow \quad \left(\frac{x}{c_2}\right)^{-v} = \frac{p + \sqrt{1 + p^2}}{c_1}$$

$$\Rightarrow \quad p = \frac{1}{2}\left(c_1 c_2^v x^{-v} - \frac{x^v}{c_1 c_2^v}\right).$$

Let $a = c_1 c_2^v$. Then we have

$$\frac{dy}{dx} = \frac{1}{2}\left(\frac{a}{x^v} - \frac{x^v}{a}\right). \tag{3}$$

Integrating with respect to x, we find the equation for the curve traced out by the follower (assuming that $v \neq 1$):

$$y = \frac{1}{2} \left(\frac{ax^{1-v}}{1-v} - \frac{x^{1+v}}{a(1+v)} \right) + b. \tag{4}$$

The constants a and b are found from initial conditions:

$$a = x_0^{v-1} \left(y_0 - \sqrt{x_0^2 + y_0^2} \right)$$

$$b = \frac{v \left(-v y_0 + \sqrt{x_0^2 + y_0^2} \right)}{1 - v^2}.$$

Combining (2)-(4), we can write an expression for the time taken to reach a particular x and y. It is even possible to invert this expression and write $x(t)$ and $y(t)$ in parametric form. However, the result is not pretty, and how to use it to find fixed points of ϕ analytically is still an open question. Moreover, linear pursuit curves are convex, because

$$\frac{d^2 y}{dx^2} = -\frac{v x^{-(v+1)}}{2a} \left(a^2 + (x^v)^2 \right)$$

so if $x_0 > 0$ and $y_0 > 0$ then we know $a < 0$ and hence $d^2y/dx^2 > 0$. As a result, bisection on x allows us to find fixed points of ϕ and to compute the shape of associated limit cycles numerically, to any desired precision.

5.2   Each Quadrant Is an Invariant Set

Although we have not yet been able to find the fixed points of ϕ analytically, we can show that each quadrant is an invariant set.

First, consider a single linear pursuit curve, where the target moves at constant speed v for a length of time $1/v$. If $x_0 \neq 0$, then (4) implies that $y \to \infty$ as $x \to 0$, so we know that $x(t) \neq 0$ for all $t > 0$. Also, if $y_0 > 0$, then (2) implies that $dy/dx \geq 0$ (meaning that the agent is moving downward) only so long as $y \geq \eta$. As a result, $y(t)$ is bounded below by

$$y(t) \geq \frac{v}{v+1} y_0 > 0.$$

Similarly, if $y_0 < 1$, then (2) implies that $dy/dx \leq 0$ (meaning that the agent is moving upward) for no more than an interval of time $\Delta t = (1 - y_0)/v$. As a result, $y(t)$ is bounded above by

$$y(t) \leq y_0 + \frac{1 - y_0}{v} < 1$$

Now, without loss of generality, consider an agent that begins in the upper-right quadrant. The agent's trajectory from $0 \leq t \leq 1/v$ can be modeled as a linear

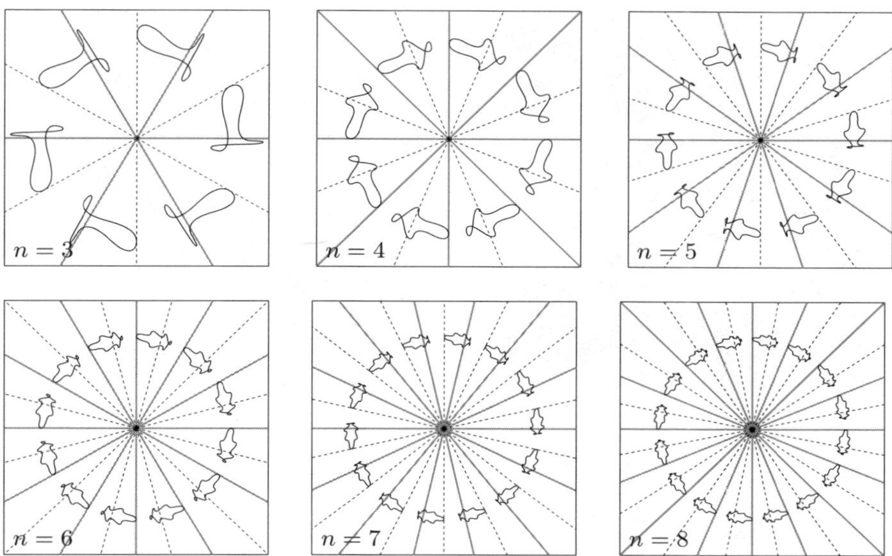

Fig. 4. Agent trajectories resulting from cyclic motion of two targets, each of which repeatedly move out and back along n rotating lines

pursuit curve for which $x_0 < 0$ and $y_0 > 0$, where we want to verify $x(t) < 0$ and $y(t) > 0$ for $t \leq 1/v$. Similarly, the agent's trajectory from $1/v < t \leq 2/v$ can be modeled as a linear pursuit curve for which $x_0 > 0$ and $y_0 < 1$, where we want to verify $x(t) > 0$ and $y(t) < 1$ for $t \leq 1/v$. But we have just proven both of these results, and the situation for $2/v < t \leq 4/v$ is symmetric. Consequently, we know that the agent remains in the upper-right quadrant for all time. An identical argument shows that each of the other quadrants is also invariant.

6 Separation Can Be Maintained between Any Number of Agents Using Two Targets

As in the previous section, we assume there are two targets. But now we conjecture that for certain trajectories $z_1(t)$ and $z_2(t)$ of these targets, it is possible to maintain separation between any number of agents (given suitable initial conditions). Although we are still unable to prove or disprove this conjecture, results in simulation lend strong support. In particular, let n be the number of agents and let $v > 1$ be the constant speed of each target. Then for any time $t > 0$ we take $s = t \mod (n/v)$ and define

$$z_1(s) = vs \cdot (\cos(\pi k/n), \sin(\pi k/n))$$
$$z_2(s) = -z_1(s)$$

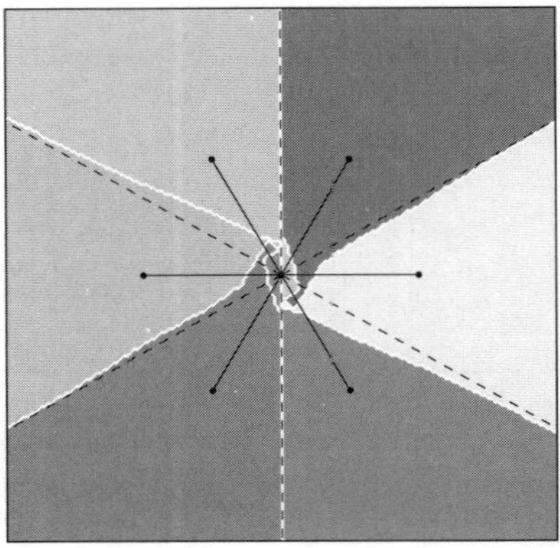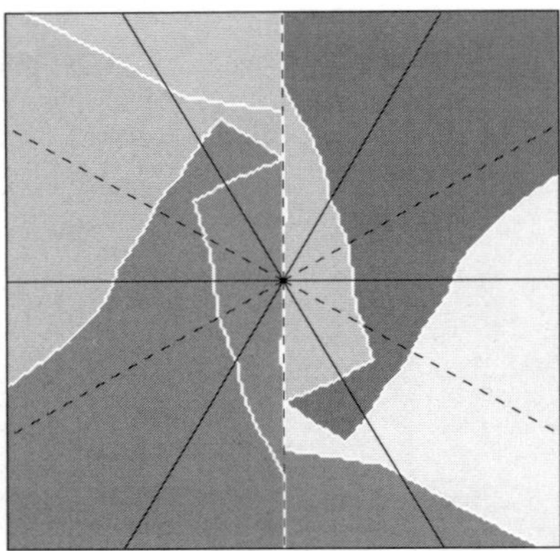

Fig. 5. Invariant sets for $n = 3$, given $v = 9.4$

on each interval

$$\frac{k}{v} < s \le \frac{k+1}{v}$$

for all $k = 0, \ldots, n - 1$. So just as in the previous section, both targets move repeatedly away from and back toward the origin. But now, the targets move along n lines (rather than just 2), rotating an angle π/n between each one.

When this system is simulated, the resulting agent trajectories fall into one of $2n$ limit cycles (see Fig. 4). Each limit cycle lies entirely in a cone, either centered on the target trajectories (when n is odd) or between them (when n is even). These limit cycles show rotational symmetry about the origin. The limit cycles seem to be passively stable as before, but the basins of attraction are not as well-defined. In particular, the cone containing each limit cycle is not an invariant set for $n > 2$ as were the quadrants for $n = 2$ (see Fig. 5 for an example). As a result, if the target speed v is too small, the limit cycles disappear. Note that it is possible to compute the minimum target speed numerically (see Table 1).

Table 1. Minimum Target Speed for Separation

N	2	3	4	5	6	7	8
$v_{\min}$	1.00	9.37	17.15	30.68	41.45	61.76	75.75

7   Conclusion

When talking about control of multi-agent systems, we usually assume that the dynamics of individual agents can be designed. In this paper, we were interested in multi-agent systems where the dynamics of each agent are fixed. Biological systems—such as groups of microorganisms, herds of cattle, or crowds of people—are perfect examples. These systems are controlled indirectly, by applying external stimuli. In general, it is not clear how to plan a sequence of stimuli that cause desired group behavior, nor even how to decide whether a given behavior is achievable. In this paper we considered a simple multi-agent system in which agents chase targets, and focused on the task of maintaining separation between the agents by specifying the target trajectories. We demonstrated in simulation that two targets are sufficient to maintain separation between any number of agents. In future work we hope to address the other questions raised in the introduction, to consider more realistic dynamic models, and to apply our work to actual biological and robotic systems.

References

1. Antonelli, G., Chiaverini, S.: Kinematic control of platoons of autonomous vehicles. IEEE Trans. Rebot. 22(6), 1285–1292 (2006)
2. Bernhart, A.: Curves of pursuit. Scripta Mathematica 20(3-4), 125–141 (1954)
3. Bernhart, A.: Curves of pursuit-II. Scripta Mathematica 23(1-4), 49–65 (1957)
4. Bernhart, A.: Curves of general pursuit. Scripta Mathematica 24(3), 189–206 (1959)
5. Bernhart, A.: Polygons of pursuit. Scripta Mathematica 24(1), 23–50 (1959)
6. Bruckstein, A.: Why the ant trails look so straight and nice. Mathematical Intelligencer 15(2), 59–62 (1993)
7. Butler, Z., Corke, P., Peterson, R., Rus, D.: From robots to animals: virtual fences for controlling cattle. Int. J. Rob. Res. 25(5-6), 485–508 (2006)
8. Caprari, G., Colot, A., Siegwart, R., Halloy, J., Deneubourg, J.-L.: Insbot: Design of an autonomous mini mobile robot able to interact with cockroaches. In: Int. Conf. Rob. Aut. (2004)
9. Cortés, J., Martínez, S., Bullo, F.: Robust rendezvous for mobile autonomous agents via proximity graphs in arbitrary dimensions. IEEE Trans. Automat. Contr. 51(8), 1289–1296 (2006)
10. Davis, H.T.: Introduction to Nonlinear Differential and Integral Equations. Dover Publications, Inc., New York (1962)
11. Freeman, R.A., Yang, P., Lynch, K.M.: Distributed estimation and control of swarm formation statistics. In: American Control Conference, Minneapolis, MN, pp. 749–755 (June 2006)
12. Gazi, V., Passino, K.M.: Stability analysis of swarms. IEEE Trans. Automat. Contr. 48(4), 692–697 (2003)
13. Jadbabaie, A., Lin, J., Morse, A.S.: Coordination of groups of mobile autonomous agents using nearest neighbor rules. IEEE Trans. Automat. Contr. 48(6), 988–1001 (2003)
14. Leonard, N.E., Fiorelli, E.: Virtual leaders, artificial potentials and coordinated control of groups. In: IEEE Conf. Dec. Cont., Orlando, FL, pp. 2968–2973 (December 2001)
15. Low, D.J.: Statistical physics: Following the crowd. Nature 407, 465–466 (2000)
16. Marshall, J.A., Broucke, M.E., Francis, B.A.: Formations of vehicles in cyclic pursuit. IEEE Trans. Automat. Contr. 49(11), 1963–1974 (2004)
17. Ogawa, N., Oku, H., Hashimoto, K., Ishikawa, M.: Microrobotic visual control of motile cells using high-speed tracking system. IEEE Trans. Rebot. 21(3), 704–712 (2005)

18. Ogawa, N., Oku, H., Hasimoto, K., Ishikawa, M.: A physical model for galvanotaxis of paramecium cell. Journal of Theoretical Biology 242, 314–328 (2006)
19. Pan, X., Han, C.S., Dauber, K., Law, K.H.: Human and social behavior in computational modeling and analysis of egress. Automation in Construction 15(4), 448–461 (2006)
20. Reif, J.H., Wang, H.: Social potential fields: A distributed behavioral control for autonomous robots. Robotics and Autonomous Systems 27, 171–194 (1999)
21. Robinson, K.R.: The responses of cells to electrical fields: A review. Journal of Cell Biology 101(6), 2023–2027 (1985)
22. Takahashi, K., Ogawa, N., Oku, H., Hashimoto, K.: Organized motion control of a lot of microorganisms using visual feedback. In: IEEE Int. Conf. Rob. Aut., Orlando, FL, pp. 1408–1413 (May 2006)
23. Tanner, H.G., Pappas, G.J., Kumar, V.: Leader-to-formation stability. IEEE Trans. Robot. Automat. 20(3), 443–455 (2004)
24. Vaughan, R., Sumpter, N., Frost, A., Cameron, S.: Robot sheepdog project achieves automatic flock control. In: Int. Conf. on the Simulation of Adaptive Behaviour (1998)
25. Vaughan, R., Sumpter, N., Henderson, J., Frost, A., Cameron, S.: Experiments in automatic flock control. Robotics and Autonomous Systems 31, 109–117 (2000)

Improved Mapping and Image Segmentation by Using Semantic Information to Link Aerial Images and Ground-Level Information

Martin Persson[1], Tom Duckett[2], and Achim Lilienthal[1]

[1] Centre of Applied Autonomous Sensor Systems, Örebro University, Sweden
   martin.persson@tech.oru.se, achim@lilienthals.de
[2] Department of Computing and Informatics, University of Lincoln, UK
   tduckett@lincoln.ac.uk

Summary. This paper investigates the use of semantic information to link ground-level occupancy maps and aerial images. A ground-level semantic map is obtained by a mobile robot equipped with an omnidirectional camera, differential GPS and a laser range finder. The mobile robot uses a virtual sensor for building detection (based on omnidirectional images) to compute the ground-level semantic map, which indicates the probability of the cells being occupied by the wall of a building. These wall estimates from a ground perspective are then matched with edges detected in an aerial image. The result is used to direct a region- and boundary-based segmentation algorithm for building detection in the aerial image. This approach addresses two difficulties simultaneously: 1) the range limitation of mobile robot sensors and 2) the difficulty of detecting buildings in monocular aerial images. With the suggested method building outlines can be detected faster than the mobile robot can explore the area by itself, giving the robot an ability to "see" around corners. At the same time, the approach can compensate for the absence of elevation data in segmentation of aerial images. Our experiments demonstrate that ground-level semantic information (wall estimates) allows to focus the segmentation of the aerial image to find buildings and produce a ground-level semantic map that covers a larger area than can be built using the onboard sensors.

1 Introduction

A mobile robot has a limited view of its environment. Mapping of the operational area is one way of enhancing this view for visited locations. In this paper we explore the possibility to use information extracted from aerial images to further improve the mapping process. Semantic information (classification of buildings versus non-buildings) is used as the link between the ground level information and the aerial image. The method allows to speed up exploration or planning in areas unknown to the robot.

Colour image segmentation is often used to extract information about buildings from an aerial image. However, it is hard to perform automatic detection of buildings in monocular aerial images without elevation information [15]. Buildings can not easily be separated from other man-made structures such as driveways, tennis courts, etc. due to the resemblance in colour and shape. We show

S. Lee, I.H. Suh, and M.S. Kim (Eds.): Recent Progress in Robotics, LNCIS 370, pp. 157–169, 2008.

that wall estimates found by a mobile robot can compensate for the absence of elevation data. In the approach proposed in this paper, wall estimates detected by a mobile robot are matched with edges extracted from an aerial image. A virtual sensor[1] for building detection is used to identify parts of an occupancy map that belong to buildings (wall estimate). To determine potential matches we use geo-referenced aerial images and an absolute positioning system on board of the robot. The matched lines are then used in region- and boundary-based segmentation of the aerial image for detection of buildings. The purpose is to detect building outlines faster than the mobile robot can explore the area by itself. Using a method like this, the robot can estimate the size of found buildings and using the building outline it can "see" around one or several corners without actually visiting the area. The method does not assume a perfectly up-to-date aerial image, in the sense that buildings may be missing although they are present in the aerial image, and vice versa. It is therefore possible to use globally available[2] geo-referenced images.

1.1   Related Work

Overhead images in combination with ground vehicles have been used in a number of applications. Oh *et al.* [10] used map data to bias a robot motion model in a Bayesian filter to areas with higher probability of robot presence. Mobile robot trajectories are more likely to follow paths in the map and using the map priors, GPS position errors due to reflections from buildings were compensated. This work assumed that the probable paths were known in the map. Pictorial information captured from a global perspective has been used for registration of sub-maps and subsequent loop-closing in SLAM [2].

Silver *et al.* [14] discuss registration of heterogeneous data (e.g. data recorded with different sampling density) from aerial surveys and the use of these data in classification of ground surface. Cost maps are produced that can be used in long range vehicle navigation. Scrapper *et al.* [13] used heterogeneous data from, e.g., maps and aerial surveys to construct a world model with semantic labels. This model was compared with vehicle sensor views providing a fast scene interpretation.

For detection of man-made objects in aerial images, lines and edges together with elevation data are the features that are used most often. Building detection in single monocular aerial images is very hard without additional elevation data [15]. Mayer's survey [8] describes some existing systems for building detection and concludes that scale, context and 3D structure were the three most important features to consider for object extraction, e.g., buildings, roads and vegetation, in aerial images. Fusion of SAR (Synthetic Aperture Radar) and aerial images has been employed for detection of building outlines [15]. The

---

[1] A virtual sensor is understood as one or several physical sensors with a dedicated signal processing unit for recognition of real world concepts.

[2] E.g. Google Earth, Microsoft Virtual Earth, and satellite images from IKONOS and its successors.

building location was established in the overhead SAR image, where walls from one side of buildings can be detected. The complete building outline was then found using edge detection in the aerial image. Parallel and perpendicular edges were considered and the method belongs to edge-only segmentation approaches. The main difference to our work regarding building detection is the use of a mobile robot on the ground and the additional roof homogeneity condition.

Combination of edge and region information for segmentation of aerial images has been suggested in several publications. Mueller *et al.* [9] presented a method to detect agricultural fields in satellite images. First, the most relevant edges were detected. These were then used to guide both the smoothing of the image and the following segmentation in the form of region growing. Freixenet *et al.* [4] investigated different methods for integrating region- and boundary-based segmentation, and also claim that this combination is the best approach.

1.2   Outline and Overview

The presentation of our proposed system is divided into three main parts. The first part, Sect. 2, concerns the estimation of walls by the mobile robot and edge detection in the aerial image. The wall estimates are extracted from a probabilistic semantic map. This map is basically an occupancy map built from range data and labelled using a virtual sensor for building detection [11] mounted on the mobile robot. The second part describes the matching of wall estimates from the mobile robot with the edges found in the aerial image. This procedure is described in Sect. 3. The third part presents the segmentation of an aerial image based on the matched lines, see Sect. 4. Details of the mobile robot, the experiments performed and the obtained result are found in Sect. 5. Finally, the paper is concluded in Sect. 6 and some suggestions for future work are given.

2   Wall Estimation

A major problem for building detection in aerial images is to decide which of the edges in the aerial image correspond to building outlines. The idea of our approach, to increase the probability that correct segmentation is performed, is to match wall estimates extracted from two perspectives. In this section we describe the process of extracting wall candidates, first from the mobile robot's perspective and then from aerial images.

2.1   Wall Candidates from Ground Perspective

The wall candidates from the ground perspective are extracted from a semantic map acquired by a mobile robot. The semantic map we use is a probabilistic occupancy grid map with two classes: buildings and non-buildings [12]. The probabilistic semantic map is produced using an algorithm that fuses different sensor modalities. In this paper, a range sensor is used to build an occupancy

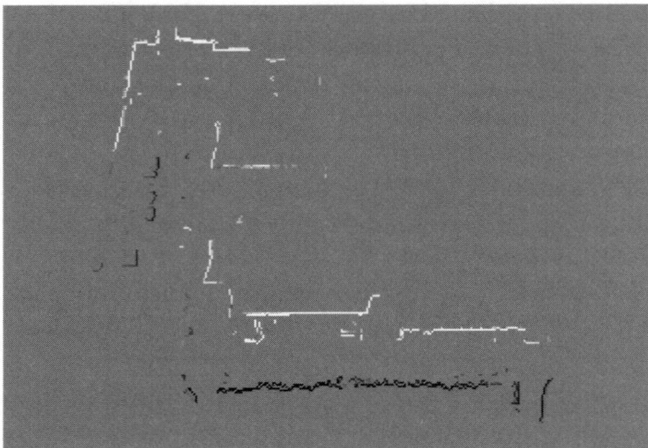

Fig. 1. An example of a semantic map where white lines denote high probability of walls and dark lines show outlines of non-building entities

map, which is converted into a probabilistic semantic map using the output of a virtual sensor for building detection based on an omnidirectional camera.

The algorithm consists of two parts. First, a local semantic map is built using the occupancy map and the output from the virtual sensor. The virtual sensor uses the AdaBoost algorithm [5] to train a classifier that classifies close range monocular grey scale images taken by the mobile robot as buildings or non-buildings. The method combines different types of features such as edge orientation, grey level clustering, and corners into a system with high classification rate [11]. The classification by the virtual sensor is made for a whole image. However, the image may also contain parts that do not belong to the detected class, e.g., an image of a building might also include some vegetation such as a tree. Probabilities are assigned to the occupied cells that are within a sector representing the view of the virtual sensor. The size of the cell formations within the sector affects the probability values. Higher probabilities are given to larger parts of the view, assuming that larger parts are more likely to have caused the view's classification [12].

In the second step the local maps are used to update a global map using a Bayesian method. The result is a global semantic map that distinguishes between buildings and non-buildings. An example of a semantic map is given in Fig. 1. From the global semantic map, lines representing probable building outlines are extracted. An example of the extracted lines is given in Fig. 2.

2.2 Wall Candidates in Aerial Images

Edges extracted from an aerial image are used as potential building outlines. We limit the wall candidates used for matching in Sect. 3 to straight lines extracted from a colour aerial image taken from a nadir view. We use an output fusion method for the colour edge detection. The edge detection is performed separately

Fig. 2. Illustration of the wall estimates (*black lines*) calculated from the semantic map. The grey areas illustrate building and nature objects (manually extracted from Fig. 3). The semantic map in Fig. 1 belongs to the upper left part of this figure.

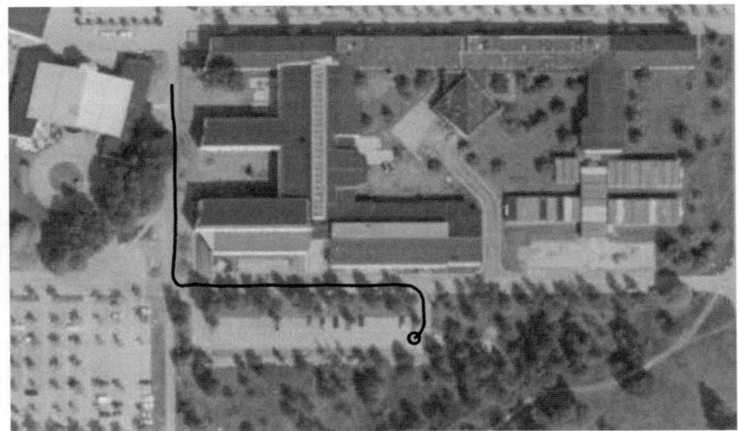

Fig. 3. The trajectory of the mobile robot (*black line*) and a grey scale version of the used aerial image

on the three RGB-components using Canny's edge detector [1]. The resulting edge image I_e is calculated by fusing the three binary images obtained for the three colour components with a logical OR-function. Finally a thinning operation is performed to remove points that occur when edges appear slightly shifted in the different components. For line extraction in I_e an implementation by Peter Kovesi[3] was used. The lines extracted from the edges detected in the aerial image in Fig. 3, are shown in Fig. 4.

---

[3] http://www.csse.uwa.edu.au/~pk/Research/MatlabFns/, University of Western Australia, Sep 2005

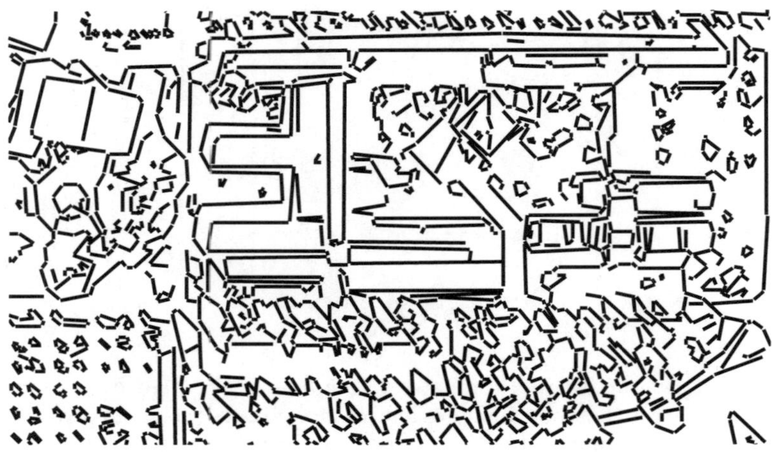

Fig. 4. The lines extracted from the edge version of the aerial image

3   Wall Matching

The purpose of the wall matching step is to relate a wall estimate, obtained at ground-level with the mobile robot to the edges detected in the aerial image. In both cases the line segments represent the wall estimates. We denote a wall estimate found by the mobile robot as L_g and the N lines representing the edges found in the aerial image by L_a^i with $i \in \{1, \ldots, N\}$. Both line types are geo-referenced in the same Cartesian coordinate system.

The lines from both the aerial image and the semantic map may be erroneous, especially concerning the line endpoints, due to occlusion, errors in the semantic map, different sensor coverage, etc. We therefore need a metric for line-to-line distances that can handle partially occluded lines. We do not consider the length of the lines and restrict the line matching to the line directions and the distance between two points, one point on each line. The line matching calculations are performed in two sequential steps: 1) decide which points on the lines are to be matched, and 2) calculate a distance measure to find the best matches.

3.1   Finding the Closest Point

In this section we define which points on the lines are to be matched. For L_g we use the line midpoint, P_g. Due to the possible errors described above we assume that the point P_a on L_a^i that is closest to P_g is the best candidate to be used in our 'line distance metric'.

To calculate P_a, let e_n be the orthogonal line to L_a^i that intersects L_g in P_g, see Fig. 5. We denote the intersection between e_n and L_a^i as ϕ where $\phi = e_n \times L_a^i$ (using homogenous coordinates). The intersection ϕ may be outside the line segment L_a^i, see right part of Fig. 5. We therefore need to check if ϕ is within

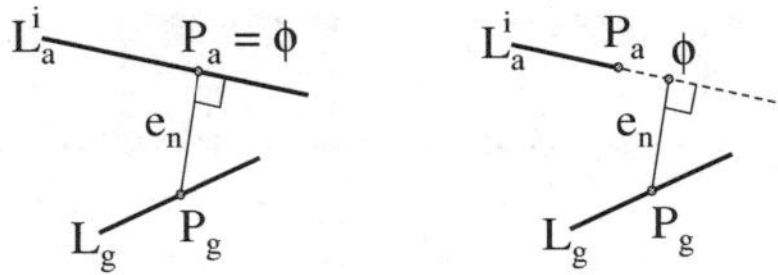

Fig. 5. The line L_g with its midpoint $P_g = (x_m, y_m)$, the line L_a^i, and the normal to L_a^i, e_n. To the left, $P_g = \phi$ since ϕ is on L_a^i and to the right, P_g is the endpoint of L_a^i since ϕ is not on L_a^i.

the endpoints and then set $P_a = \phi$. If ϕ is not within the endpoints, then P_a is set to the closest endpoint on L_a.

3.2 Distance Measure

The calculation of a distance measure is inspired by [7], which describes geometric line matching in images for stereo matching. We have reduced the complexity in those calculations to have fewer parameters that need to be determined and to exclude the line lengths. Matching is performed using L_g's midpoint P_g, the closest point P_a on L_a^i and the line directions, θ_i. First, a difference vector is calculated as

$$\mathbf{r_g} = [P_{g_x} - P_{a_x}, P_{g_y} - P_{a_y}, \theta_g - \theta_a]^T. \tag{1}$$

Second, the similarity is measured as the Mahalanobis distance

$$\mathbf{d_g} = \mathbf{r_g}^T \mathbf{R}^{-1} \mathbf{r_g} \tag{2}$$

where the diagonal covariance matrix $\mathbf{R}$ is defined as

$$\mathbf{R} = \begin{bmatrix} \sigma_{Rx}^2 & 0 & 0 \\ 0 & \sigma_{Ry}^2 & 0 \\ 0 & 0 & \sigma_{R\theta}^2 \end{bmatrix} \tag{3}$$

with σ_{Rx}, σ_{Ry}, and $\sigma_{R\theta}$ being the expected standard deviation of the errors between the ground-based and aerial-based wall estimates.

4 Aerial Image Segmentation

This section describes how local segmentation of the colour aerial image is performed. Segmentation methods can be divided into two groups; discontinuity- and similarity-based [6]. In our case we combine the two groups by first performing an edge based segmentation for detection of closed areas and then colour segmentation based on a small training area to confirm the areas' homogeneity. The following is a short description of the sequence that is performed for each line L_g:

1. Sort L_a^N based on $\mathbf{d_g}$ from (2) in increasing order and set $i = 0$.
2. Set $i = i + 1$.

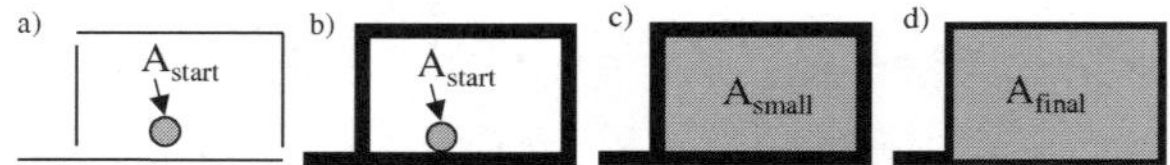

Fig. 6. Illustration of the edge-based algorithm. (a) shows a small part of I_e and A_{start}. In (b) I_e has been dilated and in (c) A_{small} has been found. (d) shows A_{final} as the dilation of A_{small}.

3. Define a start area A_{start} on the side of L_a^i that is opposite to the robot (this will be in or closest to the unknown part of the occupancy grid map).
4. Check if A_{start} includes edge points (parts of edges in I_e). If yes, return to step 2.
5. Perform edge controlled segmentation.
6. Perform homogeneity test.

The segmentation based on L_g is stopped when a region has been found. Step 4 makes sure that the regions have a minimum width. Steps 5 and 6 are elaborated in the following paragraphs.

4.1   Edge Controlled Segmentation

Based on the edge image I_e constructed from the aerial image, we search for a closed area. Since there might be gaps in the edges bottlenecks need to be found [9]. We use morphological operations, with a 3×3 structuring element, to first dilate the interesting part of the edge image in order to close gaps and then search for a closed area on the side of the matched line that is opposite to the mobile robot. When this area has been found the area is dilated in order to compensate for the previous dilation of the edge image. The algorithm is illustrated in Fig. 6.

4.2   Homogeneity Test

Classical region growing allows neighbouring pixels with properties according to the model to be added to the region. The model of the region can be continuously updated as the region grows. We started our implementation in this way but it turned out that the computation time of the method was quite high. Instead we use the initial starting area A_{start} as a training sample and evaluate the rest of the region based on the corresponding colour model. This means that the colour model does not gradually adapt to the growing region, but instead requires a homogeneous region on the complete roof part that is under investigation. Regions that gradually change colour or intensity, such as curved roofs, might then be rejected.

Gaussian Mixture Models, GMM, are popular for colour segmentation. Like Dahlkamp *et al.* [3] we tested both GMM and a model described by the mean and the covariance matrix in RGB colour space. We selected the mean/covariance model since it is faster and we noted that the mean/covariance model performs approximately equally well as the GMM in our case.

5 Experiments

5.1 Data Collection

The above presented algorithms have been implemented in Matlab for evaluation and currently work off-line. Data were collected with a mobile robot, a Pioneer P3-AT from ActivMedia, equipped with differential GPS, laser range scanner, cameras and odometry. The robot is equipped with two different types of cameras; an ordinary camera mounted on a PT-head and an omni-directional camera. The omni-directional camera gives a 360° view of the surroundings in one single shot. The camera itself is a standard consumer-grade SLR digital camera (Canon EOS350D, 8 megapixels). On top of the lens, a curved mirror from 0-360.com is mounted. From each omni-image we compute 8 (every 45°) planar views or sub-images with a horizontal field-of-view of 56°. These sub-images are the input to the virtual sensor. The images were taken with ca. 1.5 m interval and were stored together with the corresponding robot's pose, estimated from GPS and odometry. The trajectory of the mobile robot is shown in Fig. 3.

5.2 Tests

The occupancy map in Fig. 7 was built using the horizontally mounted laser range scanner. The occupied cells in this map (marked in black) were labelled by the virtual sensor giving the semantic map presented in Fig. 1. The semantic map contains two classes: buildings (values above 0.5) and non-buildings (values below 0.5). From this semantic map we extracted the grid cells with a high probability of being a building (above 0.9) and converted them to the lines L_g^M presented in Fig. 2. Matching of these lines with the lines extracted from the aerial image L_a^N, see Fig. 4, was then performed. Finally, based on best line matches the segmentation was performed according to the description in Sect. 4.

In the experiments, the three parameters in $\mathbf{R}$ (3) were set to $\sigma_{Rx} = 1$ m, $\sigma_{Ry} = 1$ m, and $\sigma_{R\theta} = 0.2$ rad. Note that it is only the relation between the parameters that influences the line matching.

We have performed two different types of tests. *Tests 1-3* are the nominal cases when the collected data are used as they are. The tests intend to show the influence of a changed relation between σ_{Rx}, σ_{Ry} and $\sigma_{R\theta}$ by varying $\sigma_{R\theta}$. In *Test 2* $\sigma_{R\theta}$ is decreased by a factor of 2 and in *Test 3* $\sigma_{R\theta}$ is increased by a factor of 2. In *Tests 4* and *5* additional uncertainty (in addition to the uncertainty already present in L_g^M and L_a^N) was introduced. This uncertainty is in the form of Gaussian noise added to the midpoints (σ_x and σ_y) and directions σ_θ of L_g^M. The tests are defined in Table 1.

5.3 Quality Measure

We introduce two quality measures to be able to compare different algorithms or sets of parameters in an objective way. For this, four sets (A-D) are defined: A is the ground truth, a set of cells/points that has been manually classified as

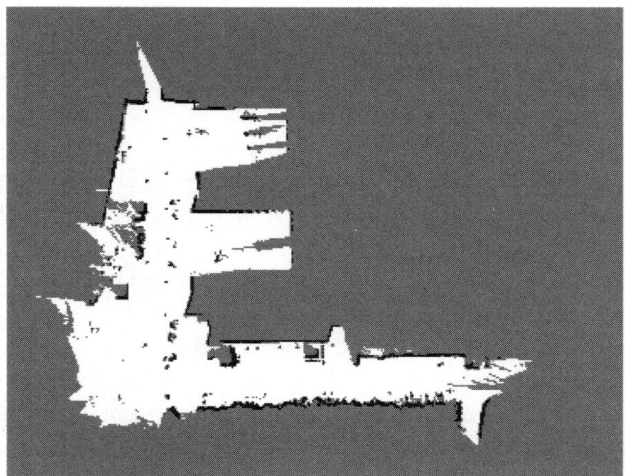

Fig. 7. Occupancy map used to build the semantic map presented in Fig. 1

Table 1. Definition of tests and the used parameters

Test	σ_x [m]	σ_y [m]	σ_θ [rad]	$\sigma_{R\theta}$ [rad]	N_{run}
1	0	0	0	0.2	1
2	0	0	0	0.1	1
3	0	0	0	0.4	1
4	1	1	0.1	0.2	20
5	2	2	0.2	0.2	20

building; B is the set of cells that has been classified as building by the algorithm; C is the set of false positives, $C = B \setminus A$, the cells that have been classified as building B but do not belong to ground truth A; and D are the true positives, $D = B \cap A$, the cells that have been classified as building B and belong to ground truth A. Using these sets, two quality measures are calculated as:

- The true positive rate, $\Phi_{TP} = \#D/\#B$.
- The false positive rate, $\Phi_{FP} = \#C/\#B$.

where $\#D$ denotes the number of cells in D, etc.

5.4 Result

The results of *Test 1* show a high detection rate (96.5%) and a low false positive rate (3.5%), see Table 2. The resulting segmentation is presented in Fig. 8. Four deviations from an ideal result can be noted. At a and b tree tops are obstructing the wall edges in the aerial image, a white wall causes a gap between two regions at c, and a false area, to the left of b, originates from an error in the semantic map (a low hedge was marked as building).

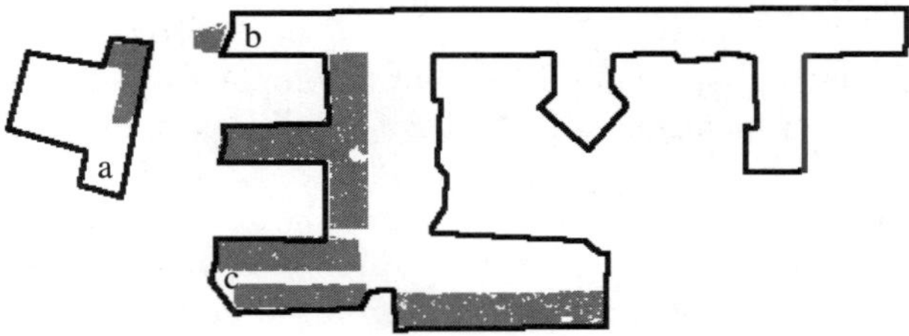

Fig. 8. The result of segmentation of the aerial image using the wall estimates in Fig. 2 (*grey*) and the ground truth building outlines (*black lines*)

The results of *Test 1-3* are very similar which indicate that the algorithm in this case was not specifically sensitive to the changes in $\sigma_{R\theta}$. In *Test 4* and *5* the scenario of *Test 1* was repeated using a Monte Carlo simulation with introduced pose uncertainty. The result is presented in Table 2. One can note that the difference between the nominal case and *Test 4* is very small. In *Test 5* where the additional uncertainties are higher the detection rate has decreased slightly.

Table 2. Results for the tests. The results of Test 4 and 5 are presented with the corresponding standard deviation.

Test	Φ_{TP} [%]	Φ_{FP} [%]
1	96.5	3.5
2	97.0	3.0
3	96.5	3.5
4	96.8 ± 0.2	3.2 ± 0.2
5	95.9 ± 1.7	4.1 ± 1.7

6   Conclusions and Future Work

This paper discusses how semantic information obtained with a virtual sensor for building detection on a mobile robot can be used to link ground-level information to aerial images. This approach addresses two difficulties simultaneously: 1) buildings are hard to detect in aerial images without elevation data and 2) the range limitation of the sensors of mobile robots. Concerning the first difficulty the high classification rate obtained shows that the semantic information can be used to compensate for the absence of elevation data in aerial image segmentation. The benefit from the extended range of the robot's view can clearly be noted in the presented example. Although the roof structure in the example is quite complicated, the outline of large building parts can be extracted even though the mobile robot has only seen a minor part of the surrounding walls.

There are a few issues that should be noted:

- It turns out that we can seldom segment a complete building outline due to, e.g., different roof materials, different roof inclinations and additions on the roof.
- It is important to check several lines from the aerial image since the edges are not always as exact as expected. Roofs can have extensions in other colours and not only roofs and ground are usually seen in the aerial image. In addition, when the nadir view is not perfect, walls appear in the image in conjunction with the roof outline. Such a wall will produce two edges in the aerial image, one where ground and wall meet and one where wall and roof meet.

6.1 Future Work

An extension to this work is to use the building estimates as training areas for colour segmentation in order to make a global search for buildings within the aerial image. Found regions would then have a lower probability until the mobile robot actually confirms that the region is a building outline.

The presented solution performs a local segmentation of the aerial image after each new line match. An alternative solution would be to first segment the whole aerial image and then confirm or reject the regions as the mobile robot finds new wall estimates.

As can be seen in the result, the building estimates can be parts of large buildings. It could therefore be advantageous to merge these regions. Another improvement would be to introduce a verification step that could include criteria such as:

- The building area should not cover ground that the outdoor robot has traversed.
- The size of the building estimate should exceed a minimum value (in relation to a minimum roof part).
- The found area should be checked using shadow detection to eliminate false building estimates.

References

1. Canny, J.: A computational approach for edge detection. IEEE Transactions on Pattern Analysis and Machine Intelligence 8(2), 279–298 (1986)
2. Chen, C., Wang, H.: Large-scale loop-closing with pictorial matching. In: Proceedings of the 2006 IEEE International Conference on Robotics and Automation, Orlando, Florida, pp. 1194–1199 (May 2006)
3. Dahlkamp, H., Kaehler, A., Stavens, D., Thrun, S., Bradski, G.: Self-supervised monocular road detection in desert terrain. In: Proceedings of Robotics: Science and Systems, Cambridge, USA (June 2006)

4. Freixenet, J., Munoz, X., Raba, D., Marti, J., Cufi, X.: Yet another survey on image segmentation: Region and boundary information integration. In: European Conference on Computer Vision, Copenhagen, Denmark, vol. III, pp. 408–422 (May 2002)
5. Freund, Y., Schapire, R.E.: A decision-theoretic generalization of on-line learning and an application to boosting. Journal of Computer and System Sciences 55(1), 119–139 (1997)
6. Gonzales, R.C., Woods, R.E.: Digital Image Processing. Prentice-Hall, Englewood Cliffs (2002)
7. Guerrero, J., Sagüés, C.: Robust line matching and estimate of homographies simultaneously. In: IbPRIA 2003. Pattern Recognition and Image Analysis: First Iberian Conference, Puerto de Andratx, Mallorca, Spain, pp. 297–307 (2003)
8. Mayer, H.: Automatic object extraction from aerial imagery – a survey focusing on buildings. Computer vision and image understanding 74(2), 138–149 (1999)
9. Mueller, M., Segl, K., Kaufmann, H.: Edge- and region-based segmentation technique for the extraction of large, man-made objects in high-resolution satellite imagery. Pattern Recognition 37, 1621–1628 (2004)
10. Oh, S.M., Tariq, S., Walker, B.N., Dellaert, F.: Map-based priors for localization. In: IEEE/RSJ 2004 International Conference on Intelligent Robotics and Systems, Sendai, Japan, pp. 2179–2184 (2004)
11. Persson, M., Duckett, T., Lilienthal, A.: Virtual sensor for building detection by an outdoor mobile robot. In: Proceedings of the IROS 2006 workshop: From Sensors to Human Spatial Concepts, Beijing, China, pp. 21–26 (October 2006)
12. Persson, M., Duckett, T., Valgren, C., Lilienthal, A.: Probabilistic semantic mapping with a virtual sensor for building/nature detection. In: CIRA 2007. The 7th IEEE International Symposium on Computational Intelligence in Robotics and Automation (June 21-24, 2007)
13. Scrapper, C., Takeuchi, A., Chang, T., Hong, T.H., Shneier, M.: Using a priori data for prediction and object recognition in an autonomous mobile vehicle. In: Gerhart, G.R., Shoemaker, C.M., Gage, D.W. (eds.) Proceedings of the SPIE Unmanned Ground Vehicle Technology, vol. 5083, pp. 414–418 (September 2003)
14. Silver, D., Sofman, B., Vandapel, N., Bagnell, J.A., Stentz, A.: Experimental analysis of overhead data processing to support long range navigation. In: Proceedings of the 2006 IEEE/RSJ International Conference on Intelligent Robots and Systems, Beijing, China, pp. 2443–2450 (October 9-15, 2006)
15. Tupin, F., Roux, M.: Detection of building outlines based on the fusion of SAR and optical features. ISPRS Journal of Photogrammetry & Remote Sensing 58, 71–82 (2003)

Active Visual Search by a Humanoid Robot

Francois Saidi, Olivier Stasse, and Kazuhito Yokoi

ISRI/AIST-STIC/CNRS Joint Japanese-French Robotics Laboratory (JRL)
{francois.saidi,olivier.stasse,kazuhito.yokoi}@aist.go.jp

1 Introduction

1.1 The Visual Search Behavior

Object search is a very common task we perform each time we need an object. Humanoid robots are multipurpose platforms and will need to use generic tools to extend their capacities. It must thus be able to look for objects, to localize and use them. A search behavior would be a great improvement in humanoid autonomy and a step forward toward their rise outside laboratories.

Before starting a search behavior, the robot needs a model of the desired object. This model could be provided by an external mechanism, but a humanoid has all the required abilities to build that model by its own. An undergoing project in our laboratory, called the "Treasure hunting" aim at integrating in a unique cycle, the model building of an unknown object, and the search for that object in an unknown environment. With such a combined skill, the robot may incrementally build a knowledge of its surrounding environment and the object it has to manipulate without any a-priori models. Latter the robot would be able to find and recognize that object. The time constraint is crucial, as a reasonable limit has to be set on the time an end user can wait the robot to achieve its mission. This paper will focus on the search behavior and we assume that the object model is already created.

1.2 Problem Statement and Contributions

Object search is a sensor planning problem which is proven to be NP-complete [1] thus a heuristic strategy is needed to overcome that task. Because of the limited field of view, the limited depth, the lighting condition, the recognition algorithm limitation, and possible occlusion, many images from different point of view are necessary to detect and locate a given object. The knowledge of the target position is represented by a discrete presence probability [2]. A rating function to evaluate the interest of a potential next view must be created and optimized at each sensing step. The rating function will analyze the theoretical field of view for a given configuration according to various criteria defined further in

S. Lee, I.H. Suh, and M.S. Kim (Eds.): Recent Progress in Robotics, LNCIS 370, pp. 171–184, 2008.
springerlink.com

this paper. Such a function is costly and thus must be used as less as possible to evaluate a configuration interest.

In [3], we introduce the concept of *Visibility Map* a statistical accumulator in the sensor configuration space which takes into account the characteristics of the recognition system to constrain the sensor configuration space and avoid unnecessary call to the rating function. The present paper proposes an extention of the visibility map and exposes a process to retrieve interesting configuration out of that map (section 2.4).

1.3   Related Works

Few works on active 3D object search are available, fortunately the sensor planning research field provides us with some hints.

Wixon [4] uses the idea of indirect search (in which one first finds an object that commonly has a spatial relationship with the target, and then restrict the search in the spatial area defined by that relationship) he proposes a mathematical model of search efficiency, which shows that indirect search can improve the search.

Works done by Ye and Tsotsos [2] tackle the field of sensor planning for 3D object search. The search agent's knowledge of object location is encoded as a discrete probability density which is updated after each sensing action performed by the detection function. The detection function uses a simple recognition algorithm, and all factors which influence the detection ability such as imaging parameters, lighting condition, complexity of the background, occlusions etc. are included in the detection function value by averaging experimental results done under various conditions. The vision system uses one pan tilt zoom camera and a laser range finder to build a model of the environment. The search is not really 3D as, the object is recognized using a 2D technique, and the height of the camera is fixed.

Works by Sujan [5] are not focused on object search but on accurate mapping of unknown environment by the mean of sensor planning. The author proposes a model based on iterative planning, driven by an evaluation function based on Shannon's information theory. The camera parameter space is explored and each configuration is evaluated according to the evaluation function. No computational timing tests are provided, but the algorithm seems to focus on configurations which are close to obstacles or to unknown areas to improve the algorithm efficiency, this latter constraint will be formalized with the notion of visibility map introduced in 2.3.

The operational research community [6] has extensively studied the problem of optimal search, they came up with interesting theoretical results on search effort allocation which served as a basis for Tsotsos's work.

The Next Best View (NBV) research field [7] studied the sensor planning problem mainly for C.A.D. model building. These works, although sharing some commom aspects with the present topic, rely on the fundamental assumption that the object is always in the sensor field.

1.4   Problem Overview

2   Constraint on the Camera Parameters Space

2.1   Specificities of Humanoid Approach

Specificities of the HRP-2 humanoid robot must be taken early into account in the search behavior analysis.

The walking pattern generator provided by [8] constrain the waist motion on a plane, as a consequence the head is also restricted on a plane called the visual plane. During the walk, the robot point of view oscillate around that plane with an amplitude of 2cm which falls inside the resolution used by environment model. This constrain will be removed in a future work as a new pattern generator is available [9] which accepts large perturbations on the waist height.

Unlike [5], the visual sensor, which is located in the head of the robot, is subjected to stability constraint. In this work we don't consider robot postures in which the head of the robot goes over obstacles, thus the sensors configuration space is restricted by the 2D projection of obstacles on the visual plane. Moreover, we introduce a safety margin around obstacles in which sensor placement will not be evaluated.

These remarks on humanoids specificities provide natural constraints on the sensor configuration space. Other constraints due to the stereoscopic sensor and the recognition algorithm will be discussed in 2.3.

2.2   Model of the Recognition System

All recognition algorithms have some restrictions regarding the imaging condition (lighting, occlusion, scale...). One of the main assumption that can be easily controlled by active vision is the scale limitation: the smallest scale at which the object can still be recognized constitute a maximum distance limit for the detection algorithm (R_{max}). It is also suitable to have a sensor configuration in which the whole object is projected inside the image in order to maximize the number of imaged features, this imposes a lower limit for the sensor distance to the object (R_{min}). Without any loss of generality regarding the recognition algorithm, we can assume that these bounding values (R_{min} and R_{max}) are determined theoritically or experimentally during the model building and are stored with the object model. These limit values will be used to further constrain the sensor parameters to improve optimization time.

2.3   The Visibility Map

To take into account the limitation of the recognition algorithm, and to restrict the optimization to area of interest, we use the concept of visibility sphere which represents the configuration set of the stereoscopic head in which a particular 3D point can be well recognized by a given recognition algorithm. This sphere is created using R_{min} and R_{max} defined in 2.2. Figure 1 shows a 2D representation of the visibility sphere when a unique solid point is considered.

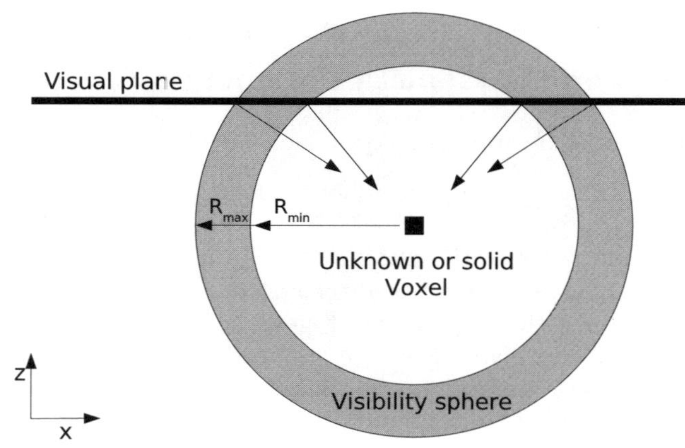

Fig. 1. Visibility sphere for a given 3D point

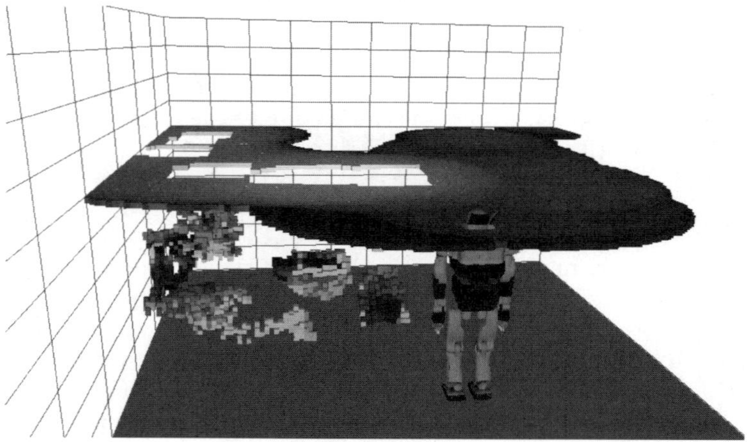

Fig. 2. This visibility map is only computed for reconstructed solid points (gray points under the plane). Each point is creating a visibility sphere around it. Lighter area on the plane represent configurations in which the solid points can be well imaged.

The configuration space of the stereoscopic head has initially 6 DOF but because the robot motion is constrained on the z axis, and the roll parameter (rotation around the line of sight) has a small influence on the visible area, only 4 DOF are considered. The sensor configuration space parameters are discretisized using the same resolution as the occupancy grid for x and y (5 cm). Whereas for pan and tilt, a resolution of half the stereoscopic field of view value, which is 33 degrees horizontally and vertically, is used.

For each solid or unknown point, the visibility sphere according to R_{min} and R_{max} values is computed and the contributions of all solid and unknown points are summed up in an accumulation map. The visibility map is then constrained on the z axis by computing its intersection with the visual plane. The figure 2 shows a 2D projection of the 4D visibility map.

In a previous work, the visibility map was computed on a 2.5D projection of the environment, this solution although computationally efficient, did not take into account an important part of the potentially visible points of the environment. Moreover, this technique did introduce a skew in the visibility map creating false interesting configurations. In the current paper, we now compute the visibility map for all boundary points (unknown or solid voxels with an empty neighbor). This new approach increases the computation time of the visibility map but takes into account all the visible 3D surface made of unknown or solid points of the environment. This computational overload can be reduced by some algorithmic improvement discussed in 4.2.

2.4 Local Maxima Extraction

The visibility map can be seen as a 4D, gray values map:

- The value of each configuration in the visibility map is called the visibility of the configuration. A candidate is a configuration which has a non zero visibility.
- The set of candidate which have the same x and y parameter is called a cluster (the cluster visibility is the sum of all its candidates visibility). Figure 2 shows in fact the clusters of the visibility map.

In order not to introduce unuseful candidates, the visibility map is only computed in the reachable area (area of the visual plan which is connected to the current sensor position). Nevertheless, a pretreatment of the visibility map is necessary to reduce the number of configurations to send to the rating function.

The basic idea of the treatment is to provide the evaluation function with configurations which respect certain criteria:

- For each configuration, a certain amount of points of interest must be visible
- Points of interest must be seen under imaging condition which allow a reliable recognition
- Configuration must have a low coupling (their view field must weakly intercept)
- The set of all configurations must partition the visible space

The coupling inside the same cluster is low because a change in the pan tilt parameter will bring a lot of new information in the field of view. On the other hand, a change in the x,y parameters will most likely produce a small change in the field of view. A local maxima extraction of the visibility map based on a window with different size for the rotation and translation parameters will output the 'locally best' configurations for which a reasonable amount of point is visible. A small size is used for the pan and tilt parameter, reflecting the fact that configurations with close orientation value are weakly coupled. A larger window size is used on the translation parameters. In this paper we use a window of size 3 for rotation and 9 for translation in the discreet parameter space.

The greedy exploration of sensor's parameter space is constrained to the local maxima of the visibility map. An interesting feature of the visibility map comes from the fact that solid and unknown points are treated the same way, and generate their visibility sphere, thus suitable configurations for exploring unknown areas are also created.

Next section will present the overall algorithm.

3   Algorithm

3.1   Overview

The flowchart of the next best view selection process is depicted in figure 3. When a new world model is available, the corresponding visibility map is computed and the local maxima extraction is performed providing a candidate list. The following sections describe the different steps of the next view selection as well as the formulation of the rating function, more details can be found in [3].

3.2   The Probability World Map

A discrete occupancy grid is generated by the stereoscopic sensor of the robot (figure 4). Localization will be done through a SLAM process [10] which merges odometric information provided by the walking pattern generator and visual information to provide accurate positioning. The target presence is represented by a discrete probability distribution function p. Since this probability will be updated after each recognition action, it is a function of both position and time. $p(v_i, t)$ represents the probability that the voxel v_i is a part of the target. For a given camera configuration c,

$$P(c) = \sum_{\Psi} p(v_i, t), \tag{1}$$

represents the probability that the object is inside the current field of view Ψ. The field of view takes into account occlusions for already mapped obstacles as well as the depth of field.

3.3   The Rating Function

The rating function must evaluate the interest of a given configuration according to different criteria:

1. the probability of detecting the object: the detection probability (DP),
2. the new area of the environment that will be seen: the new information (NI),
3. the cost in time/energy to reach that configuration: the motion cost (MC).

The DP, NI and MC are combined in the rating function:

$$RF = \lambda_{DP} \cdot DP + \lambda_{NI} \cdot NI - \lambda_{MC} \cdot MC, \tag{2}$$

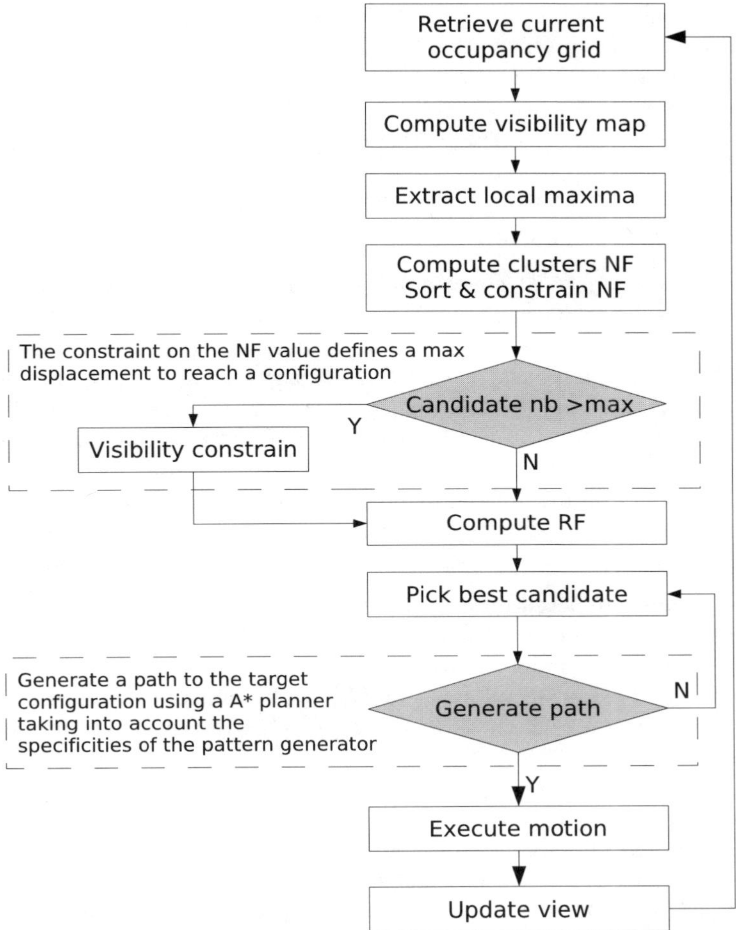

Fig. 3. Flowchart of the next view selection

where λ_{DP}, λ_{NI} and λ_{MC} are scaling factor to balance the contribution of each member of the rating function. This function will be optimized to select the next view.

The weights selection depends on the current strategy of the search:

- a high λ_{NI} will support a wide exploration of the environment,
- a high λ_{DP} will support a deep search of each potential target.

The following sections will describe the different part of the rating function.

3.4   The Detection Probability

Resolution studies done by [11] provide a characterization of the stereoscopic sensor of the robot. The resolution factor $\rho(v_i)$, which gives the resolution at which each voxel is perceived, is used to modulate the recognition likelihood. This function is defined on the field of view Ψ and has 3 parameters (θ, δ, l).

From equation 1 we define the detection probability (DP) for a given camera parameter c as:

$$DP(c) = \sum_{\Psi} p(v_i, t)\rho(v_i). \tag{3}$$

3.5 The New Information

The new information (NI) concept already introduced by [12] and [5] is also used in the overall configuration rating process but with a different formulation. In these works, the expected information evaluation for a given sensor configuration did not take into consideration the occlusion problem. The only occlusion that was considered is the one created for already known obstacles. In [3] we proposed a novel formulation of the information measurement which integrates an occlusion prediction. With such a formulation we could maximize the expected information while minimizing the likelihood of occlusion.

In order to have a measurement on the possible occlusion in unmapped areas, we evaluate both the minimum and maximum expected information:

- The minimum predicted information (I_{min}), in which all unknown voxels are expected to be solid and thus causes high occlusion which, in return, will decrease the available information.
- The maximum expected information (I_{max}), in which all voxels are expected to be empty and for which all unknown voxels will reveal information.

$$NI = \alpha_{avg} \cdot \frac{I_{max} + I_{min}}{2 \cdot N} + \alpha_{err} \cdot \frac{I_{min}}{I_{max}}, \tag{4}$$

where N is the total number of voxel in the field of view when there is no occlusion, α_{avg} and α_{err} are the coefficient for the expected average and error $(I_{min} \leq I_{max})$ and $NI = 0$ when $I_{max} = 0$. With this formulation maximizing NI, will on one hand, maximize the average expected information $\frac{I_{max}+I_{min}}{2 \cdot N}$, while on the other hand, minimize the error on the prediction $\frac{I_{min}}{I_{max}}$.

3.6 The Motion Cost

In addition to maximizing the NI and DP, it is also interesting to minimize the distance to travel to reach the configuration. An Euclidean metric in the configuration space of the sensor with individual weights on each DOF, is used to define the motion cost (MC). Moreover to take into account obstacles, we integrate a navigation function based on a 2D projection of the occupancy grid to evaluate the motion cost on the x and y parameters of the sensor.

$$MC = \alpha_{NF} \cdot NF(x, y) + \sqrt{\alpha_p (p' - p)^2 + \alpha_t (t' - t)^2}, \tag{5}$$

In this paper, the pan-tilt (p,t) parameters have a low weight (α_p, α_t) whereas x and y have a higher weight (α_{NF}) reflecting the fact that a change of x and y is achieved by moving the whole robot which takes more time and energy than moving only the head.

Next section presents the optimization of this rating function in order to determine the next sensor configuration.

3.7   Candidates Examination

The local maxima extraction presented in section 2.4 provides us with a list of candidates. This candidates list could directly be sent to the rating function, but for efficiency reasons the different parts of the rating function are evaluated separately starting with the less computationally expensive part, the motion cost. The navigation function (section 3.6) $NF(x, y)$ is computed for all positions. A distance criteria is first applied to constrain the candidates inside a neighborhood around the current robot position (a typical value is 2m, which guaranties that the next view will be within a 2m distance).

If the candidates are still too numerous, a visibility constrain is applied and the best candidates are taken (i.e. candidates wich recived to maximum amount of votes). The number of candidates that can be sent to the rating function depends on the reaction time we want to achieve an on the state of the robot (i.e. when the robot is moving, the threshold will be higher than when the robot is standing and waiting for a decision). Typically we set a limit of 1000 candidates to rate. The actual implementation of the rating function takes (initially) 3 ms per candidate (section 4.2 gives some timing results for each step of the process), thus in the worst case, it takes up to 3 sec to plan the next view. These steps are depicted in figure 3.

Moreover, the examination process could select the weight of the rating function linear combination depending on the current strategy. When the examination process comes out with a candidate, the existence of a path to the target is then checked using an A^* 2D planner. This simple path planner, takes into account the bounding box of the robot while walking. The planning is done only for the robot body, and the residual head motion is then executed to reach the target sensor configuration.

3.8   The Recognition Function and the Update Process

A simulation of the recognition system has been implemented. Although the simulation is simple, it has the main characteristics of a real recognition system. A random function creates false target that adds some noise in the probability map. Few assumptions are made on the underlaying recognition system and the output of the recognition is a list of object pose with their associated likelihood. The recognition system is a color detector based on a normalized color histogram. The 3D position of the center of the color region detected in an pair of image is computed using the camera calibration information. The matching score is proportional to the size of the segmented color region, the closer this size is to real object size, the higher the matching score will be.

Each object pose is then converted into the corresponding voxel set and their probabilities are merged with the target presence probability map through the update process.

4   Experiments

4.1   Object Search and Exploration Behavior

Preliminary experiments were done to validate the algorithm. Two simulations where performed: one in which the target object is not present and another one in which the object is present but not hidden.

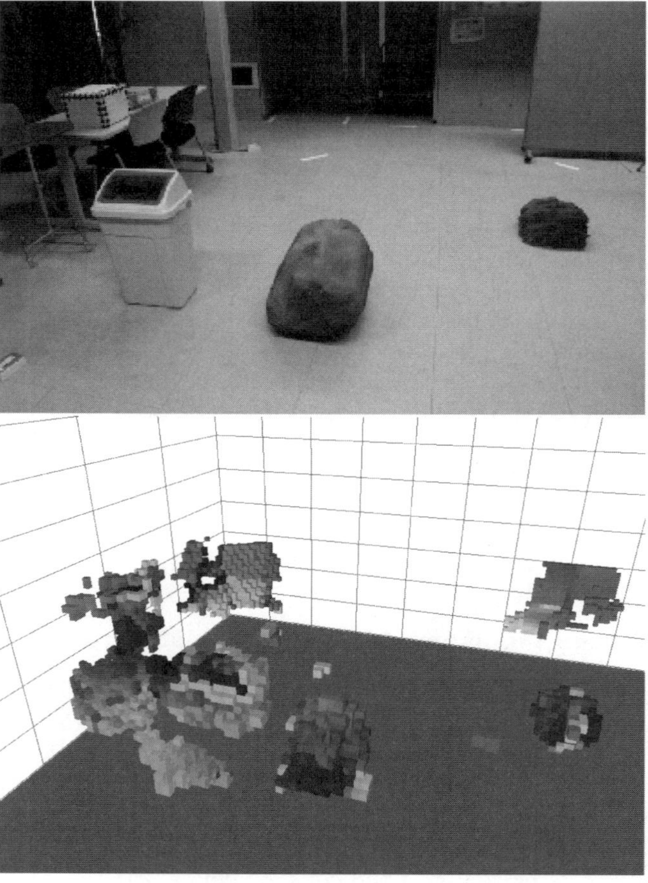

Fig. 4. Real view of the experiment environment and the corresponding 3D occupancy grid generated by the robot

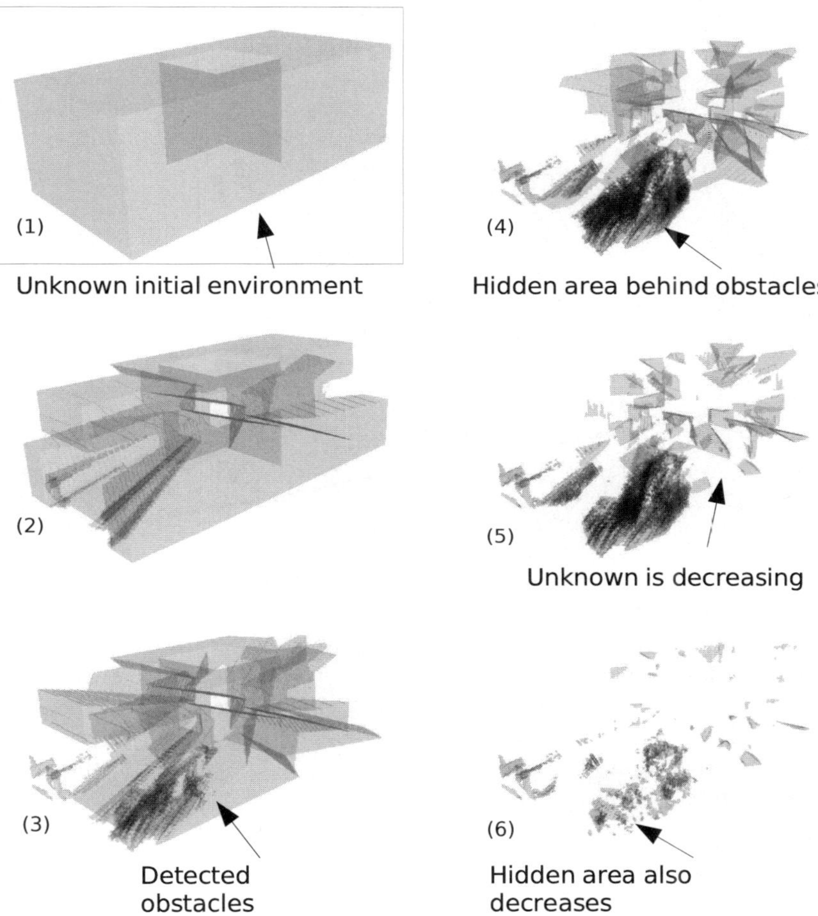

Fig. 5. A screenshots sequence of the exploration behavior performed in simulation. The environment is a 12x6x4 meter box initially unknown, the robot starts from the center surrounded by a fully known safety area in order to bootstrap the algorithm.

In the first experiment, the robot mainly driven by the NI explore the full environment (figure 5). The complete exploration is done in 100 hundred views and lasts 5 minutes (the displacement time is not taken into account). The motion cost weight is very low, thus the system was focusing on retrieving the maximum information at each step whatever displacement it needs ($\lambda_{NI} = 1000$, $\lambda_{MC} = 0.1$, $\alpha_{NF} = 1$, $\alpha_p = \alpha_t = 0$). The table below gives the total distance traveled by the robot for 50 different views, and the remaining unknown voxels in the environment for different values of λ_{MC} ($\lambda_{NI} = 1000$).

λ_{MC}	0.01	0.1	0.2	0.5	2	3
Total distance (m)	91.3	71.4	56.3	45.7	21	16
Unknown (%)	13.8	13.7	13.7	16	21	19

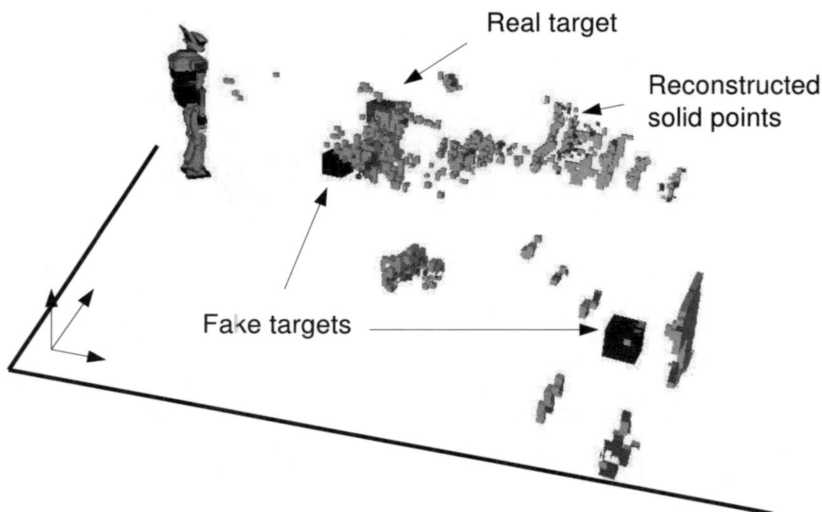

Fig. 6. A screen capture of the simulator at the end of the search behavior

In the second experiment the robot finds the target after 45 views (figure 6). Depending on the settings (the $\lambda_{NI}/\lambda_{DP}$ ratio) the robot will lock the target after the first view or will do some remaining exploration before focusing its attention on the target. An online video[1] shows the complete search sequence.

Next section gives some implementation details and benchmark results on the different parts of the algorithm.

4.2   Implementation Notes

The whole design and implementation were done while targeting a fast and reactive behavior of the robot, thus time constraints are crucial and have guided the project. The table below shows some benchmark results done with a 5cm resolution of a 12x6x4 meter environment, using a 3GHz bi-Xeon workstation with Hyper-Threading[TM].

Many improvement of the initial code were performed. Concerning the visibility map, the visibility sphere of a point is precomputed according to the R_{min}, R_{max} values and stored in a look-up-table (LUT). Then, the map update is done incrementally, which means that only points which have a change in their state will be considered. Because it is done incrementally, the update process gets faster.

14500 points with no LUT	6s
24600 points with the LUT	3.1s
average for 50 updates with LUT	380msec

---

[1] http://staff.aist.go.jp/francois.saidi/video/HRP2SearchBehavior.avi

The constrain achieved by the visibility map drastically reduces the configurations to consider. The discretized configuration space of the robot sensor in this experiment contains 240x120x200=5.76 million configurations, the visibility map and local maxima extraction only outputs 1000 configurations.

The rating function computation is a highly parallelisable process which benefits of multi-core/cpu machines. Thus, the number of physical/logical cpu is detected at runtime and the corresponding number of threads is used to compute the score of the candidates. Once more, the visibility map update gets faster as the unknown in the environment decreases. Moreover as the environment is being mapped, the number of unknown voxels decreases quickly and the computation of the rating function gets faster. The average computation time over 50 views of the rating function using 4 threads is around 1 msec per candidate.

5   Conclusion

This paper exposed the framework for a search behavior developed for the humanoid robot HRP-2. The problem, which falls in the sensor planning field, is formulated as an optimization problem. The concept of visibility map introduced in [3] to constrain the sensor parameter space according to the detection characteristics of the recognition algorithm is used to reduce the dimension of the sensor parameter space. The rating function uses a formulation of the expected information takes into account a prediction on occlusion in the unexplored space to provide a more accurate information prediction. Simulation results of an exploration and search behavior has been presented to validate the model. Work is on progress, and experiments on the real robot to validate parts of the algorithm are already undertaken and the z axis limitation of the sensor is on the way of beeing removed.

Acknowledgment

This research was partially supported by a Post-doctoral Fellowship of Japan Society for Promotion of Science(JSPS) and JSPS Grand-in-Aid for Scientific Research.

References

1. Ye, Y., Tsotsos, J.K.: Sensor planning in 3d object search: its formulation and complexity. In: Fourth International Symposium on Artificial Intelligence and Mathematics, Florida, U.S.A (January 3-5, 1996)
2. Ye, Y., Tsotsos, J.K.: Sensor planning for 3d object search. Computer Vision and Image Understanding 73(2), 145–168 (1999)
3. Saidi, F., Stasse, O., Yokoi, K.: A visual attention framework for a visual search by a humanoid robot. In: IEEE-RAS International Conference on Humanoid Robots, Genova, Italy, pp. 346–351 (December 4-6, 2006)

4. Wixson, L.E.: Gaze selection for visual search. Ph.D. dissertation, Department of Computer Science, Univ. of Rochester (1994)
5. Sujan, V.A., Dubowsky, S.: Efficient information-based visual robotic mapping in unstructured environments. The International Journal of Robotics Research 24(4), 275–293 (2005)
6. Koopman, B.O.: Search and Screening. Pergamon Press, Oxford (1980)
7. Connolly, C.J.: The determination of next best views. In: IEEE Int. Conf. on Robotics and Automation, pp. 432–435 (1985)
8. Kajita, S., Kanehiro, F., Kaneko, K., Yokoi, K., Hirukawa, H.: The 3d linear inverted pendulum mode: A simple modeling of a biped walking pattern generation. In: International Conference on Intelligent Robots and Systems, Maui, Hawaii, USA, pp. 239–246 (November 2001)
9. Verrelst, B., Yokoi, K., Stasse, O., Arisumi, H., Vanderborght, B.: Mobility of humanoid robots: Stepping over large obstacles dynamically. In: International Conference on Mechatronics and Automation, Luoyang, China, pp. 1072–1079 (June 25-28, 2006)
10. Stasse, O., Davison, A., Sellaouti, R., Yokoi, K.: Real-time 3d slam for humanoid robot considering pattern generator information. In: International Conference on Intelligent Robots and Systems, IROS, Beijing, China, pp. 348–355 (October 9-15, 2006)
11. Telle, B., Stasse, O., Ueshiba, T., Yokoi, K., Tomita, F.: Three characterisations of 3d reconstruction uncertainty with bounded error. In: Proceedings of the 2004 IEEE International Conference on Robotics and Automation, pp. 3905–3910 (2004)
12. Makarenko, A., Williams, S., Bourgault, F., Durrant-Whyte, H.: An experiment in integrated exploration. In: IEEE/RSJ International Conference on Intelligent Robots and System, vol. 1, pp. 534–539 (2002)

Visual Control of a Micro Helicopter under Dynamic Occlusions

Yuta Yoshihata, Kei Watanabe, Yasushi Iwatani, and Koichi Hashimoto

Department of System Information Sciences, Tohoku University
Aoba-ku Aramaki Aza Aoba 6-6-01, Sendai, Japan
{yoshihata, watanabe, iwatani, koichi}@ic.is.tohoku.ac.jp

Summary. This paper proposes a switched visual feedback control method for a micro helicopter under occlusions. Two stationary cameras are placed on the ground. They track four black balls attached to rods connected to the bottom of the helicopter. The control input is computed by using the errors between the positions of the tracked objects and pre-specified reference values for them. The multi-camera configuration enables us to design a switched controller which is robust against occlusions. The proposed controller selects a camera which correctly measures all of the tracked object positions at each time. If a camera loses a tracked object, then the camera searches for the four tracked objects by using an image data captured from the other camera. The proposed controller can keep the helicopter in a stable hover even when one of the cameras loses tracked objects due to occlusions.

1 Introduction

Autonomous control of unmanned helicopters has potential for surveillance tasks in dangerous areas, including chemical or radiation spill monitoring, forest-fire reconnaissance, monitoring of volcanic activity and surveys of natural disaster areas. For vehicle navigation, the use of computer vision as a sensor is effective in unmapped areas. Visual feedback control is also suitable for the task of autonomous aircraft takeoff or landing, since vision sensors provide variable information about the helicopter position and posture relative to the launch pad or the landing pad. They have generated considerable interest in the vision based control community [2, 3, 4, 6, 7, 8, 9, 11, 12].

The authors have developed a visual control system for a micro helicopter [10]. The helicopter does not have any sensors which measure its position or posture. A camera is placed on the ground. They track four black balls attached to rods connected to the bottom of the helicopter. The differences between the current ball positions and given reference positions in the camera frame are fed to a set of PID controllers. No sensors are installed on the helicopter body. Thus we need no mechanical or electrical improvements of existing unmanned helicopters which are controlled remotely and manually.

In visual control, tracked objects have to be visible in the camera view, but tracking may fail due to occlusions. An occlusion occurs when an object moves across in front of a camera or when the background color happens to be similar to

S. Lee, I.H. Suh, and M.S. Kim (Eds.): Recent Progress in Robotics, LNCIS 370, pp. 185–197, 2008.
springerlink.com

the color of a tracked object. Multi-camera systems are suitable for designing a robust controller under occlusions. In fact, for multi-camera systems, even when a tracked object is not visible in a camera view, other cameras may track it.

This paper proposes a visual feedback control system for a helicopter using two cameras. The multi-camera configuration is redundant for helicopter control, but it enables us to design a switched controller which is robust against occlusions. We use four tracked objects, and each camera tracks the four balls. The proposed controller selects a camera which correctly measures all of the tracked object positions at each time. If a camera loses a tracked object, then the camera searches for the four tracked objects by using an image data captured from the other camera. The proposed controller can keep the helicopter in a stable hover even when one of the cameras loses tracked objects due to occlusions.

2   Experimental Setup

2.1   System Configuration

The system considered in this paper consists of a small helicopter and two stationary cameras as illustrated in Fig. 1. The helicopter does not have any sensors

Fig. 1. System configuration

Fig. 2. X.R.B. with four black balls

Table 1. Specifications of the system

Length of the helicopter,	0.40 [m].
Height of the helicopter,	0.20 [m].
Rotor length of the helicopter,	0.35 [m].
Weight of the helicopter,	0.22 [kg].
Focal length of the lens,	4.5 [mm].
Camera resolution,	640 × 480 [pixels].
Pixel size,	7.4 × 7.4 [μm^2].

which measure the position or posture. It has four small black balls, and they are attached to rods connected to the bottom of the helicopter. The two cameras are placed on the ground and they look upward. Each camera tracks the four balls.

The system requires 8.5 milli-seconds to compute control input voltages from capturing images of the balls. This follows from the use of a fast IEEE 1394 camera, Dragonfly Express, developed by Point Grey Research Inc.

The small helicopter used in experiments is X. R. B–V2–lama developed by HIROBO (see Fig. 2). It has a coaxial rotor configuration. The two rotors share the same axis, and they rotate in opposite directions. The tail is a dummy. A stabilizer is installed on the upper rotor head. It mechanically keeps the posture horizontal. Table 1 summarizes specifications of the system.

Snapshots of the helicopter captured from the two cameras can be seen in Fig. 3.

2.2 Coordinate Frames

Let Σ^g be the global reference frame and a coordinate frame Σ^b be attached to the helicopter body as illustrated in Fig. 4. The z^g axis is directed vertically downward. A coordinate frame Σ^j is attached to camera j for $j = 1, 2$. The z^j axis lies along the optical axis of camera j. The axes x^g, x^1 and x^2 are parallel. The coordinate frame $x^j y^j$ corresponds to the image frame of camera j, and it is denoted by Σ^{cj} for $j = 1, 2$.

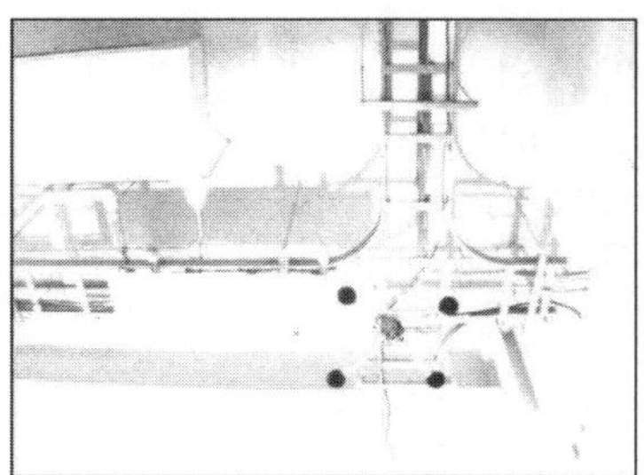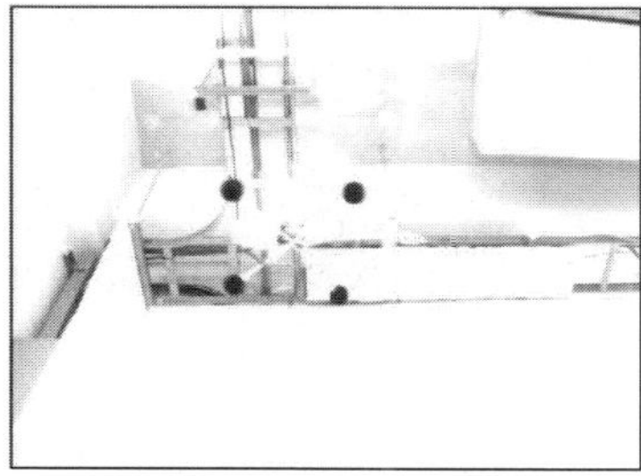

Fig. 3. Snapshots of helicopter flight. The left was captured from camera 1 and the right from camera 2. The helicopter was controlled manually.

The cameras capture images of the four black balls attached to rods connected to the bottom of the helicopter. The black balls are labeled from 1 to 4. Let $^{\mathrm{b}}\boldsymbol{p}_i \in \mathbb{R}^3$ denote the position of ball i in the frame Σ^{b} and the ball positions in the frame Σ^{b} be given by

$$^{\mathrm{b}}\boldsymbol{p}_1 = \begin{bmatrix} 0.1 & 0.1 & 0.04 \end{bmatrix}^{\top}, \tag{1}$$

$$^{\mathrm{b}}\boldsymbol{p}_2 = \begin{bmatrix} -0.1 & 0.1 & 0.04 \end{bmatrix}^{\top}, \tag{2}$$

$$^{\mathrm{b}}\boldsymbol{p}_3 = \begin{bmatrix} 0.1 & -0.1 & 0.04 \end{bmatrix}^{\top}, \tag{3}$$

$$^{\mathrm{b}}\boldsymbol{p}_4 = \begin{bmatrix} -0.1 & -0.1 & 0.04 \end{bmatrix}^{\top}. \tag{4}$$

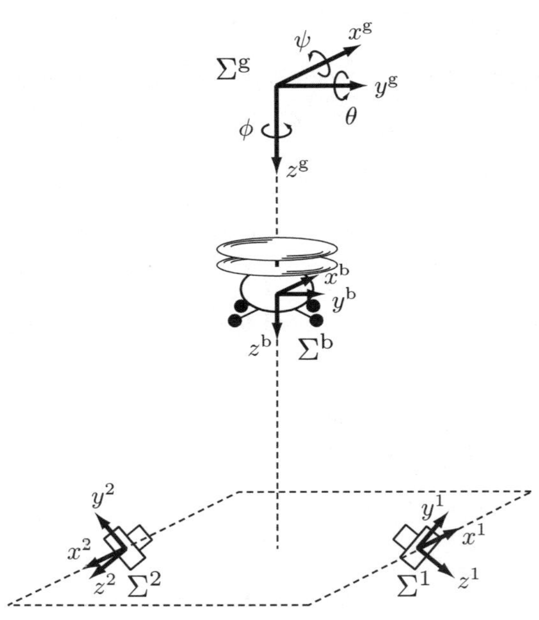

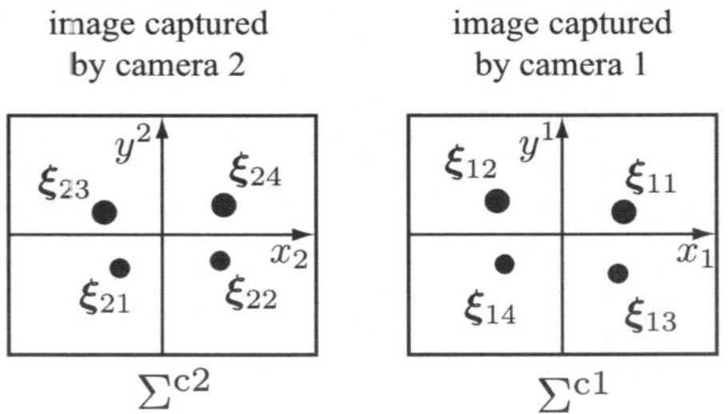

Fig. 4. Coordinate frames

The position of the center of mass of ball i in the image frame Σ^{cj} is denoted by $\boldsymbol{\xi}_{ji} = [\xi_{jix}, \xi_{jiy}] \in \mathbb{R}^2$ as illustrated in Fig. 4. We define

$$\boldsymbol{\xi}_j = \begin{bmatrix} \boldsymbol{\xi}_{j1}^\top & \boldsymbol{\xi}_{j2}^\top & \boldsymbol{\xi}_{j3}^\top & \boldsymbol{\xi}_{j4}^\top \end{bmatrix}^\top \tag{5}$$

for $j = 1, 2$.

The helicopter position relative to the global reference frame Σ^g is denoted by (x, y, z). The roll, pitch and yaw angles are denoted by ψ, θ, ϕ, respectively.

The following four variables are individually controlled by voltages supplied to a transmitter (see Fig. 5):

B : Elevator, pitch angle of the lower rotor.

A : Aileron, roll angle of the lower rotor.

T : Throttle, resultant force of the two rotor thrusts.

Q : Rudder, difference of the two torques generated by the two rotors.

The corresponding input voltages are denoted by V_B, V_A, V_T and V_Q. The state variables x, y, z and ϕ are controlled by applying V_B, V_A, V_T and V_Q, respectively.

It is assumed in this paper that

$$\theta(t) = 0 \quad \text{and} \quad \psi(t) = 0, \ \forall \, t \geq 0. \tag{6}$$

Recall that the helicopter has the horizontal-keeping stabilizer. Both the angles θ and ψ converge to zero fast enough even when the body is inclined. Thus, the assumption is not far from the truth in practice. We here define

$$\boldsymbol{r} = \begin{bmatrix} x & y & z & \phi \end{bmatrix}^\top. \tag{7}$$

Note that $\boldsymbol{r}$ means the vector of the generalized coordinates of the helicopter.

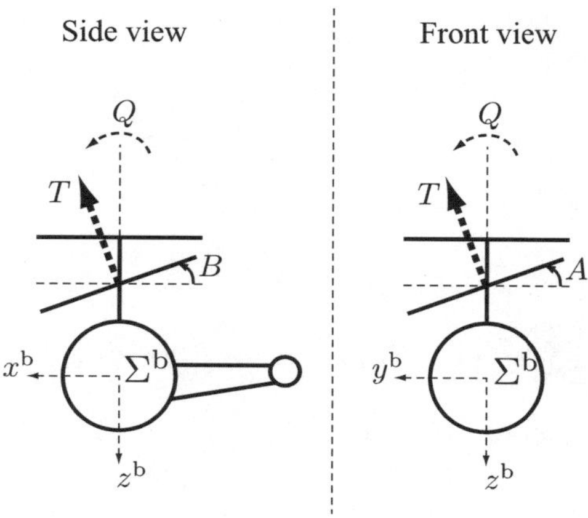

Fig. 5. The helicopter coordinate frame and input variables

3   Image Jacobians

This section derives image Jacobians, say J_1 and J_2, which gives the relationship between ξ_j and r by

$$\dot{\xi}_j = J_j(r)\dot{r} \tag{8}$$

for $j = 1, 2$, where recall that ξ_j is the vector of the image features in the image plane Σ^{cj} and r the vector of the generalized coordinates of the helicopter.

Let the position of ball i in the frame Σ^j be given by

$$^{j}p_i = \begin{bmatrix} ^{j}x_i \\ ^{j}y_i \\ ^{j}z_i \end{bmatrix} \in \mathbb{R}^3. \tag{9}$$

Then we have

$$\begin{bmatrix} ^{j}p_i \\ 1 \end{bmatrix} = {}^{j}H_{\mathrm{g}}{}^{\mathrm{g}}H_{\mathrm{b}} \begin{bmatrix} ^{\mathrm{b}}p_i \\ 1 \end{bmatrix} \tag{10}$$

where $^{j}H_{\mathrm{g}}$ and $^{\mathrm{g}}H_{\mathrm{b}}$ are the homogeneous transformation matrices from Σ^{g} to Σ^j and from Σ^{b} to Σ^{g}, respectively. It then holds that

$$^{j}z_i \begin{bmatrix} \xi_{ji} \\ 1 \end{bmatrix} = F \begin{bmatrix} r_i \\ 1 \end{bmatrix} \tag{11}$$

where

$$F = \begin{bmatrix} fs & 0 & 0 & 0 \\ 0 & fs & 0 & 0 \\ 0 & 0 & 1 & 0 \end{bmatrix} \tag{12}$$

and f is the focal length of the lens and s the length of a pixel side. It is straightforward to verify that

$$\xi_{ji} = \frac{f}{^{j}z_i} \begin{bmatrix} ^{j}x_i \\ ^{j}y_i \end{bmatrix} \tag{13}$$

$$=: \alpha_{ji}(r). \tag{14}$$

We here define

$$\alpha_j(r) = \begin{bmatrix} \alpha_{j1}^{\top}(r) & \alpha_{j2}^{\top}(r) & \alpha_{j3}^{\top}(r) & \alpha_{j4}^{\top}(r) \end{bmatrix}^{\top} \tag{15}$$

for $j = 1, 2$. Then we obtain the image Jacobians by

$$J_j(r) := \frac{\partial \alpha_j}{\partial r}. \tag{16}$$

4   Controller Design

The reference position relative to the global reference frame Σ^{g} is always set to
$\mathbf{0}$. When the reference position is changed, the global reference frame is replaced
and the reference position is set to the origin of the new global reference frame.
Then, our goal is that $\boldsymbol{r}(t) \to \mathbf{0}$ as $t \to \infty$. In this paper, $\boldsymbol{J}_j(\mathbf{0})$ is simply denoted
by $\boldsymbol{J}_j$.

This paper assumes that at least one camera can capture all of the ball positions at each time. This section proposes a switched visual feedback control
system illustrated in Fig. 6 under the assumption. The proposed controller uses
the errors between the image feature $\boldsymbol{\xi}_j(t)$ and the corresponding given reference
$\boldsymbol{\xi}_j^{\mathrm{ref}}$ to obtain the input voltages, and it is image-based visual servo control which
is robust against model uncertainties [5].

The switch in closed loop depends on whether an occlusion is detected or not.
The criterion for making the decision will be described in Section 4.2. In this
paper, camera j is labeled as "normal" at time t, if it measures all of the image
features correctly. Otherwise, camera j is labeled as "occluded" at time t.

4.1   Measurement of Image Features

An image feature $\boldsymbol{\xi}_{ji}(t)$ is given by the following manner. A binary data matrix
at time t is first obtained from an image captured by camera j, and it is denoted
by $I_j(x,y)$. The matrix $I_j(x,y)$ has values of 1 for black and 0 for white. We
then make a search window $\mathbb{S}_{ji}$ whose center is defined as follows:

Normal case: It is set at $\boldsymbol{\xi}_{ji}(t-h)$, where h denotes the sampling time.
Occluded case: We first estimate $\boldsymbol{\xi}_{ji}(t)$ by the following equation:

$$\boldsymbol{J}_j \boldsymbol{J}_{\bar{j}}^{+}(\boldsymbol{\xi}_{\bar{j}}(t) - \boldsymbol{\xi}_{\bar{j}}^{\mathrm{ref}}) + \boldsymbol{\xi}_j^{\mathrm{ref}} =: \begin{bmatrix} \tilde{\boldsymbol{\xi}}_{j1}^{\top} & \tilde{\boldsymbol{\xi}}_{j2}^{\top} & \tilde{\boldsymbol{\xi}}_{j3}^{\top} & \tilde{\boldsymbol{\xi}}_{j4}^{\top} \end{bmatrix}^{\top} \tag{17}$$

where $\boldsymbol{J}_j^{+}$ denotes the Moore-Penrose inverse of $\boldsymbol{J}_j$ and $\bar{j}$ satisfies $\bar{j} \neq j$ and
$\bar{j} \in \{1, 2\}$. The center is set at $\tilde{\boldsymbol{\xi}}_{ji}$.

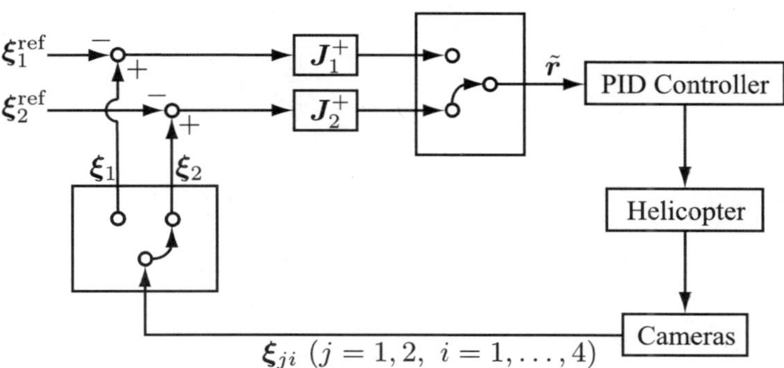

Fig. 6. Closed loop system

The size of the window $\mathbb{S}_{ji}$ is given by a constant value. We define an image data matrix by

$$\bar{I}_{ji}(x,y) = \begin{cases} I_j(x,y), & \text{for } (x,y) \in \mathbb{S}_{ji}, \\ 0, & \text{otherwise.} \end{cases}$$

The image feature $\boldsymbol{\xi}_{ji}(t)$ is the center of mass of $\bar{I}_{ji}(x,y)$.

4.2   Camera Selection

Let three constants δ, $m_{\min}$ and $m_{\max}$ be given. Let $m_{ji}(t)$ denote the area, or equivalently the zero-th order moment, of the image data $\bar{I}_{ji}(x,y)$. An occlusion is detected or cancelled for camera j in the following manner:

Normal case: If $m_{\min} \leq m_{ji}(t) \leq m_{\max}$ holds for all i, then camera j is labeled as "normal" again. Otherwise, it is labeled as "occluded".

Occluded case: If it holds that $m_{\min} \leq m_{ji}(t) \leq m_{\max}$ and

$$\|\boldsymbol{J}_1^+(\boldsymbol{\xi}_1(t) - \boldsymbol{\xi}_1^{\mathrm{ref}}) - \boldsymbol{J}_2^+(\boldsymbol{\xi}_2(t) - \boldsymbol{\xi}_2^{\mathrm{ref}})\| < \delta \tag{18}$$

for all i, then camera j is labeled as "normal". Otherwise, it is labeled as "occluded" again.

If camera j is used at time $t - h$ and it is normal at time t, then camera j is used at time t. Otherwise, camera $\bar{j}$ is used at time t, where recall that $\bar{j} \in \{1,2\}$ and $\bar{j} \neq j$.

4.3   Control Input Voltages

We compute

$$\tilde{\boldsymbol{r}}(t) = \begin{bmatrix} \tilde{x}(t) & \tilde{y}(t) & \tilde{z}(t) & \tilde{\phi}(t) \end{bmatrix}^{\top} \tag{19}$$

$$:= \boldsymbol{J}_j^+(\boldsymbol{\xi}_j(t) - \boldsymbol{\xi}_j^{\mathrm{ref}}) \tag{20}$$

for j selected in the previous subsection. The input voltages are given by PID controllers of the form

$$V_B(t) = b_1 - P_1\tilde{x} - I_1 \int_0^t \tilde{x}\,\mathrm{d}t - D_1\dot{\tilde{x}}, \tag{21}$$

$$V_A(t) = b_2 - P_2\tilde{y} - I_2 \int_0^t \tilde{y}\,\mathrm{d}t - D_2\dot{\tilde{y}}, \tag{22}$$

$$V_T(t) = b_3 - P_3\tilde{z} - I_3 \int_0^t \tilde{z}\,\mathrm{d}t - D_3\dot{\tilde{z}}, \tag{23}$$

$$V_Q(t) = b_4 - P_4\tilde{\phi} - I_4 \int_0^t \tilde{\phi}\,\mathrm{d}t - D_4\dot{\tilde{\phi}}, \tag{24}$$

where b_i, P_i, I_i and D_i are constants for $i = 1, \ldots, 4$.

Table 2. PID gains

	b_i	P_i	I_i	D_i
V_B	3.37	3.30	0.05	2.60
V_A	3.72	3.30	0.05	2.60
V_T	2.70	1.90	0.05	0.80
V_Q	2.09	3.00	0.05	0.08

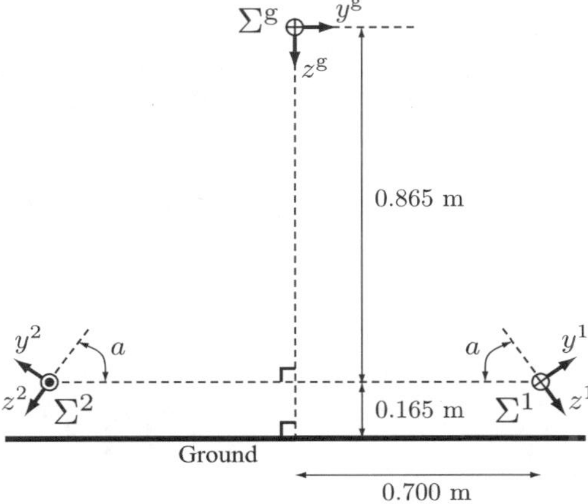

Fig. 7. Locations of two global reference frames Σ^g and the camera frames Σ^1 and Σ^2. The angle a is set to $a = 11\pi/36$.

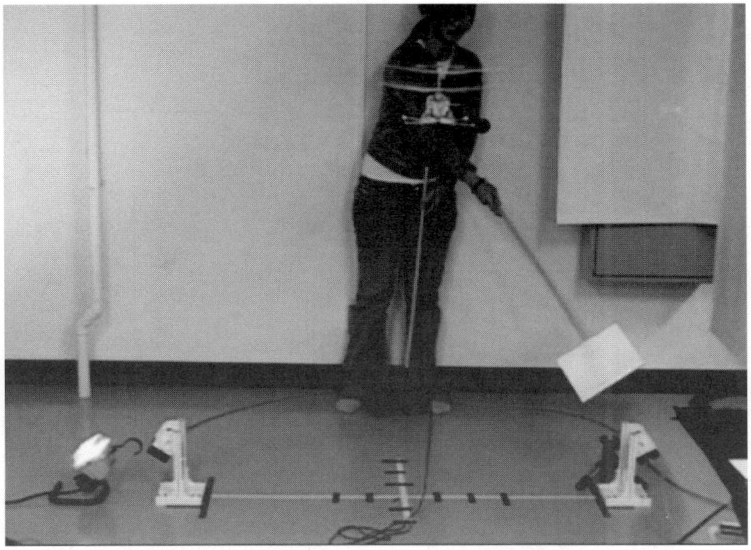

Fig. 8. A snapshot of helicopter flight under an occlusion. Camera 1 was labeled "occluded" at this time.

5   Experiment and Result

The global reference frame Σ^{g} and the camera frames Σ^1 and Σ^2 are located as shown in Fig. 7. The controller gains are tuned to the values in Table 2. The image references are set to

$$\boldsymbol{\xi}_1^{\mathrm{ref}} = \begin{bmatrix} 77.1 & -32.5 & -30.0 & -24.0 & 74.7 & -120.6 & -32.6 & -123.7 \end{bmatrix}^{\top}, \quad \text{(pixels)}, \quad (25)$$

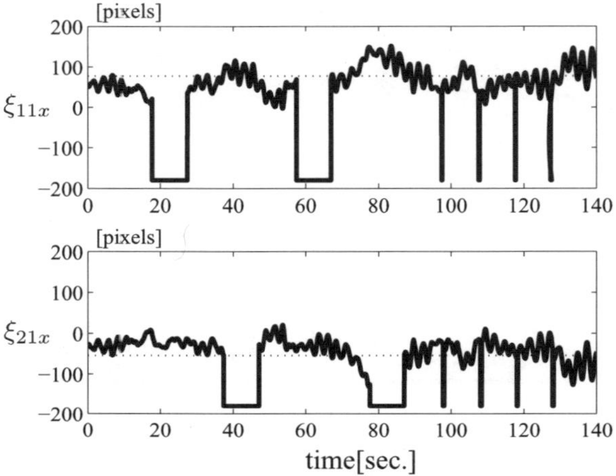

Fig. 9. Experimental result. Solid lines: Time profiles of the positions of image features. When an occlusion is detected, the value is set to -180. Dotted lines: given references.

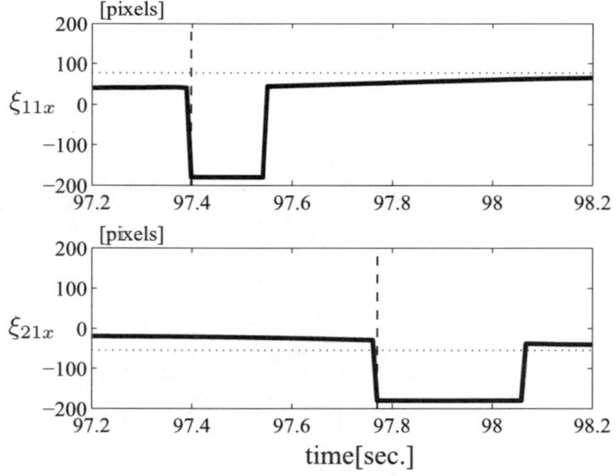

Fig. 10. Experimental result: Time profiles of the positions of image features. This is a closeup of Fig. 9 between 97.2 and 98.2 seconds. Camera 1 was used until 97.40 seconds. Camera 2 was used from 97.40 to 97.77 seconds. After that, camera 1 was selected again. Dashed lines imply time instances when the selected camera is changed.

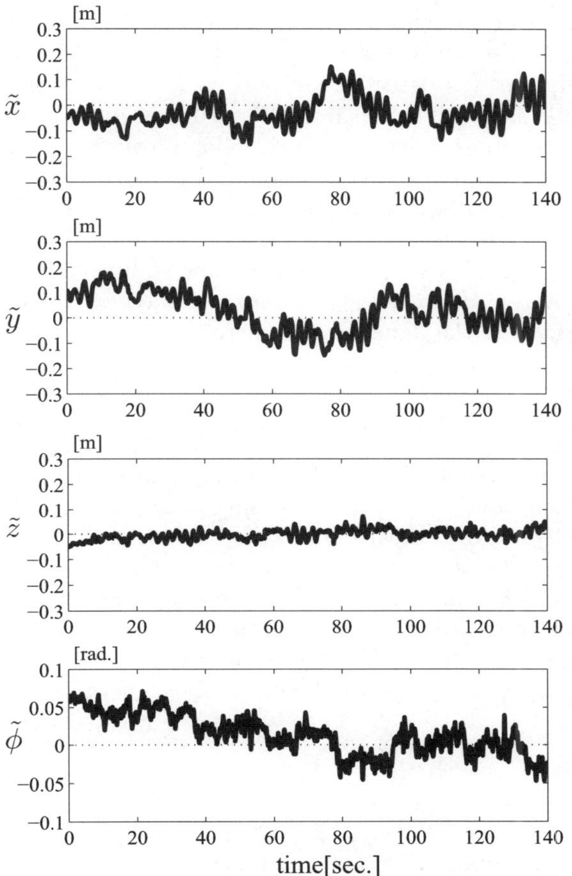

Fig. 11. Experimental result: Time profile of the estimated position $\tilde{\boldsymbol{r}}$

$$\boldsymbol{\xi}_2^{\text{ref}} = \begin{bmatrix} 54.4 & -83.2 & 47.5 & -92.4 & -60.9 & 5.1 & 62.6 & 8.5 \end{bmatrix}^\top, \quad \text{(pixels)}. \tag{26}$$

They were obtained by an actual measurement.

Camera 1 or 2 was occluded intentionally and manually. A snapshot of helicopter flight under an occlusion can be seen at Fig. 8. Long time occlusions for around 10 seconds were presented twice for each camera. Short time occlusions were done four times for each camera, and they were successively done from camera 1 to 2.

Fig. 9 shows the x positions of ball 1 in the camera frames $\Sigma^{\text{c}1}$ and $\Sigma^{\text{c}2}$. When an occlusion is detected, the value is set to -180 to make the plot easy to read. For example, camera 1 was labeled "occluded" from 17.7 to 27.5 seconds. It is seen that the number of occlusion detection is equivalent to the number of intentional occlusions.

Fig. 10 illustrates a closeup of Fig. 9 between 97.20 and 98.20 seconds. An occlusion was detected for camera 1 from 97.40 to 97.55 seconds. After 0.2 seconds, an occlusion was detected for camera 2. Our system deals with such rapid change. It can be also seen from Fig. 10 that the positions of balls are measured precisely after occlusions, which follows from the criterion (18).

Fig. 11 shows the estimated 3D position and posture $\tilde{r}$ defined by (20). It is seen that the helicopter hovered in a neighborhood of the reference position. In particular, the z position was within 8 [cm] for all time.

Several movies can be seen at [1]. They show stability, convergence and robustness of the system in an easy-to-understand way, while the properties may not be seen easily from the figures shown here.

6 Conclusion

This paper has presented a visual control system that enables a small helicopter to hover under occlusions. Two stationary and upward-looking cameras track four black balls attached to rods connected to the bottom of the helicopter. The proposed controller selects a camera which can measure all of the four tracked objects. If a camera loses tracked objects, then the camera searches for the four tracked objects by using an image data obtained from the other camera. The system can keep the helicopter in a stable hover. Several movies can be seen at [1].

The control algorithm proposed here can be generalized to multi-camera systems.

References

1. http://www.ic.is.tohoku.ac.jp/E/research/helicopter/index.html
2. Altug, E., Ostrowski, J.P., Taylor, C.J.: Control of a quadrotor helicopter using dual camera visual feedback. International Journal of Robotics Research 24(5), 329–341 (2005)
3. Amidi, O., Kanade, T., Fujita, K.: A visual odometer for autonomous helicopter flight. Robotics and Autonomous Systems 28, 185–193 (1999)
4. Ettinger, S.M., Nechyba, M.C., Ifju, P.G., Waszak, M.: Vision-guided flight stability and control for micro air vehicles. In: Proc. IEEE/RSJ International Conference on Intelligent Robots and Systems (2002)
5. Hashimoto, K.: A review on vision-based control of robot manipulators. Advanced Robotics 17(10), 969–991 (2003)
6. Mahony, R., Hamel, T.: Image-based visual servo control of aerial robotic systems using linear image features. IEEE Trans. on Robotics 21(2), 227–239 (2005)
7. Mejias, L.O., Saripalli, S., Cervera, P., Sukhatme, G.S.: Visual servoing of an autonomous helicopter in urban areas using feature tracking. Journal of Field Robotics 23(3), 185–199 (2006)
8. Saripalli, S., Montgomery, J.F., Sukhatme, G.S.: Visually-guided landing of an unmanned aerial vehicle. IEEE Trans. on Robotics and Automation 19(3), 371–381 (2003)

9. Shakernia, O., Sharp, C.S., Vidal, R., Shim, D.H., Ma, Y., Sastry, S.: Multiple view motion estimation and control for landing an unmanned aerial vehicle. In: IEEE International Conference on Robotics and Automation (2002)
10. Watanabe, K., Yoshihata, Y., Iwatani, Y., Hashimoto, K.: Image-based visual PID control of a micro helicopter using a stationary camera. Advanced Robotics (to appear)
11. Wu, A.D., Johnson, E.N., Proctor, A.A.: Vision-aided inertial navigation for flight control. In: AIAA Guidance, Navigation and Control Conference and Exhibit (2005)
12. Yu, Z., Celestino, D., Nonami, K.: Development of 3D vision enabled small-scale autonomous helicopter. In: 2006 IEEE/RSJ International Conference on Intelligent Robots and Systems (2006)

Visual Servoing from Spheres with Paracatadioptric Cameras

Romeo Tatsambon Fomena[1] and François Chaumette[2]

[1] IRISA - Université de Rennes 1, Campus de Beaulieu, 35 042 Rennes cedex, France
[2] INRIA Campus de Beaulieu, 35 042 Rennes cedex, France
`firstname.name@irisa.fr`

Summary. A paracatadioptric camera consists of the coupling of a parabolic mirror with a telecentric lens which realizes an orthographic projection to the image sensor. This type of camera provides large field of view images and has therefore potential applications for mobile and aerial robots.

This paper is concerned with visual servoing using paracatadioptric cameras. A new optimal combination of visual features is proposed for visual servoing from spheres. Using this combination, a classical control law is proved to be globally stable even in the presence of modeling error. Experimental and simulation results validate the proposed theoretical results.

1 Introduction

In visual servoing, data provided by a vision sensor is used to control the motion of a dynamic system [1]. A vision sensor provides a large spectrum of potential visual features. However, the use of some visual features may lead to stability problems if the displacement that the robot has to achieve is very large [2]. Therefore, there is a need to design optimal visual features for visual servoing. By optimality the satisfaction of the following criteria is meant: local and -as far as possible- global stability of the system, robustness to calibration and to modeling errors, non-singularity, local mimima avoidance, satisfactory trajectory of the system and of the features in the image, and finally a linear link and maximal decoupling between the visual features and the degrees of freedom (DOFs) taken into account.

Several approaches have been proposed to try to reach an optimal system behaviour using only 2D data (due to lack of space, we do not recall here the properties of pose-based visual servoing [3] and 2 1/2 D visual servoing [4]). A satisfactory motion of the system in the cartesian space can be obtained by decoupling the z-axis translational and rotational motions from the other DOFs through a partitioned approach [5]. Another way around the decoupling of the optical axis motions is to use cylindrical coordinates [6]. The partitioned approach has been coupled with a potential function in a control scheme to keep the features in the image boundary. Potential functions can also be used in path planning in the image space to keep the features in the field of view [7].

S. Lee, I.H. Suh, and M.S. Kim (Eds.): Recent Progress in Robotics, LNCIS 370, pp. 199–213, 2008.
springerlink.com

Similarly, navigation functions can be combined with a global diffeomorphism from a visible set of rigid-body configurations of a special target to an image space, to construct global, dynamical visual servoing systems that guarantees the visibility of the features all times [8].

Central catadioptric systems (except perspective cameras), despite their more complex projection equations, are well suited for large field of view images. Considering feature points on such cameras, the interaction with the system (the link between the robot velocities and the image observations) has been shown to present the same singularities as classical perspective cameras [9]. Lately, a spherical projection model has been used to design a new minimal set of optimal visual features for visual servoing from spheres with any central catadioptric system [10]. These features mostly draw a straight line trajectory from the initial position to the desired position in the image space.

For paracatadioptric cameras, straight line trajectories are not always suitable in the image space because of the dead angle in the center of the image inherent to the physiscal realization of such systems. For this reason, there is a need to search for other features more suitable for such imaging systems. This paper presents a new optimal set of visual features for visual servoing from spheres specific to this type of cameras. This new set is built from the previous combination [10] using a cylindrical coordinate system which is appropriate to the motion of the measures in the image.

In the next section, we recall the general results concerning visual servoing from spheres using any central catadioptric system. The optimal visual features obtained from this generalization are then derived in the case of paracatadioptric cameras. In section III we propose a new optimal set of three features which is shown to be more appropriate to the feature motion in the image plane of such systems. For the proposed visual features, a theoretical analysis of the stability and the robustness of a classical control law with respect to modeling errors is given. In section IV, we validate experimentally on a paracatadioptric system the combination proposed for any central catadioptric system. Finally, simulation results are given in this same section to validate the new optimal combination.

2   General Visual Features

In this section, we recall the optimal visual features obtained for visual servoing from spheres using any central catadioptric system. These features are designed using a spherical projection model. Indeed, with this projection model, it is quite easy and intuitive to determine optimal features compared to omnidirectional projection models.

2.1   Spherical Projection of a Sphere and Potential Visual Features

Let $S_{(O,R)}$ be a sphere of radius R and center O with coordinates (X_O, Y_O, Z_O) in the camera frame. Let $S_p(C,1)$ be the unit sphere located at the camera optical center C. The spherical projection of $S_{(O,R)}$ onto $S_p(C,1)$ is a dome hat [8]. This

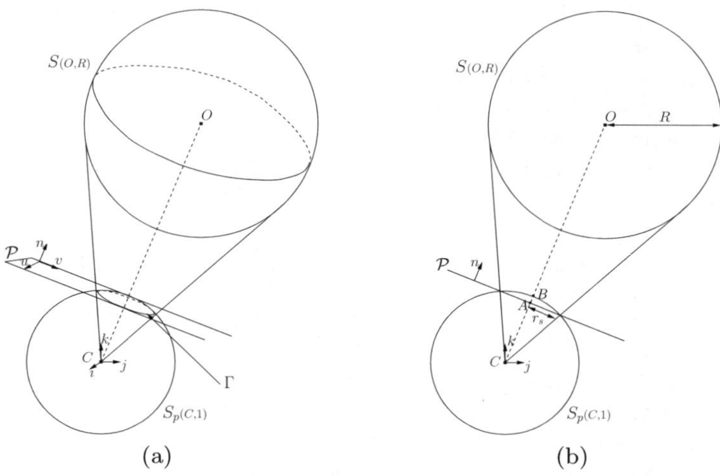

Fig. 1. Spherical projection of a sphere: **(a)** contour of the dome hat base; **(b)** cut made perpendicular to $\mathcal{P}$

dome hat can be characterized by the contour Γ of its base. This contour is pictured in Fig. 1(a). The analytical form of Γ is given by

$$\Gamma = S_{p(C,1)} \cap \mathcal{P} = \begin{cases} X_S^2 + Y_S^2 + Z_S^2 = 1 \\ X_O X_S + Y_O Y_S + Z_O Z_S = K_O, \end{cases} \tag{1}$$

where $K_O = \sqrt{X_O^2 + Y_O^2 + Z_O^2 - R^2}$. The contour Γ is therefore a circle. Let A and r_s be respectively the center and the radius of Γ (see Fig. 1(b)). After some developments we obtain in the camera frame

$$r_s = R/d_O, \tag{2}$$

$$\begin{cases} X_A = X_O \sqrt{1 - r_s^2}/d_O \\ Y_A = Y_O \sqrt{1 - r_s^2}/d_O \\ Z_A = Z_O \sqrt{1 - r_s^2}/d_O \end{cases} \tag{3}$$

where $d_O = \sqrt{X_O^2 + Y_O^2 + Z_O^2}$.

In addition to A and r_s, the dome hat summit B (see Fig. 1(b)) can also be considered as a potential visual feature. The coordinates of B in the camera frame are given by

$$\begin{cases} X_B = X_O/d_O \\ Y_B = Y_O/d_O \\ Z_B = Z_O/d_O. \end{cases} \tag{4}$$

2.2   Visual Features Selection

In this section we present the interaction matrix related to the optimal visual features selected. We recall that the interaction matrix $\mathbf{L_f}$ related to a set of features $\mathbf{f} \in \mathbb{R}^n$ is defined such that $\dot{\mathbf{f}} = \mathbf{L_f}\mathbf{v}$ where $\mathbf{v}=(\boldsymbol{v},\boldsymbol{\omega}) \in \mathfrak{se}(3)$ is the instantaneous camera velocity [11]; $\boldsymbol{v}$ and $\boldsymbol{\omega}$ are respectively the translational and the rotational velocities of the camera and $\mathfrak{se}(3) \simeq \mathbb{R}^3 \times \mathbb{R}^3$ is the Lie algebra of the Lie group of displacements $\mathbf{SE(3)}$.

Three parameters are sufficient to characterize the spherical projection of a sphere. Therefore, we need to select a combination of three visual features among $\{X_A, Y_A, Z_A, X_B, Y_B, Z_B, r_s\}$.

The combination $\mathbf{s} = (\frac{X_B}{rs}, \frac{Y_B}{r_s}, \frac{Z_B}{r_s})$ compared to the other is seductive since its interaction matrix $\mathbf{L_s}$ is simple and maximally decoupled [10]:

$$\mathbf{L_s} = \left[-\frac{1}{R}\mathbf{I_3} \ [\mathbf{s}]_\times \right]. \tag{5}$$

In addition to the decoupling property, $\mathbf{L_s}$ presents the same dynamic $(\frac{1}{R})$ in the translational velocities. Since R is a constant, there is a linear link between the visual features and the camera translational velocities. We can also see that the interaction matrix presents the passivity property, which is important to control certain under-actuated systems [12]. For these reasons, we propose the combination $\mathbf{s} = (\frac{X_B}{r_s}, \frac{Y_B}{r_s}, \frac{Z_B}{r_s})$ for visual servoing from spheres.

The only unknown 3D parameter in $\mathbf{L_s}$ is the constant R. In practice, $\widehat{R}$ (estimated value of R) is used instead. The robustness domain of a classical control law has been shown in [10] to be extremely large: $\widehat{R} \in]0, +\infty[$. Therefore, from a practical point of view, a rough estimate of R is sufficient.

We will now show how to compute this set of features using any central catadioptric system.

2.3 Visual Features Computation Using Any Central Catadioptric System

Considering a catadioptric system with (φ, ξ) as the mirror parameters, we show in this section that we can compute the visual features $\mathbf{s} = (\frac{X_B}{rs}, \frac{Y_B}{r_s}, \frac{Z_B}{r_s})$ from the catadioptric image of a sphere.

The catadioptric image of a sphere is an ellipse. Ellipse formation can be decomposed in two steps (see Fig. 2(a)) considering the unified model of catadioptric image formation [13]. From Fig. 2, note that the unique viewpoint is V and the camera optical center is C.

The first step is the spherical projection of $S_{(O,R)}$ onto $S_p(V,1)$. This result has been presented in section 2.1. Since $S_{(O,R)}$ is described in the virtual frame centered in V, we obtain

$$\Gamma = \begin{cases} X_S^{V2} + Y_S^{V2} + Z_S^{V2} = 1 \\ X_O X_S^V + Y_O Y_S^V + Z_O Z_S^V = K_O. \end{cases} \tag{6}$$

Γ is then expressed in the camera frame and projected onto the catadioptric image plane $Z = \varphi - 2\xi$. Γ is therefore the intersection of the sphere

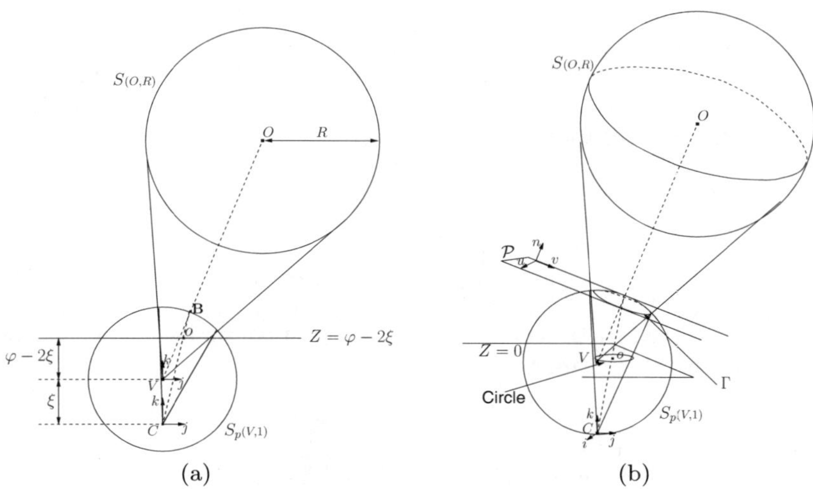

Fig. 2. Central catadioptric image of a sphere: **(a)** general case; **(b)** paracatadioptric projection $(\xi = 1)$

$$X_S{}^2 + Y_S{}^2 + (Z_S - \xi)^2 = 1 \tag{7}$$

with the plane

$$X_O X_S + Y_O Y_S + Z_O Z_S = K_O + \xi Z_O. \tag{8}$$

The equations of projection onto the catadioptric image plane are nothing but

$$\begin{cases} x = \dfrac{X_S}{Z_S} \\ y = \dfrac{Y_S}{Z_S}. \end{cases} \tag{9}$$

Plugging (9) in (8) gives

$$\frac{1}{Z_S} = \frac{X_O x + Y_O y + Z_O}{K_O + \xi Z_O} \tag{10}$$

and (9) in (7) gives

$$x^2 + y^2 + 1 - 2\frac{\xi}{Z_S} + \frac{\xi^2 - 1}{Z_S^2} = 0. \tag{11}$$

Finally, injecting (10) in (11) leads to the ellipse equation

$$k_0 x^2 + k_1 y^2 + 2k_2 xy + 2k_3 x + 2k_4 y + k_5 = 0 \tag{12}$$

with
$$\begin{cases}
k_0 = (K_O + \xi Z_O)^2 + (\xi^2 - 1)\, X_O^2 \\
k_1 = (K_O + \xi Z_O)^2 + (\xi^2 - 1)\, Y_O^2 \\
k_2 = (\xi^2 - 1)\, X_O Y_O \\
k_3 = X_O \left((\xi^2 - 1)\, Z_O - \xi\,(K_O + \xi Z_O)\right) \\
k_4 = Y_O \left((\xi^2 - 1)\, Z_O - \xi\,(K_O + \xi Z_O)\right) \\
k_5 = (K_O + \xi Z_O)^2 + (\xi^2 - 1)\, Z_O^2 - 2\xi Z_O\,(K_O + \xi Z_O).
\end{cases}$$

Now, we show how to compute $\mathbf{s}$ using the ellipse moments $\mu = (x_g, y_g, n_{20}, n_{11}, n_{02})$ measured on the catadioptric image plane: (x_g, y_g) is the ellipse center of gravity; n_{02} and n_{20} are the ellipse axes length and n_{11} is equivalent to the ellipse orientation.

First of all, we recall that:

$$\begin{cases} \dfrac{X_B}{r_s} = \dfrac{X_O}{R} \\[1mm] \dfrac{Y_B}{r_s} = \dfrac{Y_O}{R} \\[1mm] \dfrac{Z_B}{r_s} = \dfrac{Z_O}{R} \end{cases} \tag{13}$$

From (12), the ellipse moments on the catadioptric image plane can be expressed using the 3D parameters:

$$\begin{cases} x_g = X_O H_1/H_2 \\ y_g = Y_O H_1/H_2 \\ 4n_{20} = \left(H_2 - \left(\xi^2 - 1\right) X_O^2\right) R^2/H_2^2 \\ 4n_{11} = -X_O Y_O \left(\xi^2 - 1\right) R^2/H_2^2 \\ 4n_{02} = \left(H_2 - \left(\xi^2 - 1\right) Y_O^2\right) R^2/H_2^2 \end{cases} \tag{14}$$

$$\text{with} \begin{cases} H_1 = Z_O + \xi K_O \\ H_2 = H_1^2 + \left(\xi^2 - 1\right) R^2. \end{cases}$$

After tedious computations, we obtain using (14)

$$\begin{cases} \dfrac{X_B}{r_s} = x_g \dfrac{h_2}{\sqrt{h_2 + (1 - \xi^2)}} \\[2mm] \dfrac{Y_B}{r_s} = y_g \dfrac{h_2}{\sqrt{h_2 + (1 - \xi^2)}} \end{cases} \tag{15}$$

where $h_2 = 1/f(\mu)$ with $f(\mu) = \dfrac{4n_{20} y_g^2 + 4n_{02} x_g^2 - 8n_{11} x_g y_g}{x_g^2 + y_g^2}$.

It is possible to demonstrate that $f(\mu)$ is continuous even when $x_g = y_g = 0$ in which case $f(\mu) = 4n_{20}$.

In the case of paracatadioptric systems (see Fig. 2(b)) where $h_2 = 1/\sqrt{4n_{20}}$, we also obtain:

$$\frac{Z_B}{r_s} = \frac{h_2 - \left(\frac{X_B^2}{r_s^2} + \frac{Y_B^2}{r_s^2} - 1\right)}{2\sqrt{h_2}}, \tag{16}$$

and for all other catadioptric systems ($\xi \neq 1$)

$$\frac{Z_B}{r_s} = \frac{h_1 - \xi\sqrt{h_1^2 + (1 - \xi^2)\left(\frac{X_B^2}{r_s^2} + \frac{Y_B^2}{r_s^2} - 1\right)}}{(1 - \xi^2)} \tag{17}$$

where $h_1 = \sqrt{h_2 + (1 - \xi^2)}$.

The features $\mathbf{s} = (\frac{X_B}{r_s}, \frac{Y_B}{r_s}, \frac{Z_B}{r_s})$ are intuitively proper to a cartesian image space. Therefore, for any visual servoing task, these features will mostly draw a straight line trajectory in the image plane of any catadioptric system. This is not always suitable for paracatadioptric cameras since there is a dead angle in the centre of the image. Therefore we present, in the next section, a new optimal combination for such cameras.

3   Optimal Visual Features

The new combination proposed here is shown to be more suitable with the physical realization of such cameras. In addition, the stability of the system is analysed: a sufficient condition is given for the global stability of the system with respect to modeling error.

3.1   Optimal Features Design

Let us consider a task of visual servoing from a sphere using a paracatadioptric camera where the initial and desired positions (of the center of gravity of the sphere image) are the mirror image of each other. Using the general features $\mathbf{s} = (\frac{X_B}{rs}, \frac{Y_B}{r_s}, \frac{Z_B}{r_s})$ will lead to a straight line features motion and thus to the loss of the target in the dead angle (in the center of the image) as shown in Fig. 3. Since this dead angle is inherent to the physical realization of a paracatadioptric camera, we propose to use the cylindrical coordinates of $(\frac{X_B}{rs}, \frac{Y_B}{r_s})$; this will prevent the loss of the target in the dead angle by enforcing a circular feature motion (see Fig. 3). Therefore, the new optimal visual features $\mathbf{s_p}$ computed from $\mathbf{s}$ are given by

$$\begin{cases} \rho = \sqrt{\left(\frac{X_B}{r_s}\right)^2 + \left(\frac{Y_B}{r_s}\right)^2} \\[2mm] \theta = \arctan \frac{Y_B}{X_B} \\[2mm] \frac{Z_B}{r_s} = \dfrac{h_2 - \left(\frac{X_B^2}{r_s^2} + \frac{Y_B^2}{r_s^2} - 1\right)}{2\sqrt{h_2}} \end{cases} \tag{18}$$

where $h_2 = \dfrac{1}{4n_{20}}$.

In addition to the better feature motion in the image, it is important to note that the feature ρ can never be 0 on a paracatadioptric image plane since the region where the target is visible does not include the center of the image. Thus θ is always defined.

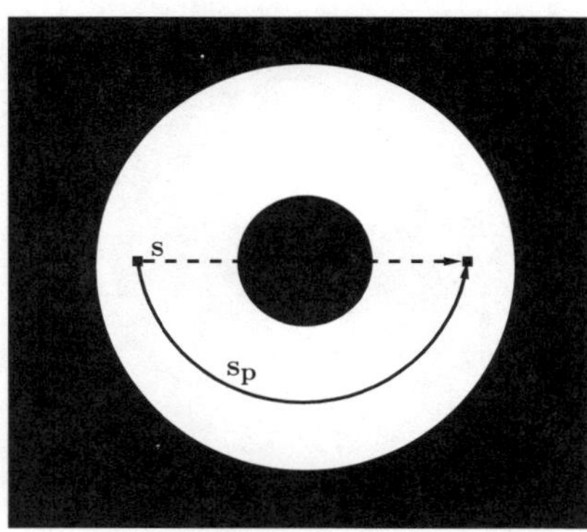

Fig. 3. Coordinate system dependence of the features motion

The interaction matrix related to $\mathbf{s_p}$ is given by

$$\mathbf{L_{s_p}} = \begin{bmatrix} \frac{-c}{R} & \frac{-s}{R} & 0 & \frac{sZ_B}{r_s} & \frac{-cZ_B}{r_s} & 0 \\ \frac{s}{\rho R} & \frac{-c}{\rho R} & 0 & \frac{cZ_B}{r_s\rho} & \frac{sZ_B}{r_s\rho} & -1 \\ 0 & 0 & -\frac{1}{R} & -\rho s & \rho c & 0 \end{bmatrix},$$

with $c = \cos\theta$ et $s = \sin\theta$.

From this interaction matrix, we can see that $\frac{Z_B}{r_s}$ is the only feature that is sensitive to the z-translation while θ is the only feature related to the rotation around the optical axis. This constrains the feature motion to avoid the dead angle. For these reasons, we propose the combination $\left(\rho, \theta, \frac{Z_B}{r_s}\right)$ for visual servoing from spheres using paracatadioptric cameras.

The only unknown 3D parameter in $\mathbf{L_{s_p}}$ is still the constant R. As before, in practice, $\widehat{R}$ (the estimated value of R) is used instead. From the stability analysis to modeling error, a robustness domain of $\widehat{R}$ will be given.

3.2   Stability Analysis to Modeling Error

Let us consider visual servoing from spheres with the combination $\mathbf{s_p} = (\rho, \theta, \frac{Z_B}{r_s})$.
We use the classical control law

$$\mathbf{v}_c = -\lambda \widehat{\mathbf{L}_{\mathbf{s_p}}}^{+}(\mathbf{s_p} - \mathbf{s_p}^*) \tag{19}$$

where $\mathbf{v}_c$ is the camera velocity sent to the low level robot controller, λ is a positive gain and $\widehat{\mathbf{L}_{\mathbf{s_p}}}^{+}$ is the pseudo-inverse of an approximation of the interaction matrix related to $\mathbf{s_p}$.

Modeling error arises from the approximation of R. In this case the closed-loop system equation can be written as:

$$\dot{\mathbf{s}}_\mathbf{p} = -\lambda \mathbf{L}_{\mathbf{s_p}} \widehat{\mathbf{L}_{\mathbf{s_p}}}^{+}(\mathbf{s_p} - \mathbf{s_p}^*) \tag{20}$$

where

$$\widehat{\mathbf{L}_{\mathbf{s_p}}}^{+} = \begin{bmatrix} \frac{-c\hat{R}(\rho^2\hat{R}^2+1)}{d} & \frac{\rho s\hat{R}}{d} & \frac{-\rho cZ_B\hat{R}^3}{dr_s} \\ \frac{-s\hat{R}(\rho^2\hat{R}^2+1)}{d} & \frac{-\rho c\hat{R}}{d} & \frac{-\rho sZ_B\hat{R}^3}{dr_s} \\ \frac{-\rho Z_B\hat{R}^3}{dr_s} & 0 & \frac{-\hat{R}\left(1+\left(\frac{Z_B}{r_s}\right)^2\hat{R}^2\right)}{d} \\ \frac{sZ_B\hat{R}^2}{dr_s} & \frac{\rho cZ_B\hat{R}^2}{dr_s} & -\frac{\rho s\hat{R}^2}{d} \\ \frac{-cZ_B\hat{R}^2}{dr_s} & \frac{\rho sZ_B\hat{R}^2}{dr_s} & \frac{\rho c\hat{R}^2}{d} \\ 0 & -\frac{\rho^2\hat{R}^2}{d} & 0 \end{bmatrix}$$

with $d = 1 + \left(\left(\frac{Z_B}{r_s}\right)^2 + \rho^2\right)\hat{R}^2$. A sufficient condition for the global asymptotic stability to modeling error is $\mathbf{L}_{\mathbf{s_p}}\widehat{\mathbf{L}_{\mathbf{s_p}}}^{+} > 0$. The eigenvalues of $\mathbf{L}_{\mathbf{s_p}}\widehat{\mathbf{L}_{\mathbf{s_p}}}^{+}$ can be computed. They are given by $\frac{\hat{R}}{R}$ and $\frac{\hat{R}}{R}\frac{r_s^2+R\hat{R}Z_B^2+R\hat{R}r_s^2\rho^2}{r_s^2+\hat{R}^2Z_B^2+\hat{R}^2r_s^2\rho^2}$ (which is a double eigenvalue). We have thus:

$$\mathbf{L_{s_p}}\widehat{\mathbf{L_{s_p}}}^{+} > 0 \iff \hat{R} > 0.$$

This condition is also necessary since if $\hat{R} \leq 0$ then $\mathbf{L_s}\widehat{\mathbf{L_s}}^{+} \leq 0$ and the system diverges. Therefore the robustness domain with respect to modeling error is: $\hat{R} \in \,]0, +\infty[$. This result is not a surprise at all since $\mathbf{s_p}$ has been computed from $\mathbf{s}$ through a bijective map. From a practical point of view, a coarse approximation of R will thus be sufficient.

4   Results

In this section we first validate the general features $\mathbf{s} = (\frac{X_B}{rs}, \frac{Y_B}{r_s}, \frac{Z_B}{r_s})$ on a real robotic system using a paracatadioptric camera. Then we show in simulation that these features, for a particular simple visual servoing task, draw a highly undesirable straight line trajectory in the image plane. We finally validate the new optimal features $\mathbf{s_p} = (\rho, \theta, \frac{Z_B}{r_s})$ in simulation.

4.1   Experimental Results

In this section, the general features $\mathbf{s}$ are validated. The experiments have been carried out with a paracatadioptric camera mounted on the end-effector of a six DOFs robotic system. The target is a $4cm$ radius polystyrene white ball. Using such a simple object allows to easily compute the ellipse moments at video rate without any image processing problem. The desired set $\mathbf{s}^*$ has been computed after moving the robot to a position corresponding to the desired image. Fig. 4 shows the desired and the initial images used for each experiment. For all the experiments, the same gain $\lambda = 0.1$ has been used.

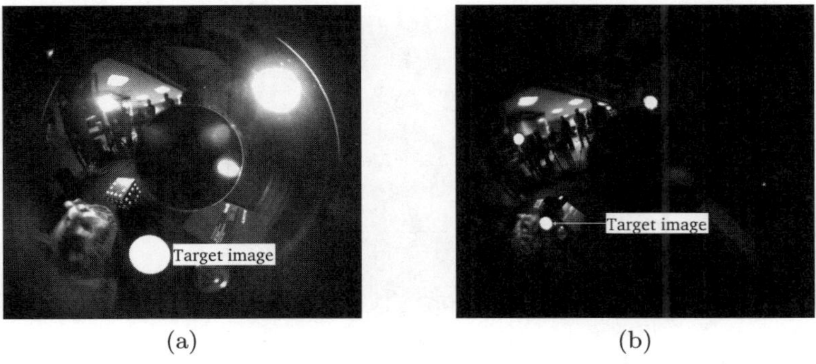

Fig. 4. (a) Desired image; (b) initial image

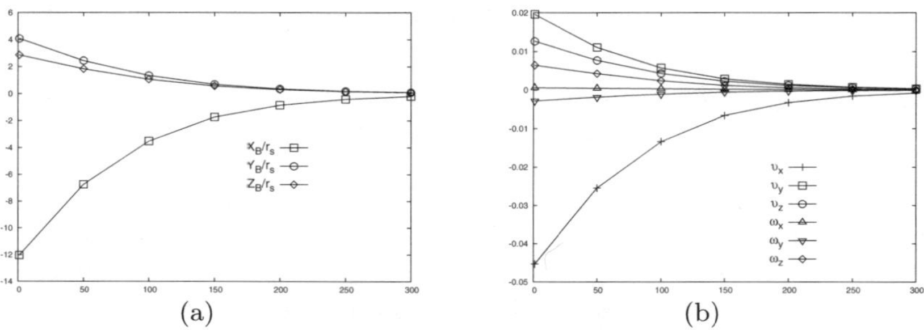

Fig. 5. Ideal case: (**a**) **s** error; (**b**) computed camera velocities (m/s and dg/s)

Ideal Case

In oder to validate the general features **s**, we first consider the ideal case where $\widehat{R} = R$. Indeed, when $\widehat{R} = R$ we have a perfect system behaviour since $\mathbf{L_s}\widehat{\mathbf{L_s}}^+ = \mathbf{I}_3$. As expected, a pure exponential decrease of the error on the visual features can be observed on Fig. 5(a) while the camera velocities are plotted on Fig. 5(b).

Modeling Error

The stability with respect to modeling error using **s** has been proved using a classical perspective camera [10]. For paracatadioptric system, we have validated this proof, with two experiments. The results in the case where $\widehat{R} = 5R$ and $\widehat{R} = 0.2R$ are depicted respectively in Fig. 6 and Fig. 7. We can note that the system still converges in both cases.

Fig. 6(b) shows a high speed on the system translational velocities while Fig. 7(b) shows a low speed on the same components. In fact, choosing an arbitrary value of $\widehat{R}$ affects the convergence speed of the system. Indeed, using the general features **s**, the velocity sent to the robot can be written as

$$\mathbf{v}_c = -\lambda\widehat{\mathbf{L_s}}^+(\mathbf{s} - \mathbf{s}^*) \tag{21}$$

where $\widehat{\mathbf{L_s}}^+$ computed from (5) is given by

$$\widehat{\mathbf{L_s}}^+ = \begin{bmatrix} -\dfrac{\widehat{R}r_s^2}{r_s^2+\widehat{R}^2}\left(\widehat{R}^2\mathbf{ss}^\top + \mathbf{I}_3\right) \\ -\dfrac{\widehat{R}^2 r_s^2}{r_s^2+\widehat{R}^2}[\mathbf{s}]_\times \end{bmatrix}.$$

After few developments we obtain from (21)

$$\begin{cases} \boldsymbol{v} = \lambda\dfrac{\widehat{R}r_s^2}{r_s^2+\widehat{R}^2}\left(\widehat{R}^2\mathbf{ss}^\top + \mathbf{I}_3\right)(\mathbf{s} - \mathbf{s}^*) \\ \boldsymbol{\omega} = \lambda\dfrac{\widehat{R}^2 r_s^2}{r_s^2+\widehat{R}^2}[\mathbf{s}]_\times(\mathbf{s} - \mathbf{s}^*). \end{cases} \tag{22}$$

When $\widehat{R}$ tends to $+\infty$, (22) tends to

$$\begin{cases} \boldsymbol{v} = \infty \\ \boldsymbol{\omega} = \lambda r_s^2 \left[\mathbf{s}\right]_\times (\mathbf{s} - \mathbf{s}^*) \end{cases}$$

which explains the fast convergence observed in Fig. 6 (100 iterations) when $\widehat{R} = 5R$. When $\widehat{R}$ tends to 0, from (22) we have: $\boldsymbol{v}$ and $\boldsymbol{\omega}$ tend to 0. This explains the slow convergence observed in Fig. 7 when $\widehat{R} = 0.2R$. In practice, the behaviour could be easily improved, by using a higher gain λ (to deal with under approximation of $\widehat{R}$) and by saturating $\mathbf{v}_c$ when needed (to deal with an over approximation of $\widehat{R}$).

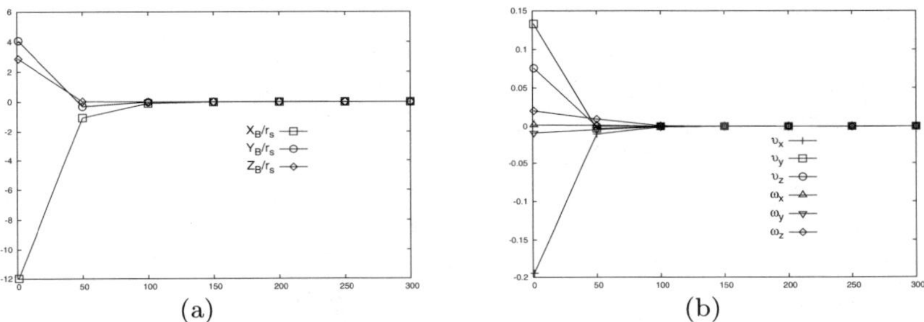

(a)   (b)

Fig. 6. Modeling error $\widehat{R} = 5R$: (**a**) $\mathbf{s}$ error; (**b**) computed camera velocities (m/s and dg/s)

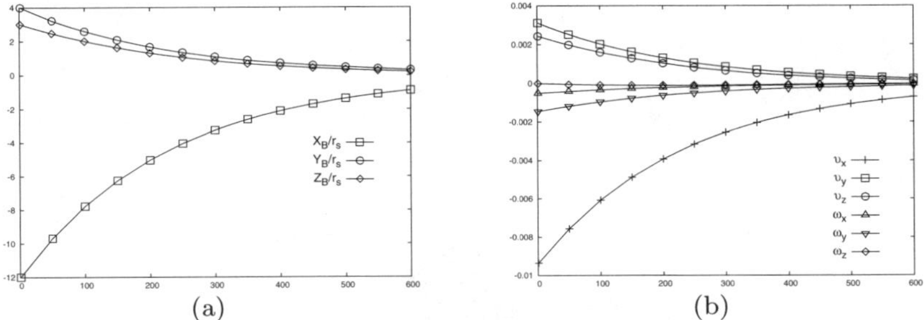

(a)   (b)

Fig. 7. Modeling error $\widehat{R} = 0.2R$: (**a**) $\mathbf{s}$ error; (**b**) computed camera velocities (m/s and dg/s)

4.2   Simulation Results

In this section, it is shown that for the general feature $\mathbf{s}$, the motion in the image plane is not suitable with paracatadioptric cameras, particularly when the initial position and desired position (in the image space) are each others mirror image. In addition, the new optimal features $\mathbf{s_p}$ specific to the paracatadioptric system are validated.

Features Motion in the Image Plane

Here we consider a visual servoing task where the initial and desired images are each the mirror image of the other (rotation of π around the z-axis). The image-plane trajectories of the center of gravity of the sphere image are drawn in Fig. 8(c). In this picture we can see that the general features $\mathbf{s}$ generate a straight line motion going through the center of the image. It means that in case of a real camera, the target would get lost in the dead angle.

Using the new features $\mathbf{s_p}$ leads to a circular trajectory as expected. It means that with a real camera, it is possible to constrain ρ to avoid the dead angle.

For all the following experiments, we consider a more complex task consisting of the previous task, a zoom and a translation in the ρ-direction.

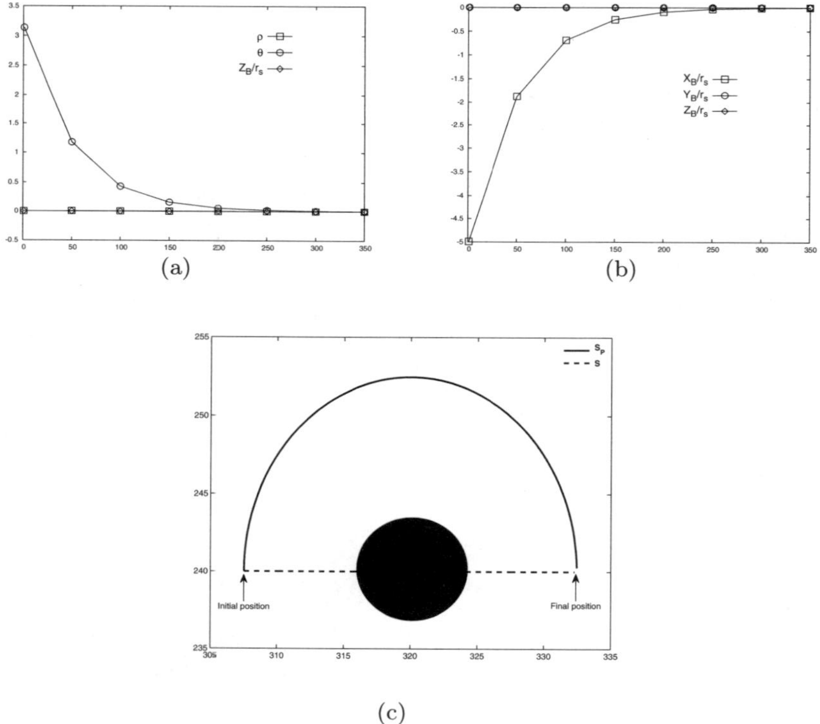

Fig. 8. Adequate features for paracatadioptric cameras: (a) $\mathbf{s_p}$ error; (b) $\mathbf{s}$ error; (c) image-plane trajectories of the center of gravity of the sphere image

Ideal Case

We first consider the case where $\widehat{R}=R$. In this case we have $\mathbf{L_{s_p}}\widehat{\mathbf{L_{s_p}}}^{+} = \mathbf{I}_3$, thus a perfect system behaviour. Fig. 9(a) plots the features error trajectory while Fig. 9(b) shows the camera velocities.

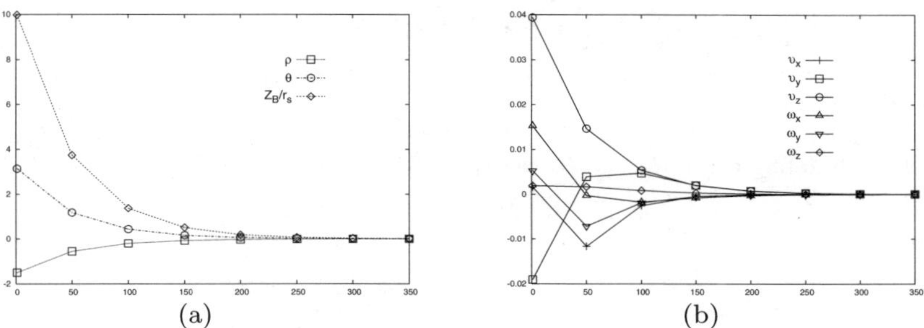

Fig. 9. Ideal case: (a) $\mathbf{s_p}$ error; (b) computed camera velocities (m/s and dg/s)

Modeling Error

The stability to modeling error has been proved in this paper. This proof is validated with two experiments. In the first case, $\widehat{R} = 5R$: Fig. 10 plots the results. In the second case, $\widehat{R} = 0.2R$: Fig. 11 shows the results. In both cases the system still converges either fastly or slowly as expected.

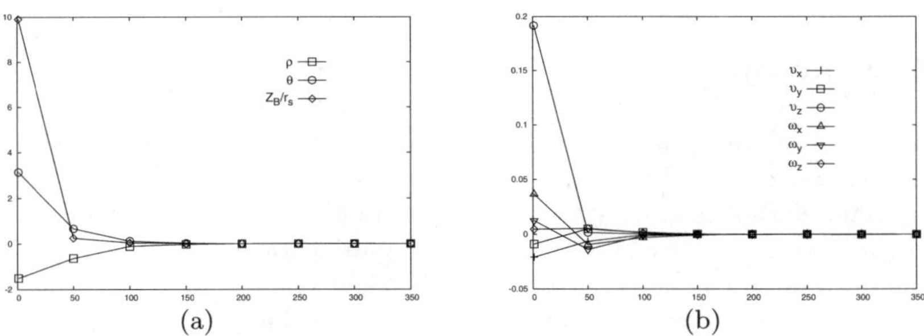

Fig. 10. Modeling error $\widehat{R} = 5R$: (a) $\mathbf{s_p}$ error; (b) computed camera velocities (m/s and dg/s)

Calibration Errors

Finally we verify the stability to calibration errors in simulation. This is done by introducing errors on the camera intrinsic parameters: $35\%f$, $-25\%u_0$ and $47\%v_0$. The results obtained are given on Fig. 12. Once again the system still converges.

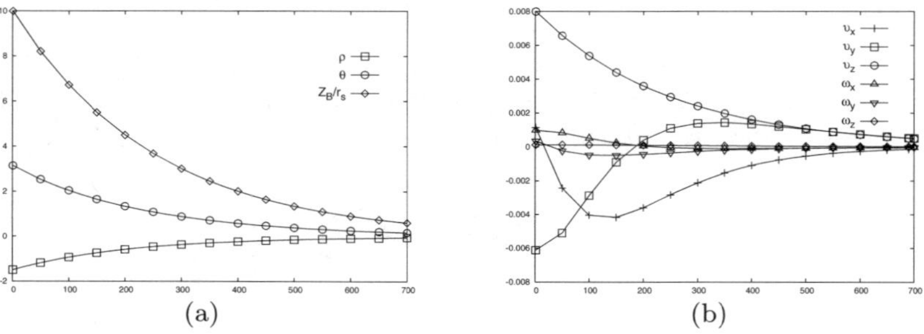

Fig. 11. Modeling error $\widehat{R} = 0.2R$: (a) $\mathbf{s_p}$ error; (b) computed camera velocities (m/s and dg/s)

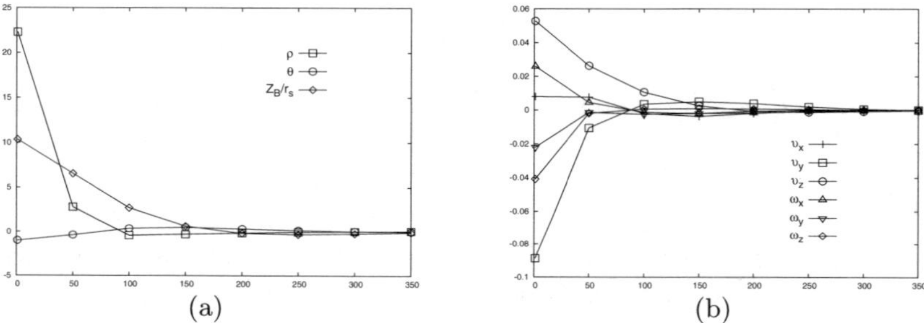

Fig. 12. Calibration errors: (a) $\mathbf{s_p}$ error; (b) computed camera velocities (m/s and dg/s)

5   Conclusions

In this paper, we have reviewed the general features designed using a spherical projection model for visual servoing from spheres with any central catadioptric system. These features usually draw a straight line trajectory in the image space which is not always suitable for paracatadioptric cameras. A new optimal combination of three visual features for visual servoing from spheres using this type of cameras has been proposed. This new set of features has been built from the previous one using a cylindrical coordinate system which enables a better feature motion in the image plane. The interaction matrix related to this new combination presents a decoupling between the rotational and the translational velocities of the optical axis. Using this new combination, a classical control law has been analytically proved to be globally stable with respect to modeling error. The general visual features have been validated experimentally with a paracatadioptric camera mounted on a robotic system and simulation results have been presented to validate the new combination.

Acknowledgment

The authors would like to thank C. Collewet, N. Mansard and S. Hutchinson for their helpful comments.

References

1. Hutchinson, S., Hager, G., Corke, P.: A tutorial on visual servo control. IEEE Trans. on Robotics and Automation 12(3), 651–670 (1996)
2. Chaumette, F.: Potential problems of stability and convergence in image-based and position-based visual servoing. In: Kriegman, D., Hager, G., Morse, A.S. (eds.) The Confluence of Vision and Control. LNCIS Series, vol. 237, pp. 66–78. Springer, Heidelberg (1998)
3. Wilson, W., Hulls, C., Bell, G.: Relative end-effector control using cartesian position-based visual servoing. IEEE Trans. on Robotics and Automation 12(5), 684–696 (1996)
4. Malis, E., Chaumette, F., Boudet, S.: 2 1/2 d visual vervoing. IEEE Trans. on Robotics and Automation 15(2), 238–250 (1999)
5. Corke, P., Hutchinson, S.: A new partitioned approach to image-based visual servo control. IEEE Trans. on Robotics and Automation 17(4), 507–515 (2001)
6. Iwatsuki, M., Okiyama, N.: A new formulation for visual servoing based on cylindrical coordinate system. IEEE Trans. on Robotics 21(2), 266–273 (2005)
7. Mezouar, Y., Chaumette, F.: Path planning for robust image-based control. IEEE Trans. on Robotics and Automation 18(4), 534–549 (2002)
8. Cowan, N., Chang, D.: Geometric visual servoing. IEEE Trans. on Robotics 21(6), 1128–1138 (2005)
9. Barreto, J., Martin, F., Horaud, R.: Visual servoing/tracking using central catadioptric images. In: Int. Symposium on Experimental Robotics, Ischia, Italy (July 2002)
10. Tatsambon Fomena, R., Chaumette, F.: Visual servoing from spheres using a spherical projection model. In: IEEE Int. Conf. on Robotics and Automation, Rome, Italy (April 2007)
11. Espiau, B., Chaumette, F., Rives, P.: A new approach to visual servoing in robotics. IEEE Trans. on Robotics and Automation 8(3), 313–326 (1992)
12. Hamel, T., Mahony, R.: Visual servoing of an under-actuated dynamic rigid-body system: an image-based approach. IEEE Trans. on Robotics and Automation 18(2), 187–198 (2002)
13. Geyer, C., Daniilidis, K.: A unifying theory for central panoramic systems and practical implications. In: European Conference on Computer Vision, vol. 29, pp. 159–179 (2000)

Dynamic Targets Detection for Robotic Applications Using Panoramic Vision System

Abedallatif Baba and Raja Chatila

LAAS-CNRS
University of Toulouse
Toulouse-France
{ababa, raja.chatila}@laas.fr

Summary. This paper presents experiments in dynamic targets detection using panoramic images, which represent rich visual sources of global scenes around a robot. Moving targets (people) are distinguished as foreground pixels in binary images detected using a modified optical flow approach where the intensity of lighting source is variable. The directions of detected targets are determined using two strategies; the first one is convenient for unfolded panoramic images; it searches most probable regions in the last binary image by calculating a histogram of foreground pixels on its columns. The second approach is applied on raw panoramic images; it regroups foreground pixels using a technique that generates a new pixel's intensity depending on the intensities of its neighbors.

1 Introduction

A robot has to be able to evaluate the occupation of its vicinity and to detect if humans are moving close to it for interacting with them. Therefore it is necessary to distinguish mobile targets, determine their number, identify them and track their movements. A panoramic vision system is probably the most adequate sensing device for this kind of purpose, thanks to its 360° field of view. It is richer than a 2D laser range finder. Our global objective is to fuse the vision detector with a laser tracker [1] to improve its reliability.This paper presents an approach for detecting and tracking moving targets from panoramic images, raw or unfolded. The approach is based on optical flow. In the next section we briefly mention related work. Section 3 discusses background extraction using the median operator. In section 4, we focus on the problem of compensating illumination changes. Two methods are explained in section 5 to determine the directions of detected targets and a comparison between them is discussed.

2 Related Work

Liu et al. [8] present a real time omnidirectional vision system for detecting and tracking multiple targets. Their detection approach is applied on colored images (RGB) which requires a long computing time, as it requires unfolded images. The problem of background objects location is not considered in their

S. Lee, I.H. Suh, and M.S. Kim (Eds.): Recent Progress in Robotics, LNCIS 370, pp. 215–227, 2008.
springerlink.com

background model, which is also not considered in other approaches [6, 10, 11, 3]. All mentioned approaches are also sensitive to illumination changes. Javed et al. [7] and Jabri et al. [5] use fusion of color and edge information for background subtraction, but our temporal constraints impose the use of gray-level images.

3   Background Extraction from a Sequenced Images Using Temporal Median Operator

Detecting dynamic targets in robot vicinity using a vision system needs to distinguish between background and foreground entities visible in it. Due to its

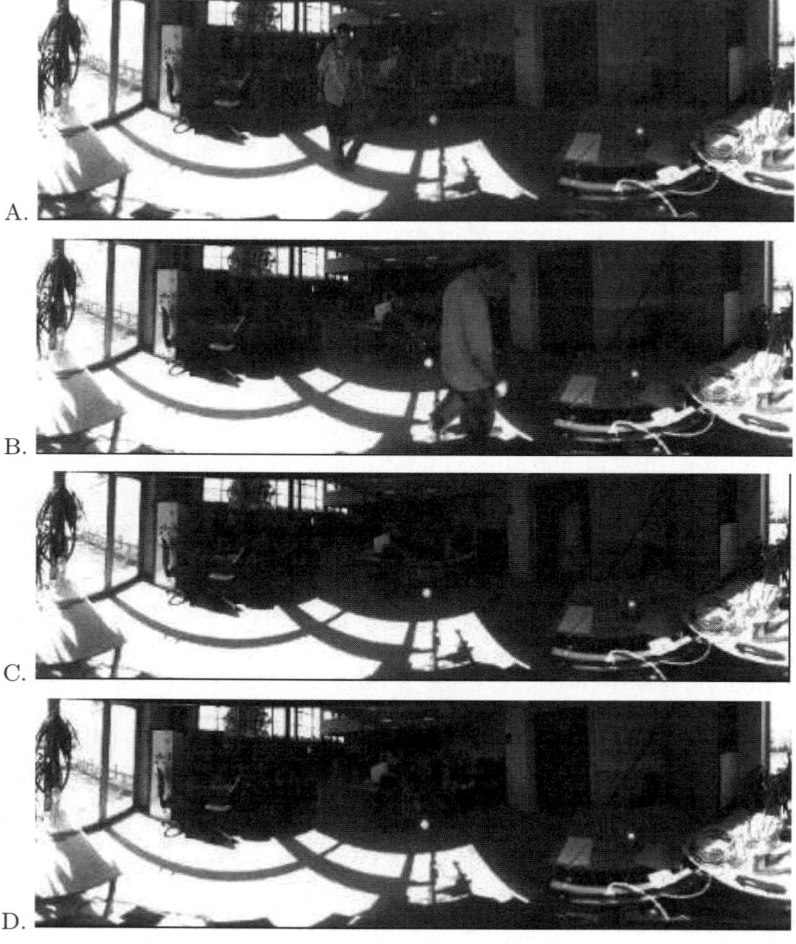

Fig. 1. Image (D) is the extracted backgrounds from three unfolded panoramic images (A, B and C)

simplicity, we use the temporal median operator [2] for background extraction. The median is the central value extracted from a series of ranked values if their number was odd, and if not, it is the average of two central values. In our application, we extract the scene's background from three images acquired from the same position with keeping a constant delay Δt among them; the more the zone of view is occupied the more Δt has to be large. Then, median operator selects only components of the global model that are correlated. For that, it is less subject to loss of detail or "bleeding" when there are objects moving in the scene (figure 1). Updating the process of background extraction continuously (or nearly continuously) using this approach solves the problem of relocation of background objects to improve the robustness of extracted background.

$$N_{x,y} = Median\left(O^0_{x,y}, O^1_{x,y}, ..., O^n_{x,y}\right) \tag{1}$$

$O^n_{x,y}$: Pixel's intensity in the nth acquired image.
$N_{x,y}$: New pixel's intensity in the extracted background.

4 Dynamic Target Detection by Calculating the Amplitudes of Pixels Velocities

From a sequence of images acquired by a camera from the same position at different moments in a dynamic environment, it is possible to detect the movement of any dynamic entity appearing in it, represented as a group of pixels moving with the temporal evolution of these images. The change of pixel intensity between a current image and the extracted background is the pixel's velocity. Thus, an image of pixels velocities represented by their amplitudes may be constructed. To calculate this image we exploited optical flow [4], but a new term related to the fact that the brightness coming from lighting sources is variable is taken in account (this is bound to happen in robotics applications). We do not assume that the brightness of a point in an image is constant. The relation between the pixels in a zone of the image can be written:

$$P(t+1)_{x+\delta x,y+\delta y} = P(t)_{x,y} + H(t)_{x,y} \tag{2}$$

Where $H(t)_{x,y}$ compensates the difference in luminous intensity between two images, and $(\delta x, \delta y)$ represents the displacement of a pixel in the interval $[t, t+1]$. Thus the intensity of a pixel in position $(x + \delta x, y + \delta y)$ equals to that of a pixel in position (x, y) with a small change which is interpreted as external physical effects (e.g. change of illumination, spectral reflection... etc). Our assumption for this application is $H(t)_{x,y} \neq 0$.

The first order of Taylor's expansion of the term $P(t+1)_{x+\delta x,y+\delta y}$ is:

$$P(t+1)_{x+\delta x,y+\delta y} = P(t)_{x,y} + \delta x\frac{\partial P(t)_{x,y}}{\partial x} + \delta y\frac{\partial P(t)_{x,y}}{\partial y} + \delta t\frac{\partial P(t)_{x,y}}{\partial t} \tag{3}$$

By replacing equation (2) in (3) we have:

$$\frac{H(t)_{x,y}}{\delta t} - \frac{\partial P(t)_{x,y}}{\partial t} = \frac{\delta x}{\delta t}\frac{\partial P(t)_{x,y}}{\partial x} + \frac{\delta y}{\delta t}\frac{\partial P(t)_{x,y}}{\partial y} \tag{4}$$

The last equation may be written as:

$$w_{x,y} - \nabla t_{x,y} = u_{x,y}\nabla x_{x,y} + v_{x,y}\nabla y_{x,y} \tag{5}$$

Where $\nabla x_{x,y}$: difference in pixel's intensity on the columns, and $\nabla y_{x,y}$: difference in pixel's intensity on the rows (In fact, these two last gradients can be helpful to extract vertical and horizontal segments in an image respectively). $\nabla t_{x,y}$: is the rate of the change of pixel's intensity provoked by the movement of a dynamic object. The term $w_{x,y} = \frac{H(t)_{x,y}}{\delta t}$ represents the rate of the difference of pixel's intensity caused by undesired reasons (as already explained), where $H(t)_{x,y}$ is a rate that interprets the change applied on a pixel between its intensities in the background and in the current image, caused by a global change that has occurred over all the pixels:

$$H(t)_{x,y} = \frac{P^{im}_{x,y} - P^{back}_{x,y}}{avrg^{im} - avrg^{back}} \tag{6}$$

Thus, the more the difference between the tow averages ($avrg^{im}$ of the current image and $avrg^{back}$ of the background) is large, the less the undesired effects will be remarkable. Horizontal and vertical components of pixel's velocity $u_{x,y}$ and $v_{x,y}$ in equation (5) have to be calculated. Here we present our solution while a full and detailed mathematical derivation may be found in [9].

$$\begin{aligned} u_{x,y} = \bar{u}_{x,y} &- \lambda\left(\frac{\nabla x_{x,y}\bar{u}_{x,y}+\nabla y_{x,y}\bar{v}_{x,y}+\nabla t_{x,y}}{1+\lambda(\nabla x^2_{x,y}+\nabla y^2_{x,y})}\right)\nabla x_{x,y} \\ &+\frac{\lambda\,w_{x,y}\,\nabla x_{x,y}}{1+\lambda(\nabla x^2_{x,y}+\nabla y^2_{x,y})} \end{aligned} \tag{7}$$

$$\begin{aligned} v_{x,y} = \bar{v}_{x,y} &- \lambda\left(\frac{\nabla x_{x,y}\bar{u}_{x,y}+\nabla y_{x,y}\bar{v}_{x,y}+\nabla t_{x,y}}{1+\lambda(\nabla x^2_{x,y}+\nabla y^2_{x,y})}\right)\nabla y_{x,y} \\ &+\frac{\lambda\,w_{x,y}\,\nabla y_{x,y}}{1+\lambda(\nabla x^2_{x,y}+\nabla y^2_{x,y})} \end{aligned} \tag{8}$$

λ : is a regularization parameter, ($\bar{u}_{x,y}$ and $\bar{v}_{x,y}$) are two calculable averages. The last added terms in equations (7) and (8) represent the new terms which we search. Now it is very simple to construct an image of amplitudes of pixels velocities:

$$V_{x,y} = \sqrt{(u_{x,y})^2 + (v_{x,y})^2} \tag{9}$$

By applying one step of thresholding (Threshold is chosen to be automatically 10 % from the maximal calculated amplitude) for the last constructed image,

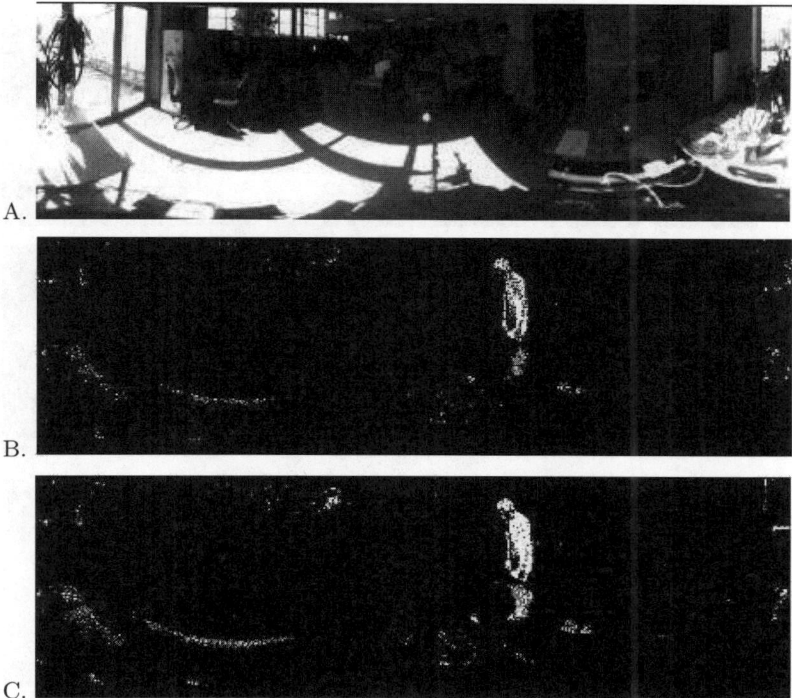

Fig. 2. Detected person, a change of illumination between the extracted background in image (1, D) and the current image (2, A). Image (B) respects our assumption; image (C) doesn't respect it.

each pixel that has a null velocity will be black, while the intensities of the other pixels which have different velocities will be unified to the level (255 or white). The corresponding extracted binary image is shown in figure (2, B), this image is less noisy than the image (2, C) which doesn't respect our assumption, where phantoms are easy to be noted because of the movement of shadows of windows frames. This movement is related to the sunlight direction changes between the moment of background extraction and the moment of taking the current image. To eliminate the noise remaining in figure (2, B), it is necessary to update the background in a repetitive way. Moreover, figure 3 shows the difference in quality for tow binary images resulting from a current image (1, B) and extracted background (1, D) but using two different methods, the first image (3, A) is constructed by applying the approach discussed in this paper, while the second one (3, B) results from a simple subtraction, and the same threshold is employed in the two cases. Image (3, A) is very less noisy than image (3, B), and the pixels that belong to the detected target are well distributed on all the parts of the target, but in image (3, B) there is a lack in the part of target legs.

Fig. 3. (A) Detected target using the discussed approach; (B) the same target is detected by a simple subtraction

5   Determining the Directions of Detected Targets

We present here two different approaches to determine the direction of one or several detected targets. And then, a comparison between them is discussed.

5.1   First Approach (for Unfolded Image)

This approach calculates a histogram on the columns of the binary image in figure (4, B). For an image of size ($NbRows$, $NbColumns$), the illustrated algorithm in (Algorithm 1) accumulates on each column the number of the foreground pixels that have maximal intensity 255, and then it calculates the degree of interest of this column represented by its probability which is proportional with the number of the foreground pixels included in it. In the histogram illustrated in figure (4, C), our approach searches zones of maximal interest (i.e. they are side by side as they have high probabilities), consequently these zones represent detected targets, and the number of columns where an interesting zone starts and ends is important to find the column that represents its mean. So, the angle of detected target is given as follows:

$$ThetaDetect = 2\pi \left(\frac{NbColumns - \mu}{NbColumns} \right) \tag{10}$$

μ : The number of column that represents the mean of interesting zone. For example, the image (4, B) has the size (450 X 1500); two targets are detected at the angles (3.39 rad and 2.27 rad) from left to right respectively.

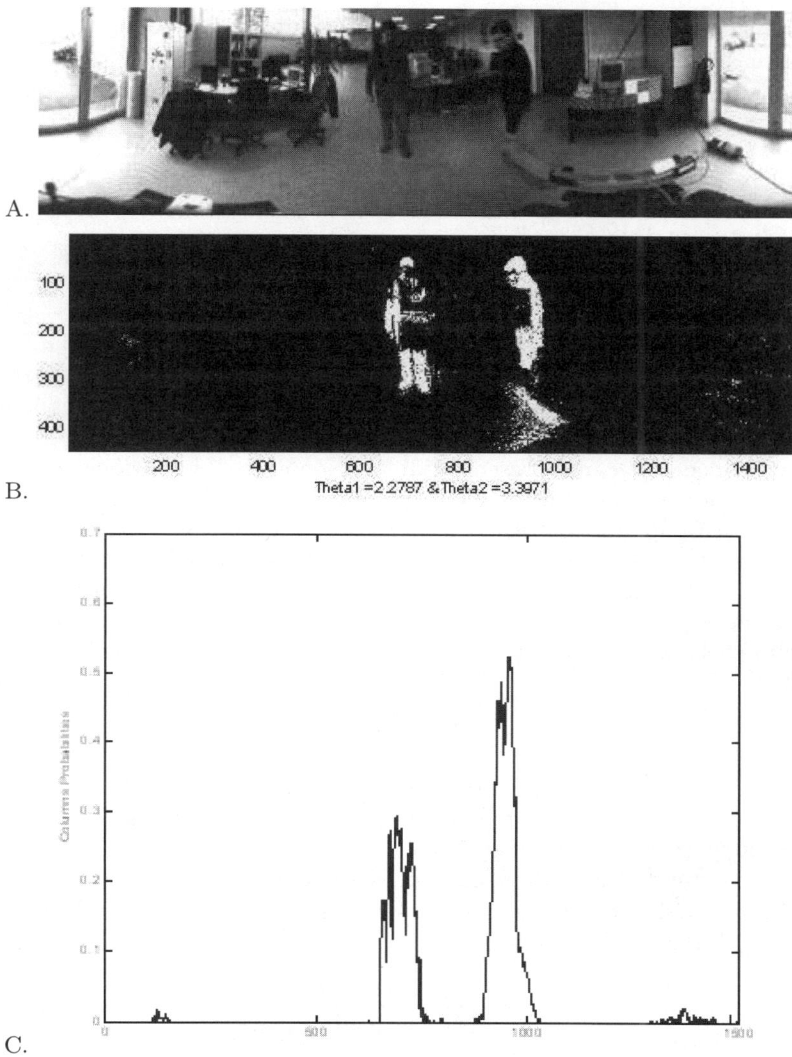

Fig. 4. (A) Two targets moving in the robot's zone of view, (B) binary image shows two detected targets, (C) histogram shows two zones of interesting columns (they have maximal probabilities

5.2   Second Approach (for Raw Panoramic Image)

The same strategies that have already employed either to extract a background (figure 5) or to produce a binary image (figure 6) are used, but in this last figure a very small threshold is applied because it will be useful that a binary

---

Algorithm 1. Algorithm that calculates the interesting probabilities of the columns in binary image, η is normalization constant

---

```
1: for (i = 0; i < NbColumns; i + +)
2:     for (j = 0; j < NbRows; j + +)
3:         if (image(j, i) == 255)
4:         ColumnCounter + +;
5:     end
6:     ColumnPro = η·ColumnCounter/NbRows
7: end
```

---

image contains a maximum of information even with much of associated noise which will be eliminated automatically in another step of treatment (grouping step) which determines simultaneously the angles of detected targets. In this case neither the columns nor the rows in the binary image (figure 6) can give any significance that can represent a target, so points which belong to detected target and don't represent a noise or a phantom have to be regrouped together. To do this a template of size (51 x 51) is centered on each pixel and a new pixel's intensity will be accumulated from the intensities of all its neighbors in the template. Therefore, a pixel that belongs to a group of pixels representing a target has a new intensity which accumulates a large value, and the summits are the intensities of pixels located in the center of the group (figure 10).

$$P_{x,y}^{acc} = \sum_{i=-N}^{N} \sum_{j=-N}^{N} P_{x+i,y+j} \tag{11}$$

$N = (TemplateDim - 1)/2 \; TemplateDim$: is a template's dimension on rows or on columns (here it is equal to 51).

After one step of normalization, pixels surrounded by neighbors are often black will be eliminated, while the pixels with summit values will take again the intensity 255, (figure 7). To normalize:

$$P_{x,y}^{new} = 255 * \frac{P_{x,y}^{acc}}{MAX(P_{x,y}^{acc})} \tag{12}$$

$P_{x,y}^{acc}$: The pixel's value after the application of the accumulation template; $MAX(P_{x,y}^{acc})$ is the maximal accumulated value; and $P_{x,y}^{new}$: The normalized pixel's intensity.

The histogram illustrated in the figure 8 shows us how the strategy which we followed has succeeded in creating a new image with grouped and quite separated levels of intensities, the pixels of summit intensities are the minority. Isolating them from phantoms will be easy by making a simple step of thresholding (at a high intensity value like 200) to obtain the final image. The result is shown in figure 9, which is well cleaned of all noise and phantoms. Two zones representing two detected targets are finally isolated in it.

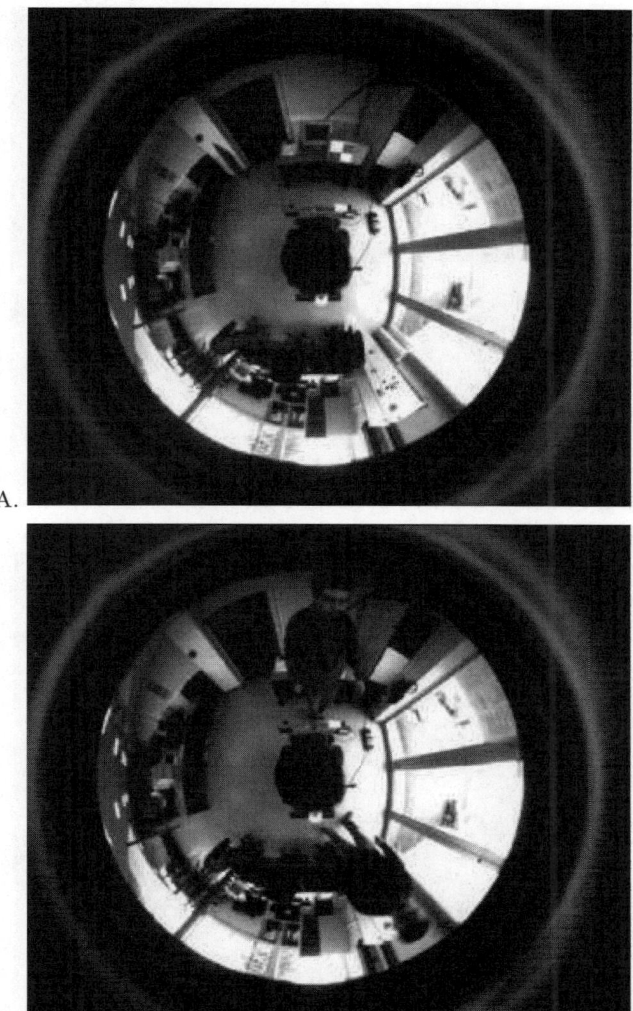

Fig. 5. (A) Panoramic background extracted from three panoramic images using temporal median operator, (B) Current panoramic image

The angle of each detected target, represented by one accumulated weight (figure 9), can be calculated simply:

$$ThetaDetect = arctg\left(\frac{y^t}{x^t}\right) \tag{13}$$

x^t And y^t are the central coordinates of the detected target, they are represented by the numbers of rows and columns occupied by the target in the binary image.

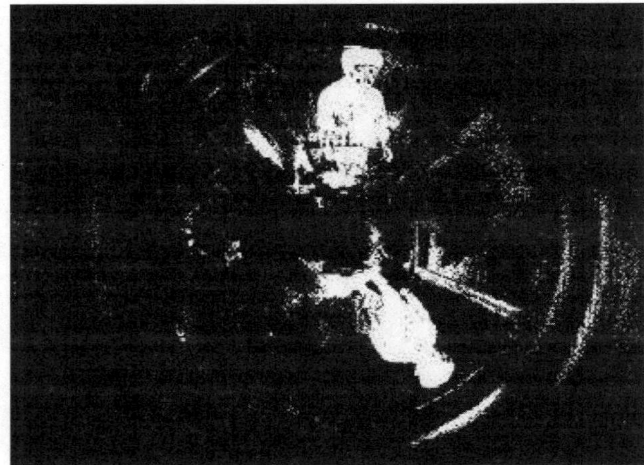

Fig. 6. Binary image generated with very small threshold

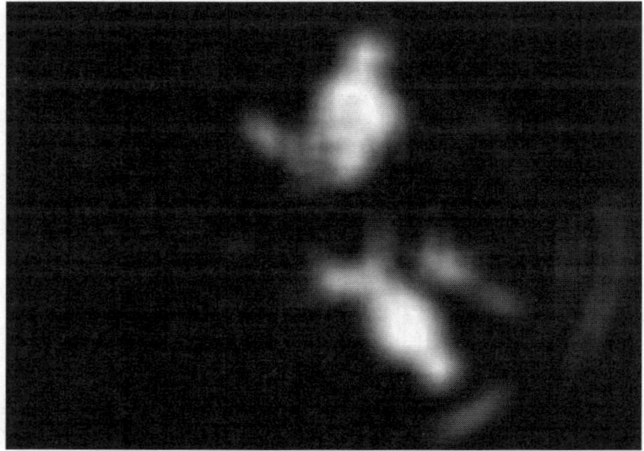

Fig. 7. The image produced after normalization

For example in figure 9 there are two detected targets at (1. 5 rad and 5. 27 rad) respectively.

5.3   Comparison between the Two Strategies

The second approach discussed in paragraph (5.2) is more reliable than the first one discussed in (5.1), but on the other hand even if it doesn't need an unfolded image, the second approach (5.2) is very expensive in computing time because of

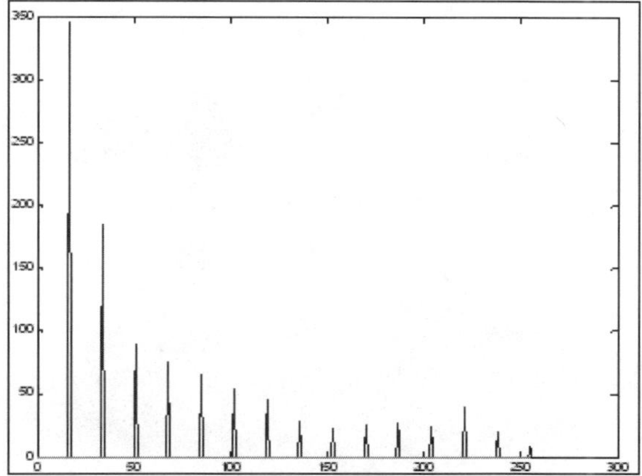

Fig. 8. Histogram of image in figure 7

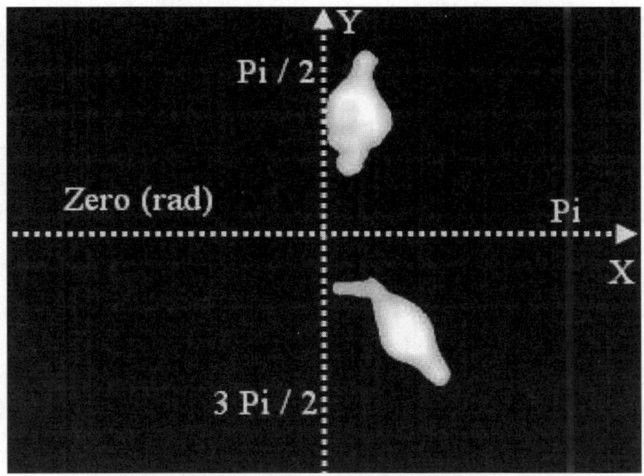

Fig. 9. Image of final step, illustrates two detected targets

the large size template. When panoramic images represent an important source of information to be fused with other sensors (like a laser), the first approach (5.1) is more convenient because it is faster, and for some small image resolutions it can work in real time (in the case of sensor fusion the matter of sensors synchronization becomes crucial) and its loss of reliability may be compensated by the other sensor. But, when a vision system is used alone for applications that need a high degree of reliability of target detection, the second approach (5.2) becomes more appropriate.

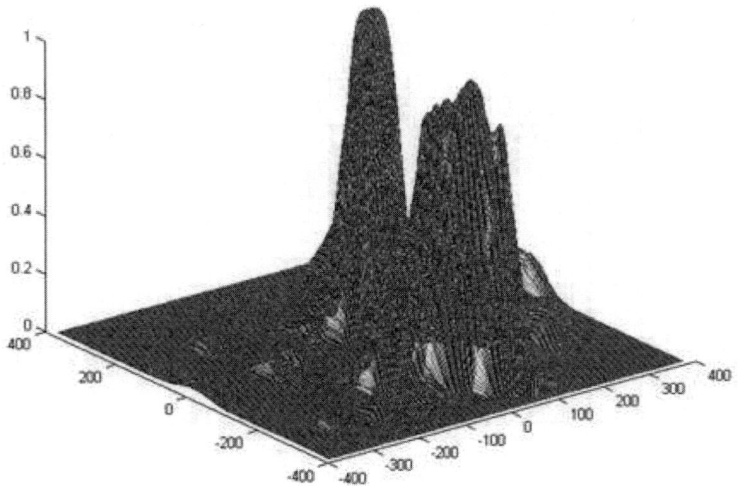

Fig. 10. 3D histogram of the image illustrated in figure 7

6   Conclusion

In this paper we have presented some experiments in dynamic target detection using panoramic images. The background is extracted repeatedly from three sequenced images with constant delay of time among them by a temporal median operator. The binary image is calculated based on a modified optical flow where a new term that compensates illumination changes is taken in account. Two strategies that determine targets directions from unfolded and raw panoramic images respectively are discussed. Future work will focus on fusion of laser and catadioptric camera to improve the reliability of dynamic target detection in one system that achieves simultaneous environment mapping and multi-target tracking.

References

1. Baba, A., Chatila, R.: Experiments with simultaneous environment mapping and multi-target tracking. In: 10th International Symposium on Experimental Robotics, Rio de Janeiro, Brazil (2006)
2. Farin, D., de With, P.H.N., Effelsberg, W.: Robust Background Estimation for Complex Video Sequences. In: International Conference on Image Processing, Barcelona, Spain (2003)
3. Haritaoglu, I., Harwood, D., Davis, L.S.: W4: real-time surveillance of people and their activities. IEEE Transactions Pattern Analysis and Machine Intelligence, 809–830 (2000)
4. Horn, B.K.P., Schunk, B.G.: Determining Optical Flow. A retrospective, Artificial Intelligence 59, 81–87 (1993)

5. Jabri, S., Duric, Z., Wechsler, H., Rosenfeld, A.: Detection and location of people using adaptive fusion of color and edge information. In: International Conference on Pattern Recognition, Vancouver, Canada (2000)
6. Jain, R., Militzer, D., Nagel, H.: Separating nonstationary from stationary scene components in a sequence of real world tv-images. In: International Joint Conferences on Artificial Intelligence, Cambridge, UK (1977)
7. Javed, O., Shafique, K., Shah, M.: A Hierarchical Approach to Robust Background Subtraction using Color and Gradient Information. In: Workshop on Motion and video computing, Orlando, Florida (2002)
8. Liu, H., Pi, W., Zha, H.: Motion Detection for Multiple Moving Targets by Using an Omnidirectional Camera. In: International Conference on Robotics, Intelligent Systems and Signal Processing, Changsha, Hunan, China (2003)
9. Nixon, M., Aguada, A.: Feature Extraction & Image Processing. Newnes (Elsevier) Linacre House, Jordan Hill, Oxford OX2 8DP 30 Corporate Drive, Burlington, MA 01803 (2002)
10. Stauffer, C., Grimson, W.E.L.: Learning patterns of activity using real-time tracking. IEEE Transactions Pattern Analysis and Machine Intelligence, 747–757 (2000)
11. Wren, C.R., Azarbayejani, A., Darrell, T., Pentland, A.P.: Pfinder, real time tracking of the human body. IEEE Transactions Pattern Analysis and Machine Intelligence, 780–785 (1997)

Vision-Based Control of the RoboTenis System

L. Angel[1], A. Traslosheros[2], J.M. Sebastian[2], L. Pari[2],
R. Carelli[3], and F. Roberti[3]

[1] Facultad de Ingeniera Electrónica Universidad Pontificia Bolivariana Bucaramanga, Colombia
langel@etsii.upm.es
[2] DISAM - ETSII Universidad Politécnica de Madrid Madrid, Spain
{atraslosheros,jsebas,lpari}@etsii.upm.es
[3] Instituto de Automática Universidad Nacional de San Juan, Argentina
{rcarelli,froberti}@inaut.unsj.edu.ar

Summary. In this paper a visual servoing architecture based on a parallel robot for the tracking of faster moving objects with unknown trajectories is proposed. The control strategy is based on the prediction of the future position and velocity of the moving object. The synthesis of the predictive control law is based on the compensation of the delay introduced by the vision system. Demonstrating by experiments, the high-speed parallel robot system has good performance in the implementation of visual control strategies with high temporary requirements.

1 Introduction

The accomplishment of robotic tasks involving dynamical environments requires lightweight yet stiff structures, actuators allowing for high acceleration and high speed, fast sensor signal processing, and sophisticated control schemes which take into account the highly nonlinear robot dynamics. As a tool for the investigation of these issues, the computer vision group of the Polytechnics University of Madrid has built the RoboTenis System, which proposes the design and construction of a high-speed parallel robot that in a future will be used to perform complex tasks, i.e. playing table tennis with the help of a vision system. The RoboTenis System is constructed with two purposes in mind. The first one is the development of a tool for use in visual servoing research. The second one is to evaluate the level of integration between a high-speed parallel manipulator and a vision system in applications with high temporary requirements.

The mechanical structure of RoboTenis System is inspired by the DELTA robot [1]. The choice of the robot is a consequence of the high requirements on the performance of the system with regard to velocity and acceleration. The kinematic analysis and the optimal design of the RoboTenis System have been presented by Angel, et al. [2]. The structure of the robot has been optimized from the view of both kinematics and dynamics respectively. The design method solves two difficulties: determining the dimensions of the parallel robot and selecting the actuators. In addition, the vision system and the control hardware have been also selected. The dynamic analysis and the preliminary control of the parallel robot

S. Lee, I.H. Suh, and M.S. Kim (Eds.): Recent Progress in Robotics, LNCIS 370, pp. 229–240, 2008.
springerlink.com

have been presented in [3], [4]. The dynamic model is based upon Lagrangian multipliers, and it uses forearms of non-negligible inertias for the development of control strategies. A nonlinear feedforward PD control has been applied and several trajectories have been programmed and tested on the prototype.

Using visual feedback to control a robot is commonly termed visual servoing. Visual features such as points, lines and regions can be used to, for example, enable the alignment of a manipulator / gripping mechanism with an object. Hence, vision is a part of a control system where it provides feedback about the state of the environment. For the tracking of fast-moving objects, several capabilities are required to a robot system, such smart sensing, motion prediction, trajectory planning, and fine sensory-motor coordination. A number of visual servo systems using model based tracking to estimate the pose of the object have been reported. Andersson presents one particular application: a ping-pong playing robot [5] [6]. The system uses a Puma robot and four video cameras. The vision system extracts the ball using simple color segmentation and a dynamic model of the ball trajectory. The system is accurately calibrated and the robot is controlled using the position-based approach. Other similar applications are: a catching robot presented in Burridge et al. [7] and a juggling robot presented by Rizzi and Koditschek [8]. Allen et al. [9] describe a system for tracking and grasping a moving object using the position-based servoing control. The object tracked is a toy-train on a circular trajectory. Buttazo et al. use a stand-alone configuration and a basket mounted at the end-effector of the robot to catch and object that moves in a plane [10]. Drummond and Cipolla present a system for the tracking of complex structures which employs the Lie algebra formalism [11]. The system is used to guide a robot into a predefined position (teach-by-showing approach). Concerning high-speed visual tracking, lots of new performing methods are appearing since a few years [12][13][14][15][16].

In this paper, we propose visual servoing architecture for the RoboTenis System. This architecture allows the 3D visual tracking of a ball at velocities of up to 1 m/s. The system uses a position-based visual servoing technique assuming the tricky problem of the 3D pose estimation of the target has been solved previously. The control law considers a prediction of the position and velocity of the ball in order to improve the performance of the movement of the robot. The synthesis of the predictive control law is based on the compensation of the delay introduced by the vision system (2 frames) and a constant acceleration motion hypothesis for the target and the robot. The presented experiments have been performed considering both predictions (position and velocity) and the position prediction only. The contributions of the paper include the use of a parallel robot in a vision-based tracking system, and the use of prediction of the movement of the target to improve tracking performance.

This paper is organized as follows. Section 2 describes the visual servo control structure. Experimental results are presented in Section 3. Finally, in Section 4 some concluding remarks are given and future work is also discussed.

2   Visual Servoing Arquitecture

This application considers an eye-in-hand configuration within dynamic look-and-move position-based scheme [6]. The task is defined as a 3D visual tracking task, keeping a constant relationship between the camera and the moving target (ball). We assume that the task is referenced with respect to a moving target that is located in the workspace of the robot and that the mobile target lays in the camera field of view so that it can always be seen as the task is executed.

The coordinate frames for the proposed visual servoing system are shown in Figure 1. $\sum_w$, $\sum_e$ and $\sum_c$ are the global, end-effector and camera coordinate frames. ^cp_b is the relative pose of camera to target object. The pose of the end-effector with respect to the global coordinate frame ^wp_e is known with $^wR_e = I$ and ^wT_e obtained from the forward kinematic model of the robot. The transformation matrix between the camera and end-effector coordinate frames (kinematics calibration), ^ep_c, is known assuming that $^cT_e = 0$.

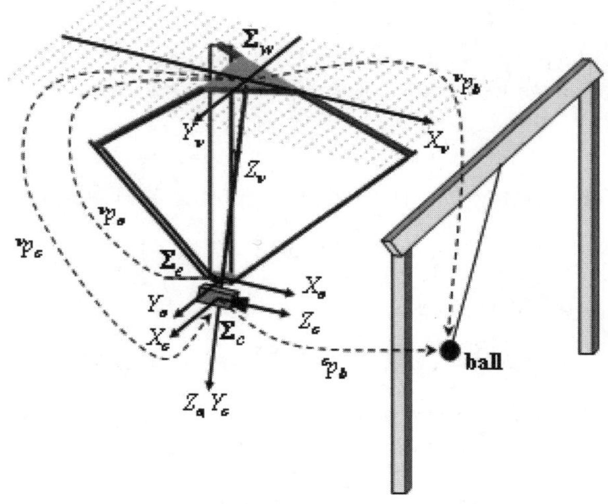

Fig. 1. Coordinate frames for the proposed visual servoing system

Fig. 2 shows a representation of the visual servo loop at an instant k. The reference position vector $^cp_b^*(k)$ of the control loop is compared to $^cp_b(k)$, this value is obtained with the vision system and the vector $^wp_c(k)$. The controller generates the control signal $^wV_e(k)$, a 3x1 vector that represents velocity references signals for each component of $^wp_e(k)$. This reference signals are expressed in the Cartesian space. So they must be converted into the joint space in order to be applied to the three joint-level velocity control loops of the robot. This transformation is computed by means of the jacobian matrix of the robot [4].

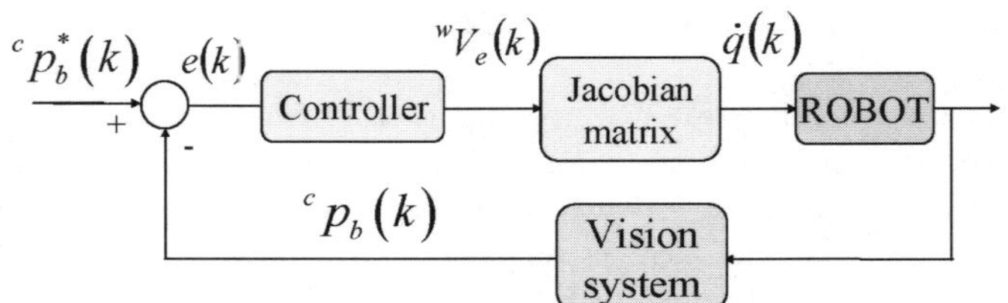

Fig. 2. Block diagram of the visual servo loop

2.1   Modeling the Visual Servoing

From Figure 2, the task error at an instant k is defined as

$$e(k) =^c p_b^*(k) -^c p_b(k) \tag{1}$$

which can be expressed by

$$e(k) =^c p_b^* -^c R_w \left(^w p_b(k) -^w p_c(k)\right) \tag{2}$$

The basic idea of control consists of trying to determine that the task error approximately behaves like a first order decoupled system, i.e.

$$\dot{e}(k) = -\lambda e(k) \tag{3}$$

with $\lambda > 0$. Differentiating (2), the following vector $\dot{e}(k)$ is obtained:

$$\dot{e}(k) = -^c R_w \left(^w v_b(k) -^w v_c(k)\right) \tag{4}$$

Using (2) and (4) in (3) it gives

$$^w v_c(k) =^w v_b(k) - \lambda^c R_w^T \left(^c p_b^* -^c p_b(k)\right) \tag{5}$$

where $^w v_c(k)$ and $^w v_b(k)$ represent the camera and ball velocities respectively. Since $^w v_e(k) =^w v_c(k)$ the control law can be written as

$$^w v_e(k) =^w v_b(k) - \lambda^c R_w^T \left(^c p_b^* -^c p_b(k)\right) \tag{6}$$

Note that (6) has two components: a component of motion prediction of the ball $^w v_b(k)$ and a component of the trajectory tracking error $(^c p_b^* -^c p_b(k))$.

A fundamental aspect in the performing of the visual servoing system is the adjustment of λ parameter. This parameter is based on the future positions of the camera and the ball. The future position of the ball according to the coordinate frame Σ_w at an instant $k + n$ can be written as

$$^w p_b(k + n) =^w \hat{p}_b(k) +^w \hat{v}_b(k) T n \tag{7}$$

where T is the sampling time. In addition, the future position of the camera according to the coordinate frame Σ_w at an instant $k + n$ is

$$^w p_c(k + n) =^w p_c(k) +^w v_c(k) T n \tag{8}$$

and λ is defined by

$$\lambda = \frac{1}{Tn} \tag{9}$$

The basic architecture of visual control is shown in Fig.3. The control law considers a prediction of the position and velocity of the ball in order to improve the performance of the movement of the robot. The synthesis of the predictive control law is based on the compensation of the delay introduced by the vision system z^{-r}

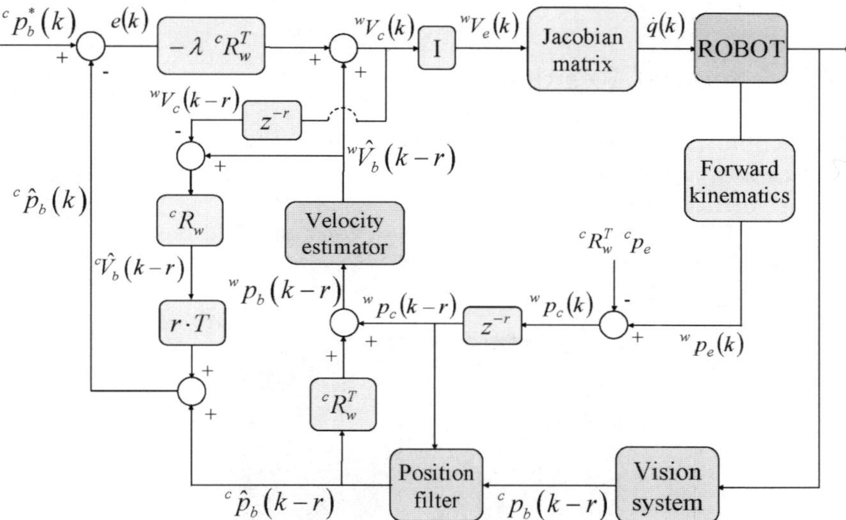

Fig. 3. Visual servoing architecture proposed for the RoboTenis System

2.2   Experimental Setup

The experimental setup is shown in Fig. 4. The control architecture of the RoboTenis System uses two control loops: one inner control loop that allows the velocity control of the robot to low level, and one external control loop that allows the 3D tracking of the ball using the information gives by vision system. The two control loops are calculated in a DSPACE card. The velocity loop is running at 0.5 ms and the vision loop at 8.33 ms. Other computer is employed for the acquisition and processing of the image in Windows 98 platform. The information given by the vision system is transmitted to the DSPACE card using a serial communication channel.

In order to minimize the delay in the vision loop, the acquisition of a new image is made in parallel to the processing of the previous image.

Image processing is simplified using a dark ball on a white background. The camera captures 120 non-interlaced images per second. At this frame rate, the resolution is limited to 640x240 pixels. Indeed, with a sampling rate of 120 Hz, the image transfer, image processing and control must no take more than 8.33 ms. In the RoboTenis System, all these tasks take about 5 ms.

A pinhole camera model performs the perspective projection of a 3D point into the image plane. The camera is pre-calibrated with known intrinsic parameters. Features extracted (centroid and diameter) together with the knowledge of the ball geometry (radius); give the pose estimation of the ball according to the camera. The position and velocity of the ball are estimated using the Kalman filter.

The control program takes the estimated position and velocity of the ball, the joints positions and, using (6) it calculates the control actions in order to be applied to the three joint-level velocity control loops of the robot.

3   Experimental Results

In this section, results related to the realization of visual tracking tasks using a parallel robot are presented. The results show the performance of the visual servoing algorithm proposed for the RoboTenis system.

The control objective consists in keeping a constant relationship between the camera and the moving target. The distance was fixed to $[600, 0, 0]T$ mm. The

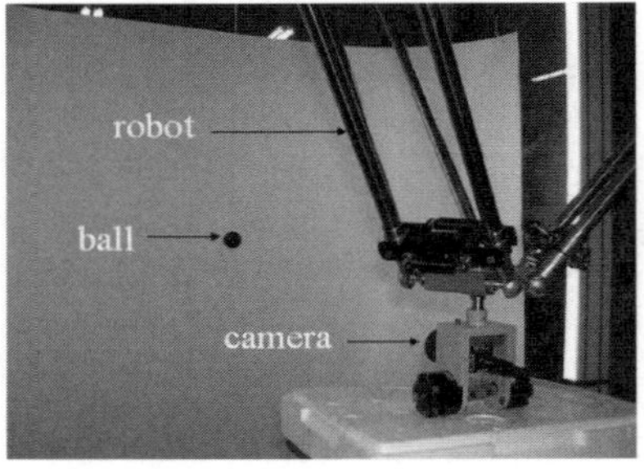

Fig. 4. Experimental setup

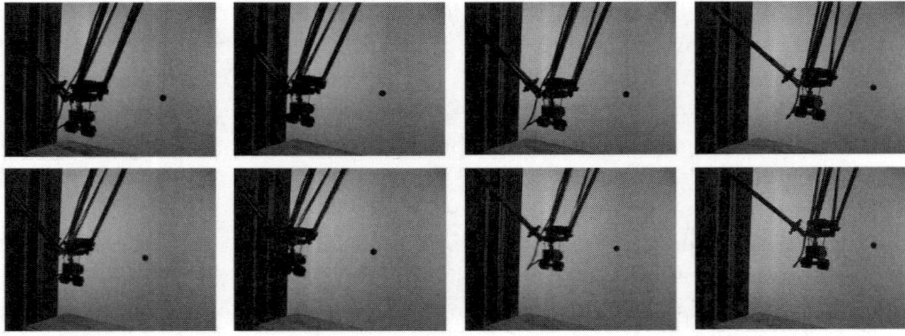

Fig. 5. 3D visual tracking of a ball using a parallel robot

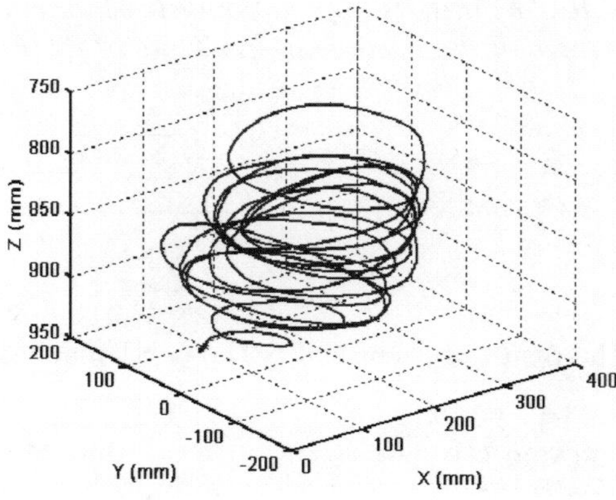

Fig. 6. Movement of the end-effector in the workspace

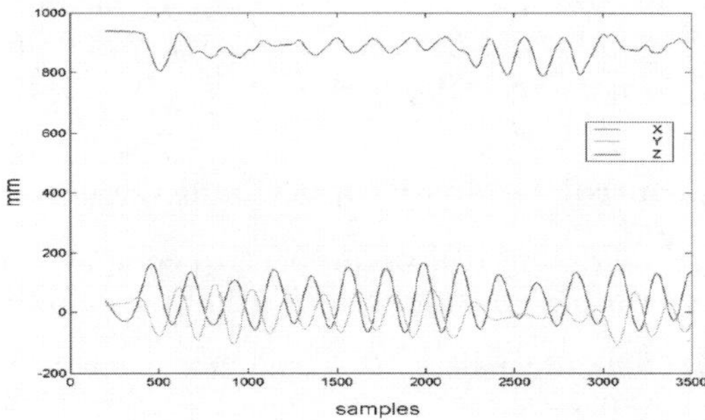

Fig. 7. Behaviour of the end-effector for the 3D visual tracking of the ball

ball is hold by a thread to the structure of the robot and it moves by means of a manual drag (Fig. 4). Different 3D trajectories have been executed. The tests have been made with speeds of up to 1000 mm/s.

For example, in Fig. 5 is represented the space evolution of the ball and the end-effector for one trial. This figure shows a sequence of eight images taken during a tracking task. Fig. 6 and Fig. 7 show the space evolution in the position of the end-effector and a time history, respectively. The nature of the motion causes appreciable variations of the velocity of the ball difficult to predict, which increases the difficulty of the tracking task.

3.1   Performance Indices

We propose two performance indices for the validation of the visual controller (6). These indices are based on the tracking error and the estimated velocity of the ball. Given the random nature of the made tests, the proposed indices are:

- *Tracking relation:It is defined as the relation between the average of modulate of the tracking error and the average of modulate of the estimated velocity of the ball, it is*

$$TrackingRelation = \frac{\frac{1}{N}\sum_{k=1}^{N} e(k)}{\frac{1}{N}\sum_{k=1}^{N} {}^{w}\hat{v}_{b}(k)} \qquad (10)$$

This index isolates the result of each trial of the particular features of motion of the ball.

- *Average of the tracking error by strips of the estimated velocity of the ball: we have defined 5 strips:*

$$\begin{aligned} &Estimated\ velocity < 200mm/s \\ &< 200mm/s < Estimated\ velocity \leq 400mm/s \\ &< 400mm/s < Estimated\ velocity \leq 600mm/s \\ &< 600mm/s < Estimated\ velocity \leq 800mm/s \\ &Estimated\ velocity > 800mm/s \end{aligned} \qquad (11)$$

3.2   Predictive Control Versus Proportional Control

With the purpose of validating (6), we propose to compare the performance of the RoboTenis System using (10) and (11) for the two following cases:

- *Predictive control law: It considers the predictive component of (6), is to say:*

$$^{w}v_{e}(k) =^{w} \hat{v}_{b}(k) - \lambda^{c}R_{w}^{T}\left(^{c}p_{b}^{*} -^{c} \hat{p}_{b}(k)\right) \qquad (12)$$

- *Proportional control law: It does not consider the predictive component of (6), is to say:*

$$^{w}v_{e}(k) = -\lambda^{c}R_{w}^{T}\left(^{c}p_{b}^{*} -^{c} \hat{p}_{b}(k)\right) \qquad (13)$$

Table 1 and Table 2 present the results obtained for the indices (10) and (11) when the control laws (12) and (13) are applied. The results present the average of 10 trials made for each algorithm of control. A high performance of the system using the predictive control algorithm is observed, given by a smaller tracking relation and a smaller error by strips.

Fig. 8 and Fig. 9 show the evolution in the tracking error and the estimated velocity of the ball when (12) is applied to the RoboTenis System. Whereas Fig. 10 and Fig. 11 show the evolution in the tracking error and the estimated velocity of

Table 1. Predictive control vs proportional control tracking relation

Algorithm	tracking relation
Proportional	40.45
Predictive	20.86

Table 2. Predictive control vs proportional control error by strips (V in mm/s)

Algorithm	$V < 200$	$200 < V < 400$	$400 < V < 600$	$600 < V < 800$	$V > 800$
Proportional	6.36	13.72	20.11	26.22	32.56
Predictive	4.21	8.19	9.50	11.38	13.54

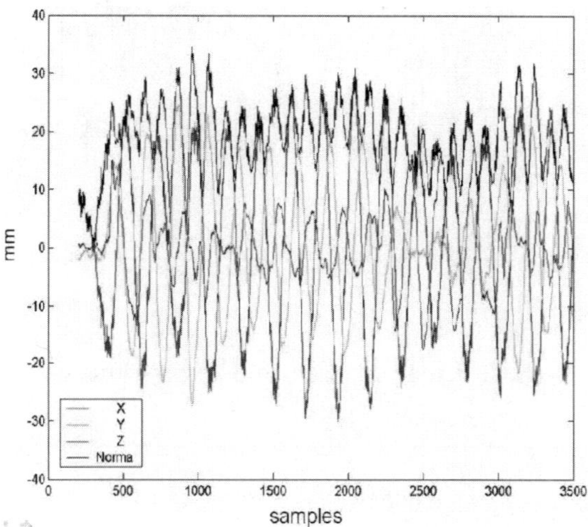

Fig. 8. Proportional Control Law: tracking error

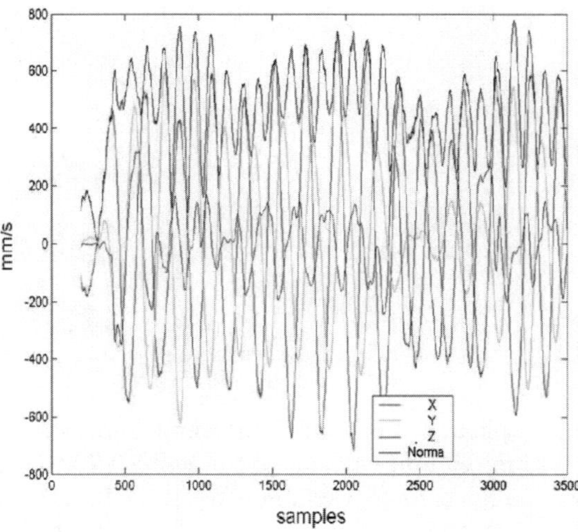

Fig. 9. Proportional Control Law: estimated velocity of the ball

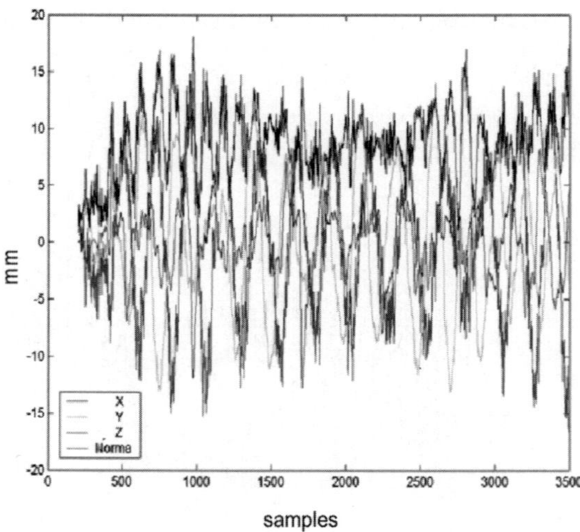

Fig. 10. Predictive Control Law: tracking error

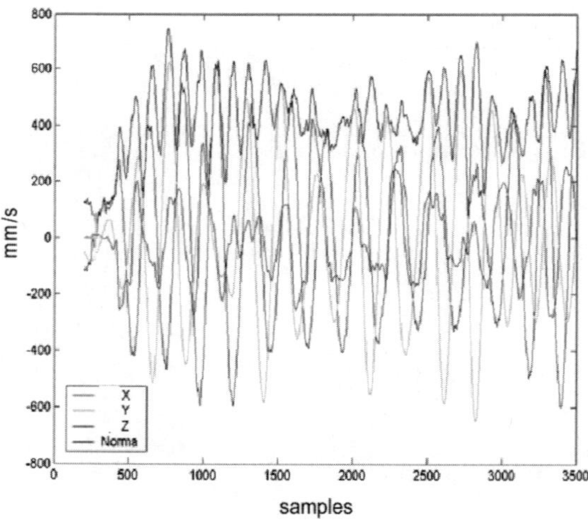

Fig. 11. Predictive Control Law: estimated velocity of the ball

the ball when (13) is applied. For proportional control law, the maximum tracking error is 34.50 mm and maximum velocity of the ball is 779.97 mm/s. For the predictive control law, the maximum tracking error is 14.10 mm and the maximum velocity of the ball is 748.16 mm/s. The error is bounded and the tracking error is reduced by introducing an estimation of the moving object velocity.

These results are no more than preliminary. Next, it will be necessary to evaluate the robustness of the control law with regard to noise in position and velocity estimation, modelling error, and particularly to the eye-in-hand calibration error.

4   Conclusion

This paper describes a position-based visual servoing system for tracking a hanging ball with a robot equipped with an attached camera. A parallel robot is used for this purpose. The ball is tracked as a single point. The control law considers a prediction of the position and velocity of the ball in order to improve the performance of the movement of the robot. The presented experiments have been performed considering both predictions and the position prediction only. These results are no more than preliminary. As future work, is necessary to evaluate the robustness of the system with respect to modeling errors, and to design new visual control strategies that allow to the system tracking velocities of up to 2 m/s.

References

1. Clavel, R.: DELTA: a fast robot with parallel geometry. In: 18th International Symposium on Industrial Robot, Sidney Australia, pp. 91–100 (1988)
2. Angel, L., Sebastian, J.M., Saltaren, R., Aracil, R., SanPedro, J.: RoboTenis: Optimal design of a Parallel Robot with High Performance. In: IROS 2005. IEEE/RSJ International Conference on Intelligent Robots and Systems, Canada, pp. 2134–2139 (2005)
3. Angel, L., Sebastian, J.M., Saltaren, R., Aracil, J., Gutiérrez, R.: RoboTenis: Design, Dynamic Modeling and Preliminary Control. In: AIM 2005. IEEE/ASME International Conference on Advanced Intelligent Mechatronics, California, pp. 747–752 (2005)
4. Angel, L., Sebastian, J.M., Saltaren, R., Aracil, R.: RoboTenis System. Part II: Dynamics and Control. In: CDC-ECC 2005. 44th IEEE Conference on Decision and Control and European Control Conference, Sevilla, pp. 2030–2034 (2005)
5. Anderson, R.: Dynamic Sensing in ping-pong playing robot. IEEE Trans. on Robotics and Automation 5(6), 728–739 (1989)
6. Anderson, R: Understanding and applying a robot ping-pong player's expert controller. In: ICRA 1989. IEEE International Conference on Robotics and Automation, vol. 3, pp. 1284–1289 (1989)
7. Burridge, R., Rizzi, A., Koditschek, D.: Toward a dynamical pick and place. In: IROS 1995. IEEE/RSJ International Conference on Intelligent Robots and Systems, vol. 2, pp. 292–297 (1995)
8. Rizzi, A., Koditschek, D.: An active visual estimator for dextrous manipulation. IEEE Trans. on Robotics and Automation 12(5), 697–713 (1996)
9. Allen, P., Timcenko, A., Yoshimi, B., Michelman, P.: Automated tracking and grasping of a moving object with a robotic hand-eye system. IEEE Trans. on Robotics and Automation 9(2), 152–165 (1993)
10. Butazzo, G., Allota, B., Fanizza, F.: Mousebuster A robot system for catching fast moving objects by vision. In: ICRA 1993. IEEE International Conference on Robotics and Automation, vol. 3, pp. 932–937 (1993)
11. Drummond, T., Cipolla, R.: Real-time tracking of multiple articulated structures in multiple views. In: Vernon, D. (ed.) ECCV 2000. LNCS, vol. 1843, pp. 20–36. Springer, Heidelberg (2000)
12. Malis, E., Benhimane, S.: A unified approach to visual tracking and servoing. Robotics and Autonomous Systems. 52, 39–52 (2005)

13. Saedi, P., Lowe, D., Lawrence, P.: 3D localization and tracking in unknown environments. In: ICRA 2003. IEEE International Conference on Robotics and Automation, vol. 1, pp. 1297–1303 (2003)
14. Gangloff, J., Mathelin, M.: High speed visual servoing of a 6 DOF manipulator using MIMO predictive control. In: ICRA 2000. IEEE International Conference on Robotics and Automation, vol. 4, pp. 3751–3756 (2000)
15. Senoo, T., Namiki, A., Ishikawa, M.: High-speed batting using a multi-jointed manipulator. In: ICRA 2004. IEEE International Conference on Robotics and Automation, vol. 2, pp. 1191–1196 (2004)
16. Kaneko, M., Higashimori, M., Takenaka, R., Namiki, A., Ishikawa, M.: The 100G capturing robot-too fast to see. In: AIM 2003. IEEE/ASME International Conference on Advanced Intelligent Mechatronics, vol. 8, pp. 37–44 (2003)

Particle Filter Based Robust Recognition and Pose Estimation of 3D Objects in a Sequence of Images

Jeihun Lee, Seung-Min Baek, Changhyun Choi, and Sukhan Lee

Intelligent Systems Research Center,
School of Information and Communication Engineering, Sungkyunkwan University,
Suwon, Korea
{jeihun81,smbaek,lsh}@ece.skku.ac.kr

Summary. A particle filter framework of multiple evidence fusion and model matching in a sequence of images is presented for robust recognition and pose estimation of 3D objects. It attempts to challenge a long-standing problem in robot vision, so called, how to ensure the dependability of its performance under the large variation in visual properties of a scene due to changes in illumination, texture, occlusion, as well as camera pose. To ensure the dependability, we propose a behavioral process in vision, where possible interpretations are carried probabilistically in space and time for further investigations till they are converged to a credible decision by additional evidences. The proposed approach features 1) the automatic selection and collection of an optimal set of evidences based on in-situ monitoring of environmental variations, 2) the derivation of multiple interpretations, as particles representing possible object poses in 3D space, and the assignment of their probabilities based on matching the object model with evidences, and 3) the particle filtering of interpretations in time with the additional evidences obtained from a sequence of images. The proposed approach has been validated by the stereo-camera based experimentation of 3D object recognition and pose estimation, where a combination of photometric and geometric features are used for evidences.

1 Introduction

The object recognition has been one of the major problems in computer vision and intensively investigated for several decades. Although to recognize object have some problems, it has been developed toward real complex objects in cluttered scenes. There are several approaches to solve the problems about object recognition in real environment. One of the most common approach for recognizing object from a measured scene is a model based recognition method. It recognizes the objects by matching features extracted from the scene with stored features of the object [1, 2, 3]. There are several methods to recognize object using predefined model information. The method proposed by Fischler and Bolles [4] uses RANSAC to recognize objects. It projects points from all models to the scene and determines if projected points are close to those of detected scene. Then recognizes the object through this. This method is not so efficient because of iterative hypothesis and verification tasks. Olson [5] proposed pose clustering method for object recognition.

S. Lee, I.H. Suh, and M.S. Kim (Eds.): Recent Progress in Robotics, LNCIS 370, pp. 241–253, 2008.
springerlink.com

This method recognizes object by producing pose space discretely and finding the cluster which is including the object. As for disadvantages of this method, data size is quite big since pose space is 6-dimentional and pose cluster can be detected only when sufficient accurate pose becomes generated. In the next, David et al. [6] proposed recognition method that matching and pose estimation are solved simultaneously by minimizing energy function. But it may not be converged to minimum value by functional minimization method due to high non-linearity of cost function. In addition, Johnson and Herbert [7] proposed a spin image based recognition algorithm in cluttered 3D scenes and Andrea Frome et al. [8] compared the performance of 3D shape context with spin-image. Jean Ponce et al. [9] introduced the 3D object recognition approach using affine invariant patches. Most recently, several authors have proposed the use of descriptor in image patch [10]. Another approach to recognize the object is local shape features based method which is inspired by the shape contexts of Belongie et al. [11]. At each edge pixel in an image, a histogram, or shape context, is calculated then each bin in the histogram counts the number of edge pixels in a neighborhood near the pixel. After searching nearest neighbor and measuring histogram distance, determine correspondences between shape contexts from a test image and shape contexts from model images [12]. But this method may not be effective when the background is concerned. To solve this problem, assessing shape context matching in high cluttered scene have studied [13] recently. Except for above method, there are many of object recognition researches. However, most of these methods are working well only at the condition under accurate 3D data or fully textured environments in single scene information with limited feature. It means that 2D or 3D measurement data from real environment contains noisy and uncertain information caused by changes of illumination, amount of texture, distance to the object and etc. Therefore, in this paper, we try to find a solution for simultaneous recognition and pose estimation of 3D object in a real environment conditions. The reminder of this paper is organized as follows : Section II outlines proposed framework for object recognition and its pose estimation. Section III explains how to assign matching similarities from different features. Description of experimental data and a feasibility analysis of our proposed framework are presented in section IV. Section V concludes by discussing scalability and implementation issues along with directions for future works.

2   Proposed Framework Overview

2.1   Outline of Proposed Approach

Some of literatures about probabilistic approaches to recognize object or estimate its pose have been reported.[15, 16, 17]. The works of [17] and [18] use maximum a posteriori (MAP) estimation under a Markov random field (MRF) model. Especially the former uses MRF as a probabilistic model to capture the dependencies between features of the object model, and employs MAP estimation to find the match between the object and a scene. Schiele and Crowley [19] have developed a probabilistic object recognition technique using multidimensional receptive field histograms. Although this technique has been shown to be somewhat robust to change of rotation and scale with low cost of computation, it only computes the probability of the presence of an object. We proposed a probabilistic method based on a sequence of images to recognize an object and to estimate its pose in our previous work [20]. But the previous framework simply uses a ratio of matched features to total features when it assigns a similarity weight to each particle. The main contribution of this paper is to propose a more systematic recognition framework which considers not only matched features but also matched pose errors. The proposed method handles the object pose probabilistically. The probabilistic pose is drawn by particles and is updated by consecutive observations extracted from a sequence of images. The proposed method can recognize not only textured but also texture-less objects because the particle filtering framework of the proposed method can deal with various features such as photometric features (SIFT; Scale Invariant Feature Transform [10], color) and geometric features (line, square) [14].

Fig. 1 illustrates block diagram of the overall 3D recognition framework. First of all, the information of circumstance - density of texture, illumination and distance of expected object pose - is calculated from input image and 3D point cloud in In-situ Monitoring. Then the valid features in an input image are selected by the Cognitive Perception Engine (CPE) which perceives an environment automatically by using information offered by In-situ Monitoring and keeps the evidences of all objects for their recognition. Valid features for recognizing the object are stored in Service Robot Information System (SRIS) and CPE uses this information as a priori-knowledge. The multiple poses are generated by features extracted from a scene and 3D point cloud. And probabilistic estimation is done by particle filter using measured poses and propagated probability distribution of target object in a sequence of images.

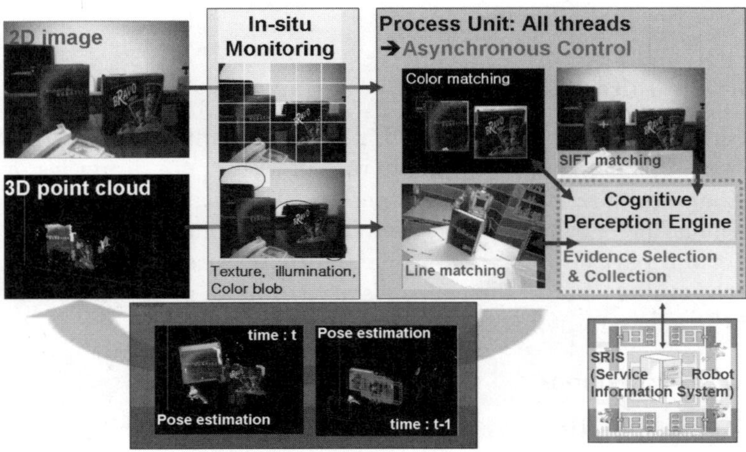

Fig. 1. Block diagram of 3D recognition framework

244   J. Lee et al.

2.2   In-Situ Monitoring

The main role of In-situ Monitoring is simply to check the changes of environment such as illumination, a mount of texture and distance between robot and target object. In this paper, we divide input image into 25 areas uniformly, 5 columns and 5 rows, and calculated values are used for selection of valid feature or feature set. The illumination means intensity information in current image that calculates not absolute value but relative such as changes of environment. Amount of texture in each block is counted pixel which is processed by Canny edge image of current frame. Lastly, we assume that existence possibility of object is high if amount of texture is abundant in particular block. So distance of each block is calculated from processed image pixel with valid 3D point cloud and average of those values.

2.3   Cognitive Perception Engine

We assume that the valid features for recognizing each object in a current scene are already defined to the CPE. In Fig. 1 this information could be delivered from SRIS. The main role of CPE is selection of proper feature or evidence set by using information from In-situ monitoring. We have 3 valid asynchronous processing pathes - color, line and SIFT - which are selected one or more automatically based on the target object and information from in-situ monitoring. For example, if the distance is far, then CPE selects color feature. On the other hand, SIFT or line features are more helpful for recognizing near or middle range object. Sometimes all three features could be used for getting maximal information from current scene. However, the processing time is longer than any other set of evidences. So, it is a trade-off problem between performance and time consumption. In this paper, strategy for feature selection is done by using distance and illumination to simplify a feasibility analysis.

2.4   Particle Filtering

Particle filtering procedure is presented in previous papers [20]. The recognized object pose is estimated by particle filtering in a sequence of images over time in order that we represent the object pose with an arbitrary distribution. We keep a formulation of Motion model and Observation model in [20] which is the most important part of proposed particle filter based framework. In this paper, we improved the way how to assign similarity weight of measured features using Bayesian theorem and probabilistic approach.

Observation Likelihood

We define the observation likelihood $p(Z_t|O_t^{[i]})$ as in previous work [20] :

$$p(Z_t|O_t^{[i]}) =$$
$$\sum_{j=1}^m w_j \ \cdot \ exp\left[\frac{-1}{2} \cdot \sum_{l=1}^4 \left\{ \times S_l^{-1} \cdot \frac{(Ob\_TP_l^j - St\_TP_l^i)^T}{(Ob\_TP_l^j - St\_TP_l^i)} \right\} \right] \qquad (1)$$

Where w_j is the similarity weight related to transformed points with $O^{[j]}$. Where m is the number of generated poses at time t. Here, we designate four points $(P1, P2, P3, P4)$ at camera frame as Fig. 2. The four points are transformed by the homogeneous transform matrix parameterized by the six spatial degrees of freedom. Fig. 2 (b) shows the transformed points $(TP1, TP2, TP3, TP4)$ with an arbitrary homogeneous transform matrix. We obtain the set of the four points $(TP1, TP2, TP3, TP4)$ transformed from $(P1, P2, P3, P4)$. Let $(Ob\_TP1[i], Ob\_TP2[i], Ob\_TP3[i], Ob\_TP4[i])$ represent the transformed points with $O^{[i]}$ while $(St\_TP1[i], St\_TP2[i], St\_TP3[i], St\_TP4[i])$ mean the transformed points with $O_t^{[i]}$.

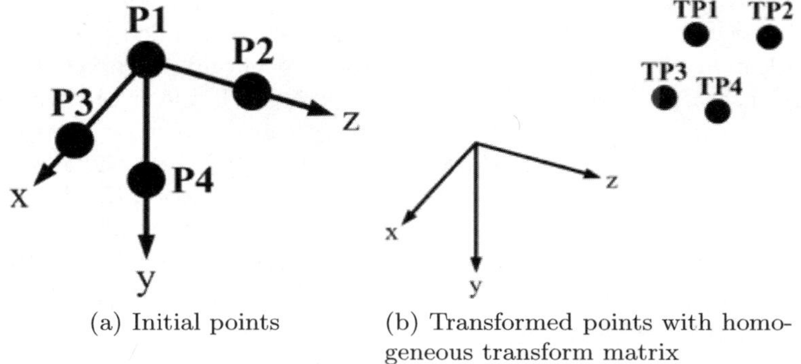

(a) Initial points

(b) Transformed points with homogeneous transform matrix

Fig. 2. The designated four points for making the observation likelihood

Similarity Assignment

To assign similarity, we consider how much correspondence exists between the recognized object and its estimated pose and real ones, respectively. In probabilistic terms the goal of proposed method is to estimate object pose which yield the best interpretation of object pose generated by multiple features in Bayesian sense. Our particle filter based probabilistic method framework approximate variant of the following posterior distribution.

$$w_j = p(O_{t,object}|E) = p(O_{t,id}, O_{t,pose}|E) \tag{2}$$

Where object O_{object} is an object to recognize, it is divided into O_{id} and O_{pose} for information of recognition and pose estimation respectively. The O_{id} means whether recognized object is correct or not and O_{pose} means precision level of estimated object pose. Where E denote the evidence, measurement, redefined $E = \{Z_1, Z_2, \cdots Z_n\}$ indicates multiple features. In other words, the O_{id} means a process of object recognition whether it is the aimed object to recognize or not. The O_{pose} is generated by accuracy rate of estimated object pose. To represent similarity weight, we assume that O_{id} and O_{pose} are independent events because object identification is considered separately as pose estimation. That means that the very well recognized object does not guarantee accurate estimation of object pose, vice versa. According to this assumption, the similarity is represented as follow :

$$p(O_{t,id}, O_{t,pose}|E) = p(O_{t,id}|E)p(O_{t,pose}|E) \qquad (3)$$

3   Object Matching Similarity from Features

3.1   Similarity Assignment from SIFT Feature

The object pose can be generated by calculating a transformation between the SIFT features [14] measured at current frame and the corresponding ones in the database. The transformation is represented by a homogeneous transform matrix. The object pose can be generated by using corresponded 3D point clouds from depth image if the matched features are 3 or more in 2D image [20]. If one scene has several candidates that have matched SIFT features, then all these candidates generate 3D poses for probabilistic fusion at particle filtering stage, as described in previous section. However, to assign similarity weight to each candidate, posterior distribution should be calculated in equation (2). For example, when an object is shown in the scene, measured average number of matched SIFT is 23 as $p(O_{t,id}|E_{SIFT})$, and average distance error is 5mm with certain variation by many trials as $p(O_{t,pose}|E_{SIFT})$. Then, the posterior distribution $p(O_{t,object}|E_{SIFT})$ can be obtained by equation (3), and the shape of probability distribution of example case is shown in Fig. 3.

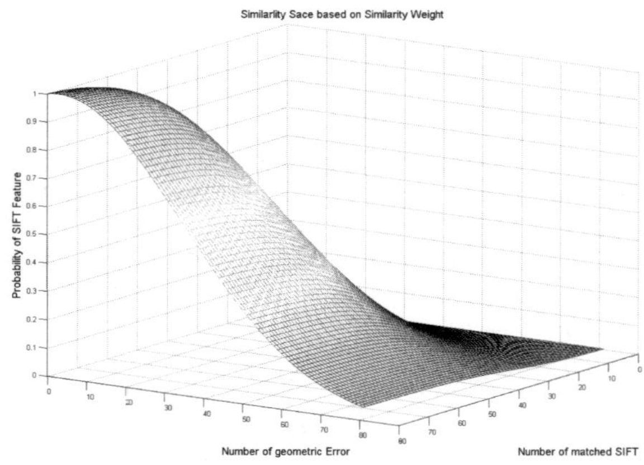

Fig. 3. Obtained $p(O_{t,object}|E_{SIFT})$ graph by experimental result

3.2   Similarity Assignment from Line Feature

Assigning similarity method of line feature is conducted the same process with SIFT. But there are two kinds of hypothesis about object identification,

$p(O_{t,id}|E_{Line})$ and pose accuracy, $p(O_{t,pose}|E_{Line})$. We define first one as a Coverage that means how many matched line with information of model line. The Coverage can be calculated by equation (4) as follow :

$$Coverage = \frac{Matched\_line\_length}{Total\_line\_length\_of\_model} \tag{4}$$

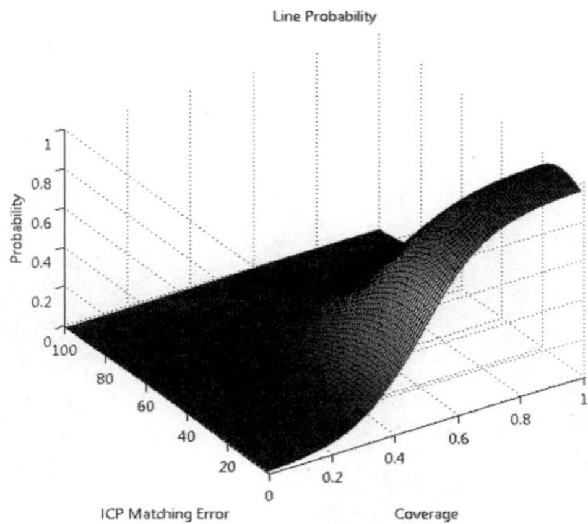

Fig. 4. Obtained $p(O_{t,object}|E_{Line})$ graph by experimental result

If the Coverage is very high, then the probability of object identification is also high. And the second one is defined as Iterative Closest Point (ICP) matching error, because we use ICP for line matching. Line matching finds several matched set like SIFT in the single scene. So, $p(O_{t,object}|E_{Line})$ can be obtained by equation (3) in each candidate and is represented as a joint probabilisty in Fig. 4.

3.3 Similarity Assignment from Color Information

The object with a particular color can be segmented by the color in the current scene. Although the segmented region can not provide an objects orientation, the objects location can be generated by using the segmented region from corresponded depth image. In homogeneous transform matrix, the rotation part is defined by an identity matrix and the translation part represents an objects location as a center of segmented area. Information of translation matrix can be approximated average of valid 3D points in segmented area. If there is no valid point in segmented area, it is not assigned as similarity. The similarity weight for jth object location, w_j, is denoted as a predefined constant with a small value in comparison with the similarity weight of the object pose generated by the other features. In particular, the color information can be combined with the other features.

4 Experimental Results

This paper focuses on simultaneous recognition of target object and estimation of its pose in a sequence of images. The proposed method is tested to recognize textured and textureless objects in various distance and illumination condition. The robot used in the experiment is a PowerBot AGV with a Videre stereo camera mounted on the pan-tile unit configuration as shown in Fig. 5. The camera motion information is calculated by the internal encoder.

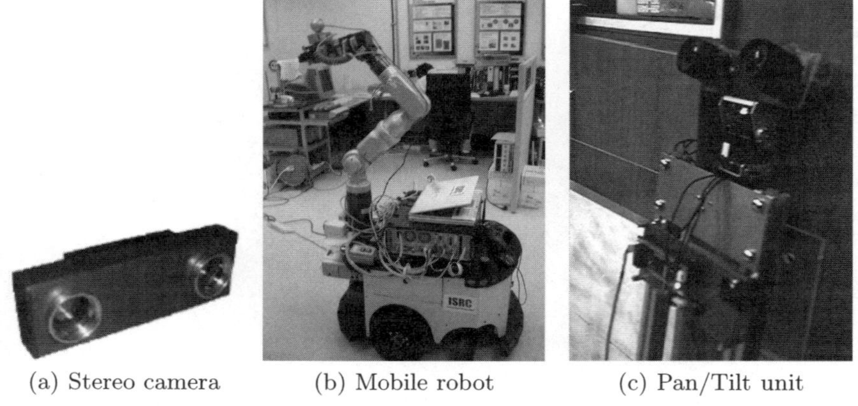

(a) Stereo camera   (b) Mobile robot   (c) Pan/Tilt unit

Fig. 5. Experimental setups

For an evaluation of the proposed method, we set up cluttered environment as Fig. 6 (a), and used illuminometer Fig. 6(b) for measuring change of illumination in environment. The target object to recognize is red circled blue book which is rectangular parallelepiped as shown in Fig. 6(a). The book has textured front side and texture-less rear side.

We made CPE strategy for selection of processing passes of three features and its combination. It means that the recognition and pose estimation of 3D object are performed by either basic features such as Color, Line and SIFT, or combined features such as Line + Color, SIFT + Color, and SIFT + Line. These features are selected automatically in accordance with illumination and distance by CPE in the proposed framework. CPE selects SIFT in the close distance to object and bright environment. If the distance is far, CPE use Color with SIFT, Line feature or Color with Line features for object recognition and its pose estimation. On the other hand, if the distance information from in-situ monitor is over the 1.0 meters, Line or Color + Line features are selected by CPE in dark environment.

Fig. 7 indicates the experimental results of object recognition under bright illumination condition. 'Mean' and 'Var.' in the table represents average distance and variation respectively. 'O' and 'X' indicates amount of detected texture from the target object. 'O' means object has enough textures, whereas 'X' means object does not have enough textures. The cross marked cells mean that the robot

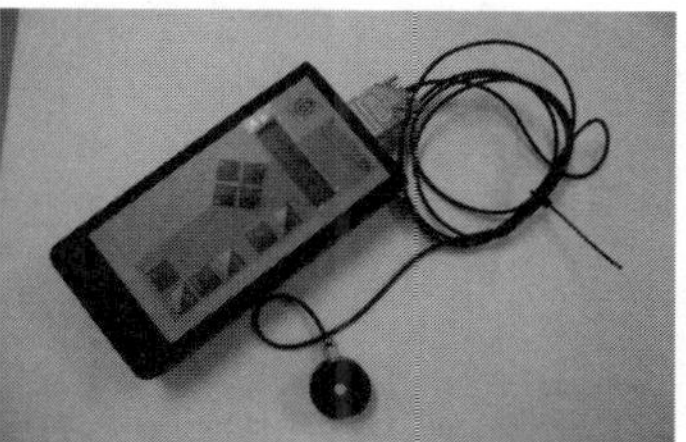

(a) Experimental environment   (b) Illuminometer

Fig. 6. Experimental environment and illuminometer

Feature		Distance	0.5		1.0		1.6	
			O	X	O	X	O	X
SIFT	Mean		0.50	0.48	1.01			
	Var.		0.002	0.02	0.002			
Color	Mean		0.46	0.50	0.99	0.94	1.55	1.53
	Var.		0.03	0.06	0.05	0.23	0.29	0.29
Line	Mean		0.39	0.44	1.44	1.04	1.68	1.59
	Var.		0.07	0.05	0.27	0.03	0.07	0.09
Line +Color	Mean		0.47	0.45	0.94	0.98	1.61	1.63
	Var.		0.05	0.04	0.06	0.03	0.02	0.02
SIFT +Color	Mean		0.51	0.49	1.02	0.97		
	Var.		0.005	0.03	0.004	0.04		
SIFT +Line	Mean		0.519	0.49	1.07		1.66	
	Var.		0.017	0.03	0.18		0.20	

*Distance Unit = meter

Fig. 7. Experimental results for object recognition under 330 lux illumination condition

is not able to recognize target object by using the selected evidence. Since the number of valid SIFT features are more extracted under the 330 lux illumination than other dark illuminations, the 330 lux can be seen as a proper illumination condition for the recognition using SIFT as a feature. Line features are also more reliably detected at 330 lux case than 120 lux case. Fig. 8 shows the experimental results under the darker illumination condition. The gray painted cells are the chosen evidences by CPE for optimal feature selection. This selection aptly shows that CPE's choices are reliable for object recognition.

First experiment is conducted as recognizing textured object, front side of the book, with changing illumination, 330 lux and 120 lux, and distances of recognition, about 0.5 meters, 1.0 meters and 1.6 meters and its result illustrates in Fig. 9. The proposed method also tested for recognizing textureless case, rear side of the book, the results are shown in Fig. 10. Means and variances of

Distance		0.5		1.0		1.6	
		Texture					
Feature		O	X	O	X	O	X
SIFT	Mean	0.51	0.48	0.99			
	Var.	0.02	0.03	0.003			
Color	Mean	0.47	0.50	0.93	1.00	1.62	1.55
	Var.	0.05	0.07	0.06	0.14	0.13	0.28
Line	Mean	0.47	0.45	0.94	0.97	1.57	1.59
	Var.	0.05	0.05	0.06	0.03	0.13	0.08
Line +Color	Mean	0.50	0.45	0.97	0.97	1.62	1.65
	Var.	0.01	0.03	0.02	0.02	0.02	0.02
SIFT +Color	Mean	0.51	0.49	1.04	0.97		
	Var.	0.03	0.03	0.07	0.04		
SIFT +Line	Mean	0.51	0.48	1.00			
	Var.	0.02	0.03	0.09			

*Distance Unit = meter

Fig. 8. Experimental results for object recognition under 120 lux illumination condition

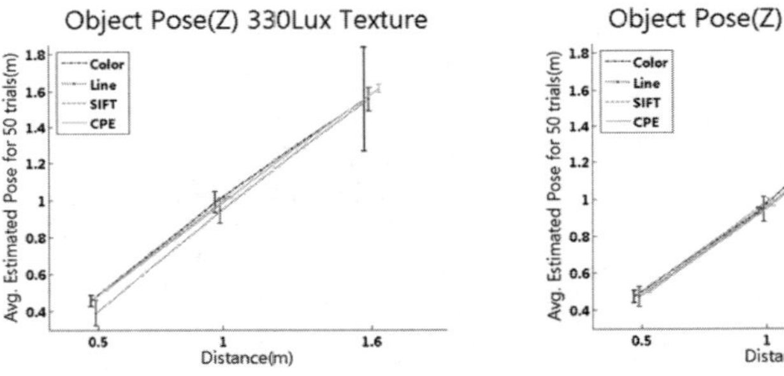

(a) Results of pose estimation of textured object in bright environment

(b) Results of pose estimation of textured object in dark environment

Fig. 9. The ARMSE for the textured object pose in accordance with illumination

estimated poses are described in each figure. So, we can see the variations of performance caused by selected set of evidences in different conditions.

Fig. 9 and Fig. 10 show that the each evidence - Color, Line and SIFT - has characteristic such as accuracy, effective distance and illumination for recognition. Color feature has the advantage for changes of not only illumination and distance but also a number of 3D point clouds. But Color feature cannot estimate object pose because it cannot identify the object. Therefore, variance of Color feature is much wider than others. Line feature has a good performance for recognition and its pose estimation with small variance. But sometimes mismatches between real object and similar ones are detected because it does not have object identification capability. Whereas other evidences, SIFT feature has very good performance to identify object. So pose estimation combined with 3D point cloud from high precision sensor is very accurate. The results from our

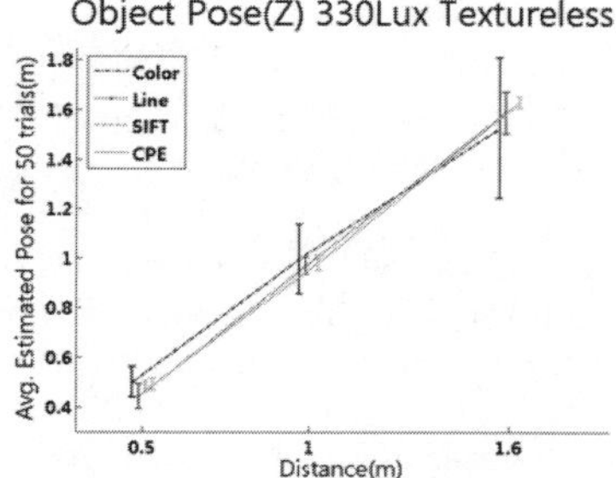

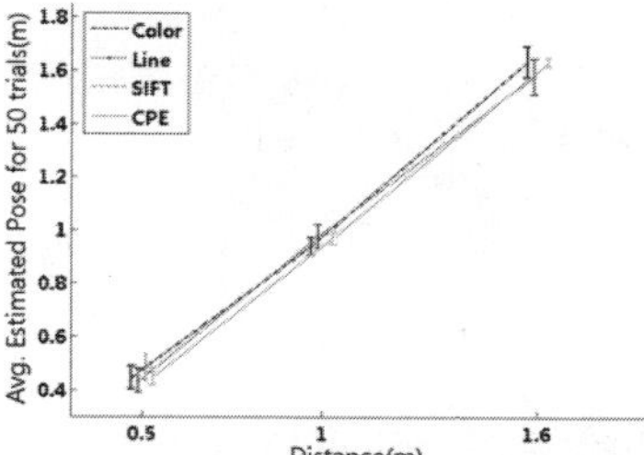

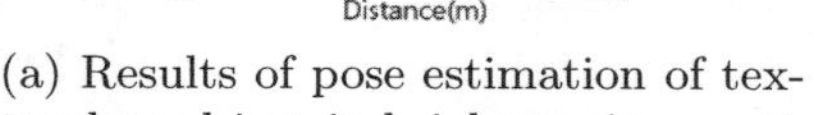

(a) Results of pose estimation of textureless object in bright environment

(b) Results of pose estimation of textureless object in dark environment

Fig. 10. The ARMSE for the textureless object pose in accordance with illumination

experiments in Fig. 9 and Fig. 10 show feasibility an effectiveness of recognition and pose estimation, despite low precision and changeable depth information of stereo camera. But some conditions such as far distance and low illumination are the fact that should be overcome in order to improve the recognition performance. The further robot is from the object, the wider variances are in the result of color and line features, as shown in Fig. 9 and Fig. 10. In dark environment is also challenging in recognition problem. Note that SIFT method cannot recognize target object when robot locates far from object and in low illuminated place in Fig. 9(b) and Fig. 10(b). The results from CPE seem to have better performance in any circumstance. Automatically selected set of features are properly achieved according to the proposed framework.

5   Conclusion

We have concentrated on developing a probabilistic method using multiple evidences based on sequence of images to recognize an object and to estimate its pose. Especially in order to design more systematic framework, we have improved the previous probabilistic method by considering both the ratio of matched features and matched pose error in assigning similarity weight. The proposed method probabilistically represents the recognized object's pose with particles to draw an arbitrary distribution. The particles are updated by consecutive observations in a sequence of images and are converged to a single pose. The proposed method can recognize various objects with individual characteristics because it can incorporate easily multiple features such as photometric features (SIFT, color) and geometric features (line, square) into the proposed filtering framework. We experiment the proposed method with a stereo camera under experimental environment including textured and texture-less objects with not only changes of illumination but also variation of distance from object. The experiment result demonstrates that the proposed method robustly recognizes various objects with individual characteristics such as textured and textureless objects in in-door environments.

References

1. Farias, M.F.S., de Carvalho, J.M.: Multi-view Technique For 3D Polyhedral Object Rocognition Using Surface Representation. Revista Controle & Automacao pp. 107–117 (1999)
2. Shirai, Y.: Three-Dimensional Computer Vision. Springer, New York (1987)
3. Ben-Arie, J., Wang, Z., Rao, R.: Iconic recognition with affine-invariant spectral. In: Proc. IAPR-IEEE International Conference on Pattern an Recognition, vol. 1, pp. 672–676 (1996)
4. Fischler, M.A., Bolles, R.C.: Random sample consensus: A paradigm for model fitting with applications to image analysis and automated cartography. Comm. Assoc. Comp. Mach. 24(6), 381–395 (1981)
5. Olson, C.F.: Efficient pose clustering using a randomized algorithm. International Journal of Computer Vision 23(2), 131–147 (1997)
6. David, P., DelMenthon, D.F., Duraiswami, R., Samet, H.: Softposit: Simultaneous pose and correspondence determination. In: Heyden, A., Sparr, G., Nielsen, M., Johansen, P. (eds.) ECCV 2002. LNCS, vol. 2352, pp. 698–703. Springer, Heidelberg (2002)
7. Johnson, A.E., Hebert, M.: Using spin images for efficient object recognition in cluttered 3D scenes. IEEE Transactions on Pattern Analysis and Machine Intelligence 21(5), 433–449 (1999)
8. Frome, A., Huber, D., Kolluri, R., Bulow, T., Malik, J.: Recognizing Objects in Range Data Using Regional Point Descriptors. In: Pajdla, T., Matas, J(G.) (eds.) ECCV 2004. LNCS, vol. 3024, Springer, Heidelberg (2004)
9. Rothganger, F., Lazebnik, S., Schmid, C., Ponce, J.: 3D Object Modeling and Recognition Using Affine- Invariant Patches and Multi-View Spatial Constraints. In: IEEE Computer Society Conference on Computer Vision and Pattrn Recognition, vol. 2, pp. 272–280 (2003)
10. Lowe, D.: Object recognition from local scale invariant features. In: ICCV 1999. Proc. 7th International Conf. Computer Vision, Kerkyra, Greece, pp. 1150–1157 (1999)
11. Belongie, S., Malik, J., Puzicha, J.: Shape matching and object recognition using shape contexts. IEEE Trans. On Pattern Analysis and Machine Intelligence 24(4), 509–522 (2002)
12. Carmichael, O., Herbert, M.: Shape-Based Recognition of Wiry Object. IEEE Transactions on Pattern Recognition and Machine Intelligence (2004)
13. Thayananthan, A., Stenger, B., Torr, P.H.S., Cipolla, R.: Shape context and chamfer matching in cluttered scenes. In: Proc. IEEE Conference On Computer Vision and Pattern Recognition (2003)
14. Lee, S., Kim, E., Park, Y.: 3D Object Recognition using Multiple Features for Robotic Manipulation. In: IEEE International Conference on Robotics and Automation, pp. 3768–3774 (2006)
15. Olson, C.F.: A probabilistic formulation for Hausdorff matching. In: IEEE Conference on Computer Vision and Pattern Recognition, pp. 150–156 (1998)
16. Subrahmonia, J., Cocper, D.B., Keren, D.: Practical reliable bayesian recognition of 2D and 3D objects using implicit polynomials and algebraic invariants. IEEE Transactions on Pattern Analysis and Machine Intelligence 18(5), 505–519 (1996)

17. Boykov, Y., Huttenlocher, D.P.: A New Bayesian Framework for Object Recognition, 517–523 (1999)
18. Li, S.Z., Hornegger, J.: A two-stage probabilistic approach for object recognition. In: Burkhardt, H., Neumann, B. (eds.) ECCV 1998. LNCS, vol. 1407, pp. 733–747. Springer, Heidelberg (1998)
19. Schiele, B., Crowley, J.L.: Probabilistic object recognition using multidimensional receptive field histograms. In: ICPR (1996)
20. Lee, S., Lee, S., Lee, J., Moon, D., Kim, E., Seo, J.: Robust Recognition and Pose Estimation of 3D Objects Based on Evidence Fusion in a Sequence of Images. In: IEEE International Conference on Robotics and Automation, pp. 3773–3779 (2007)

Preliminary Development of a Line Feature-Based Object Recognition System for Textureless Indoor Objects

Gunhee Kim[1], Martial Hebert[1], and Sung-Kee Park[2]

[1] The Robotics Institute, Carnegie Mellon University, Pittsburgh, PA, 15213, USA
{gunhee, hebert}@cs.cmu.edu
[2] Center for Cognitive Robotics Research, Korea Institute of Science and Technology,
39-1 Hawolgok-dong, Sungbuk-ku, Seoul, 136-791, Korea
skee@kist.re.kr

Summary. This paper presents preliminary results of a textureless object recognition system for an indoor mobile robot. Our approach relies on 1) segmented linear features, and 2) pairwise geometric relationships between features. This approach is motivated by the need for recognition strategies that can handle many of indoor objects that have no little or not textural information on their surfaces, but have strong geometrical consistency within the same object class. Our matching method consists of two steps. First, we find correspondence candidates between linear fragments. Second, a spectral matching algorithm is used to find the subset of correspondences which is the most consistent. Both matching methods are learnt by using logistic classifiers. We evaluated the developed recognition system with our own database, which is composed of eight indoor object classes. We also compared the performance of our line feature based recognition approach with a SIFT feature based method. Experimentally, it turned out that the line features are superior in our problem setup - the detection of textureless objects.

Index Terms: Object recognition, SIFT features, line features, spectral matching, logistic regression classifiers.

1 Introduction

In this paper, we present a recognition system for textureless objects for the navigation of an indoor mobile robot. The goal of this research is defined through the actual development of a service robot. Unfortunately, most objects which a robot encounters in indoor environments are quite bland (i.e., little visual information on their surfaces) such as a refrigerator and a microwave oven. Therefore, it is not easy to capture meaningful information such as color and texture variation, which are most popular clues to be used for object recognition. This makes it difficult to apply conventional local feature descriptors such as SIFT [1] or informative image patches [2]. Fig.1 shows some examples of SIFT [1] feature extraction on our target objects. Few features are detected on the object compared to the background.

S. Lee, I.H. Suh, and M.S. Kim (Eds.): Recent Progress in Robotics, LNCIS 370, pp. 255–268, 2008.
springerlink.com

However, one observation is that most indoor objects consist of several rigid parts and possess relatively consistent geometric relations within the same object class. For example, apparently, almost all refrigerators consist of two rectangles, in which a rectangle is located above another longer one. And, a microwave oven looks like "a big rectangle, inside another one, and some knobs and buttons in the right inside of the big rectangle."

Fig. 1. SIFT [1] feature extraction on our target objects

Based on these observations, our approach mainly uses 1) segmented line-typed features, and 2) pairwise geometric relationships between features. That is, the line type features are applied to get most information from object boundaries. Also, the pairwise geometric relationships enable us to maximally use the inherent rigidity of target objects.

Although many other approaches uses curves to model contours like such as [3][4][5], we exploit segmented lines because most of our target objects tend to contain a lot of polygonal shapes. Moreover, lines are much easier and faster to process in terms of defining geometrical relations or matching. It could be an advantage for a mobile robot which has to operate in real-time. Pairwise geometric relationships are mainly considered in the matching step by a spectral matching technique [6][7]. Basically, it finds out strongly connected (i.e., geometrically consistent) clusters between two feature sets from two images, and those clusters can be thought of matched objects.

We first find candidate correspondences by using a logistic classifier applied to a vector representing the geometric relations between lines. We expect this classifier to 1) improve the recognition performance by quickly finding correct candidates, and 2) significantly reduce the computation time by shrinking the search space of our main matching algorithm - the spectral method [6][7]. The approach described in this paper is inspired from a more general category recognition and learning procedure basic on second-order relations between simple local features (oriented points). The key difference is that we use line segments instead of oriented points, we use additional photometric features (the distribution of color and intensity between lines), and that we do not learn a single model from training images. Instead, we match the features from the input image to the entire collection of features from the training data.

Our matching is composed of two steps. First, the correspondence candidate finder will very quickly identify which features need to be considered in next matching algorithm. A spectral matching algorithm is used to find the subset of correspondences which is the most consistent.

Shape (in our case, in the form of a set of linear fragments) has been widely used as one of important cues for object recognition. Here, some closely related previous approaches are briefly reviewed. The first approach using second-order geometric relationships between features is Berg et al's method [8]. They used geometric blur descriptor as a feature, and formulated shape matching as an integer quadratic programming problem. Carmichael and Hebert [9] developed a recognition algorithm of complex-shaped objects based on edge information. A series of tree-structured classifiers separates an object from their background in a cascade way. Opelt et al [5] proposed a recognition algorithm based on "bag-of-words" approach using boundary fragment models. The object boundaries become codebook entries, and boosting is used to select discriminative combinations of boundary fragments.

Closest to our work is the approach of Leordeanu at al. [6], which uses the same spectral technique as the underlying matching engine and logistic classifiers to estimate the geometric consistency between pairs of features.

The rest of this paper is organized as follows. Section 2 illustrates how to extract line features from an image and how to define pairwise information between them. Section 3 discusses the basics of the matching algorithm. The proposed correspondence candidate finder is explained in detail in section 4. Experimental results are shown in section 5. In particular, we compared the performance of proposed line based method with that of a SIFT based approach. Finally, concluding remarks are given in section 6.

2   Line Features

Extraction of Line Features

First, we perform the Canny edge detection with hysteresis [10] on an image to obtain edge information as shown in Fig.2. The edges are regularly sampled at interval of 30 pixels, and they are approximated into line segments. The reason of sampling is that it is more robust to uncertainties of the edge detection. Moreover, in the matching step, many small lines are more desirable than a single long line because it has more chance that at least parts of segments can be matched.

In order to make a line feature more informative, we compute luminance, color, and texture descriptors in a small region around each line. The luminance descriptor is a histogram of L* values, and the color descriptor is a histogram of a*b* components in L*a*b* color space. The texture descriptor is a 32 dimensional texture histogram based on Leung-Malik filter banks [11]. That is, we make 32 universal textons from training images, and generate normalized histogram of those textons. Experiments showed that the texture descriptor is more discriminative than the luminance and color descriptors.

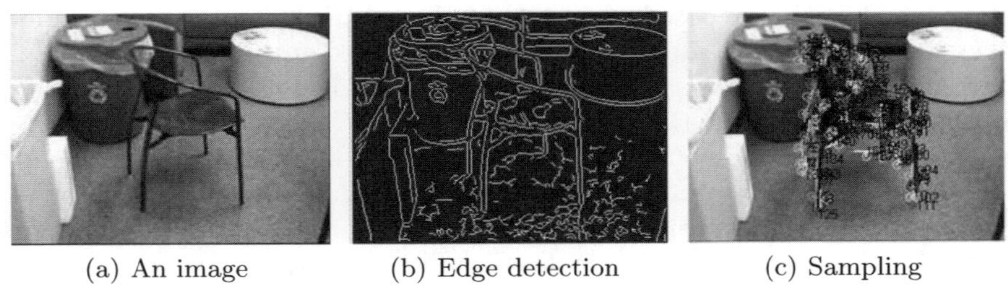

(a) An image   (b) Edge detection   (c) Sampling

Fig. 2. Line feature extraction

Pairwise Information between Lines

Fig.3 shows the pairwise geometric features used in our approach. It consists of an over-complete set of pairwise parameters since, experimentally, over-completeness gives more discriminative power [6].

We also use regional information along the segment d_{ij} between a line pair such as $H_{l_{ij}}$, $H_{ab_{ij}}$, $H_{t_{ij}}$, $V_{l_{ij}}$, and $V_{ab_{ij}}$. The basic assumption is that if a pair of line is on the object of interest, the region along d_{ij} would be located inside the object. Otherwise, most of the region along the d_{ij} would contain background, which is highly likely to show random color and texture information. As an example, Fig.4 compares variations of $V_{l_{ij}}$, that is, the luminance along d_{ij}, positive and negative samples. A positive sample represents two matched pairs of lines from different images such as Fig. 4(a), whereas a negative sample shows

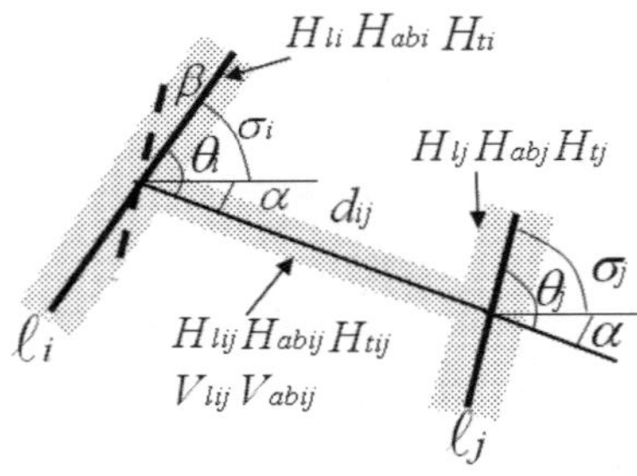

α : the absolute angles of a center line wrt. the absolute horizontal axis
β : the difference between the angles of the line pair
σ : the angles btw. the line pair and a center line
θ : the angles of the line pair wrt. the absolute horizontal axis
ℓ : the lengths of the line pair,
d_{ij} : the distance between centers of the line pair(s : small constant)
$H_{l_i}/H_{ab_i}/H_{t_i}$: the luminance/color/texture descriptors around the line pair
$H_{l_{ij}}/H_{ab_{ij}}/H_{t_{ij}}$: the luminance/color/texture descriptors around the center line
$V_{l_{ij}}/V_{ab_{ij}}$: the luminance/color variation along the center line

Fig. 3. Pairwise relations between a pair of lines

two unmatched pairs like Fig. 4(b). The right picture in Fig. 4(a) shows that the values of $V_{l_{ij}}$ are quite similar across all the positive samples. On the other hand, the variance across a set of negative examples in Fig. 4(b) are random (i.e., highly likely mismatched). That means that $V_{l_{ij}}$ could be a useful piece of information to decide whether the pairs are matched or not.

Since it is computationally burdensome to calculate the regional information along d_{ij} for all possible pairs, we use an approximation method using image segmentation. Given an image, we first over-segment it into tens of regions by using Normalized cuts [12]. For each patch, we generate representative luminance, color, and texture histograms beforehand. Then, when a pair is given, the regional information along d_{ij} such as $H_{l_{ij}}$, $H_{ab_{ij}}$, $H_{t_{ij}}$, $V_{l_{ij}}$, and $V_{ab_{ij}}$ is computed by linear combination of pre-computed color and texture histograms. They are proportionally weighted according to the fraction of the segment d_{ij} occupied by each region.

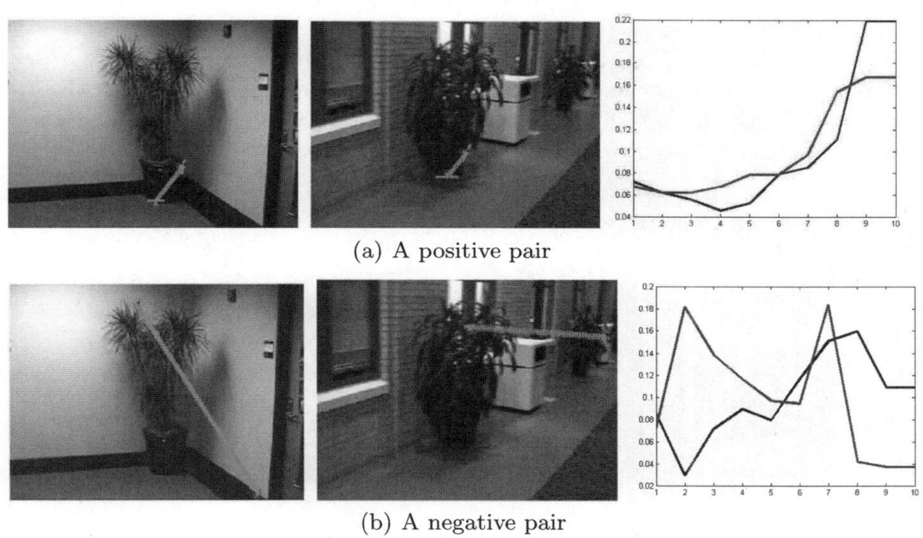

(a) A positive pair

(b) A negative pair

Fig. 4. Luminance variation along the center line between a pair of lines

3   Learning and Matching

Learning Using a Logistic Classifier

The leaning step computes which geometrical information is more important for a classification task using a set of training samples. First, we define a 22 dimensional vector of X in Eq.(1), in which each element describes a piece of geometrical information shown in Fig.3. We learn a pairwise potential G_{ij}, which will be used in the next section to construct the affinity matrix used in the spectral matching. The pairwise potentials G_{ij} are represented in the form of a logistic regression classifier (Eq.(2)). Specifically, the learning step computes the weight vector W in Eq.(2) using positive and negative samples from training images.

$$X = [\, 1\ |\alpha_1 - \alpha_2|\ |\beta_1 - \beta_2|\ |\sigma_{i,1} - \sigma_{i,2}|\ |\sigma_{j,1} - \sigma_{j,2}|\ |\theta_{i,1} - \theta_{i,2}|\ |\theta_{j,1} - \theta_{j,2}|$$
$$|l_{i,1} - l_{i,2}|\ |l_{j,1} - l_{j,2}|\ |d_{ij,1} - d_{ij,2}|\ |1 - (d_{ij,1} + s)/(d_{ij,2} + s)|$$
$$|H_{l_{i,1}} - H_{l_{i,2}}|\ |H_{ab_{i,1}} - H_{ab_{i,2}}|\ |H_{l_{j,1}} - H_{l_{j,2}}|\ |H_{ab_{j,1}} - H_{ab_{j,2}}|$$
$$|H_{t_{i,1}} - H_{t_{i,2}}|\ |H_{t_{j,1}} - H_{t_{j,2}}|\ |H_{l_{ij,1}} - H_{l_{ij,2}}|\ |H_{ab_{ij,2}} - H_{ab_{ij,2}}|$$
$$|H_{t_{ij,1}} - H_{t_{ij,2}}|\ |V_{l_{ij,1}} - V_{l_{ij,2}}|\ |V_{ab_{ij,1}} - V_{ab_{ij,2}}|\,]\tag{1}$$

$$G_{ij} = \frac{1}{1 + \exp(w_0 + \sum_{i=1}^{n} w_i X_i)}\tag{2}$$

We generate positive and negative samples as follows; within each object class, two training images are randomly picked. Then, we manually assigned a pair of corresponding features which represent the same part of the object. These pairs become positive samples, and randomly picked pairs of features are negative samples (Fig.5). That is, it is highly likely that the negative samples represent different parts of the scene. This pairwise classifier is very similar with the one used in [6], with the key difference that we use explicitly the local appearance (luminance, color) in our approach.

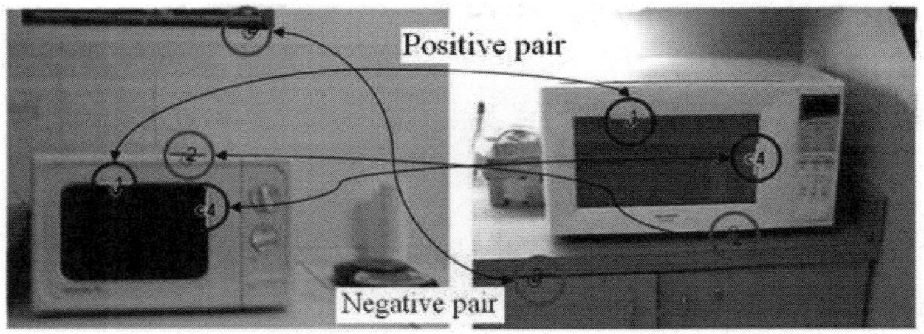

Fig. 5. Generating positive and negative samples

Matching Using a Spectral Technique

Our matching method is based on a spectral technique [6], which finds geometrically consistent correspondences between two sets of features from two images. To apply this method, we need to build an affinity matrix which consists of pairwise potentials G_{ij} for all pairs of features between two images. The diagonal elements of the affinity matrix, G_{ii} is set to 0, and off-diagonal elements are computed by Eq.(2). When an affinity matrix is given, we can easily find the matched cluster (i.e., a subset of geometrically consistent features). The details are introduced in [7].

One of the critical issues in constructing the affinity matrix is how to find corresponding candidates for each feature in a test image. In order to solve this problem, we propose a fast generator of candidate correspondences, which will be discussed in the next section. For sparseness of the affinity matrix, we use a threshold for each element of the vector X. That is, even if one of X_i $(i=0,..21)$ is lower than its corresponding threshold, G_{ij} is just set to 0. The example of this scheme will be illustrated in Fig.8.

4 Fining Candidate Correspondences

The main idea of the candidate correspondence finder is that we use not only the information of a single feature itself but also its local arrangement of neighbors. This strategy is also motivated by one of our key observations for our target objects - consistent geometric relations within the same object class. That is, our underlying assumption is that if a feature in an image is matched to a feature in a different image of the same object class, they should have similar neighborhood as shown in Fig.6. Two local structures in Fig.6 should be similar since they are from the same part of the object class, a microwave oven.

Therefore, the finding candidate correspondences can be formulated as follows: given a pair of features, we need to decide whether these two have a similar local

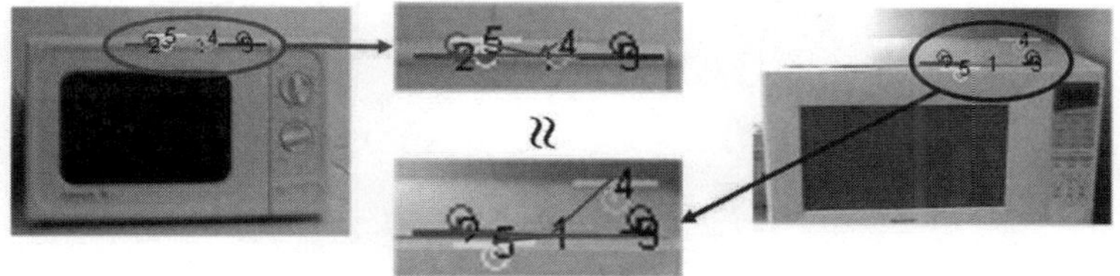

Fig. 6. Using local structure for finding correspondence candidates

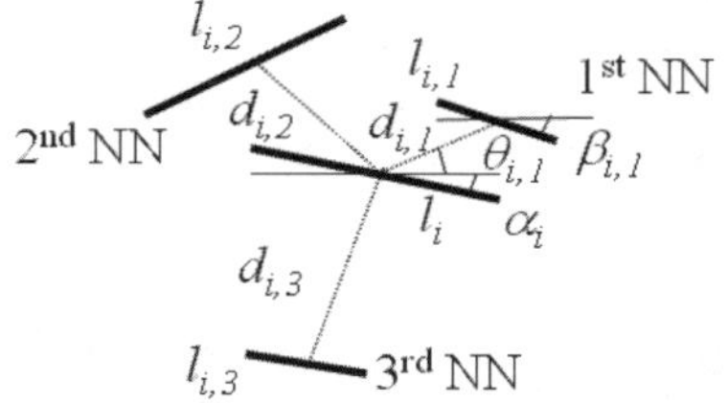

$$\begin{cases} \ell_i : \text{the length of } i\text{-th line feature} \\ \ell_{i,k} : \text{the length of } k\text{-th NN of the } i\text{-th feature} \\ d_{i,k} : \text{the distance between } i\text{-th feature and its } k\text{-th NN} \\ \alpha_i : \text{the absolute angle of the } i\text{-th feature} \\ \beta_{i,k} : \text{the absolute angle of the } k\text{-th NN of the } i\text{-th feature} \\ \theta_{i,k} : \text{the relative angle between } i\text{-th feature and its } k\text{-th NN} \end{cases}$$

Fig. 7. The local geometric structure of a feature

structure or not. It is a binary decision based on several pieces of geometric information derived from the local configuration of the features. Therefore, this setup can be a typical problem to be solved by a logistic classifier (i.e., binary decision based on several continuous random variables.) Since the correspondence finder uses a logistic classifier, its training and classifying framework are very similar to those of the spectral matching algorithm discussed in the previous section. The main difference is that the correspondence finder does not use pairwise information with all other features but only local structural information using its n-nearest neighbors (n is set to 5) in order to guarantee fast processing. It leads to different definition of input regression vector X (see Fig.7). A kd-tree is constructed for each image beforehand in order to easily find the nearest neighbors.

$$X = [\, 1 \; ratio(l_{i,1}, l_{i,2}) \; |\alpha_{i,1} - \alpha_{i,2}| \; \sum_{i=1}^{n} ratio(l_{i,k,1}, l_{i,k,2})/n$$

$$\sum_{i=1}^{n} |\beta_{i,k,1} - \beta_{i,k,2}|/n \; \sum_{i=1}^{n} ratio(d_{i,k,1}, d_{i,k,2})/n \; \sum_{i=1}^{n} |\theta_{i,k,1} - \theta_{i,k,2}|/n$$

$$|H_{l_{i,1}} - H_{l_{i,2}}| \; \sum_{i=1}^{n} |H_{l_{i,k,1}} - H_{l_{i,k,2}}|/n$$

$$|H_{t_{i,1}} - H_{t_{i,2}}| \; \sum_{i=1}^{n} |H_{t_{i,k,1}} - H_{t_{i,k,2}}|/n \,] \tag{3}$$

$$P(Y = 1|X) = \frac{1}{1 + \exp(w_0 + \sum_{i=1}^{n} w_i X_i)}$$

$$P(Y = 0|X) = \frac{\exp(w_0 + \sum_{i=1}^{n} w_i X_i)}{1 + \exp(w_0 + \sum_{i=1}^{n} w_i X_i)} \tag{4}$$

In Eq.(3) subscript 1 indicates the structure from the first image and 2 means the one from the second image. Note that we consider a pair of local structure from a pair of image. $ratio(A, B)$ means A/B if $A > B$, otherwise B/A. H_l in eighth and ninth terms represents the luminance descriptor from small regions around each feature. Similarly, H_t indicates the texture descriptor. They are exactly the same as in Eq.(1).

In learning step, using positive and negative samples, we can compute the parameter vector W in Eq.(4), which constitutes a decision surface to classify them as we did for the learning of the spectral matching.

The basic task of this classifier in the testing step is to find candidate correspondences in the training image for each feature in a novel test image. Thus, we scan all features in training image one by one to see if it can be a candidate or not. We compute Eq.(4) for each pair of features (i.e. one from a test image and the other from a training image), and select it as a candidate correspondence if $P(Y_i = 1|X) > 0.5$.

5 Experiments

Dataset

Our dataset consists of eight object classes, which are *refrigerators, microwave ovens, monitors, phones, printers, sofas, wall clocks, and wiry chairs*. The images are taken in kitchens and hallways at CMU. The database is composed of 40 labeled training images for eight object classes (i.e., five images each object class), and 145 test images.

Training Results

For training, we used 2,335 positive samples and the same number of negative samples. Using the learned parameter vector W of the logistic classifier, training errors of positive and negative samples was 1.4561%, and 3.3409%, respectively. That is, 34 positive samples and 78 negative samples among 2,334 each are misclassified by the learnt classifier.

Fig.8 shows two examples of the distributions of X_6 and X_{16} for both positive and negative samples. X_6 is an example of geometrical pairwise information and X_{16} is an example of regional information around a line feature. The red graphs, which represent negative samples, show a uniform distribution in Fig. 8(a) and a normal distribution in Fig. 8(b). This data shows that negative samples are distributed randomly. The distributions of positive samples are clearly separated from those of the negative samples. The tendency of X_i is that the values of positive samples are a little smaller than those of negative samples. The overlapped region of the two distributions for a single X_i is substantial, but we have 21 dimensions to distinguish them. Each piece of information could be hardly distinguishable, but combining them makes the decision easy. These graphs can give us the information on which X_i has more discriminative power in our training vector X. It is related to value of w_i in that more discriminative X_i has higher absolute value of w_i.

From these graphs, we also derive thresholds for each X_i for sparseness of the affinity matrix. The threshold is set so that 95% of positive samples have the value of X_i below the threshold. In the example of Fig.8, for X_6, the threshold is assigned to 0.6998. During the testing step, if X_6 of a certain pair of features is larger than this threshold, it is rejected irrespective of the value of the other X_i's.

Results of the Candidate Correspondences Finder

Fig.9 shows a result of candidate correspondence finding. The left image is a novel test image, and the right one is a training image. For a feature in a test image, we find all features in a training image whose $P(Y_i = 1|X)$ is larger than 0.5. In this example, we found six correspondence candidates. The values of $P(Y_i = 1|X)$ are [0.9689, 0.9128, 0.9426, 0.9772, 0.9804, 0.9636]. As shown Fig.9, the fourth best correspondence feature (i.e., the cyan colored line in the right image) is the actual correspondence feature to the query in the left image. This

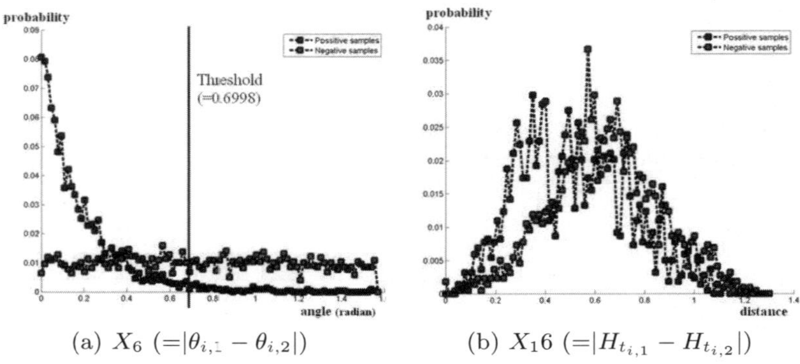

(a) X_6 $(=|\theta_{i,1} - \theta_{i,2}|)$ (b) X_16 $(=|H_{t_{i,1}} - H_{t_{i,2}}|)$

Fig. 8. Distributions of X_i of positive and negative samples

Fig. 9. An example of correspondence candidates

case is successful since the candidates include the true correspondence feature. Even though the other five features are false positives, the spectral matching can easily reject them. In conclusion, we can find not only the correct correspondence candidate but also reduce more searching space than the previous scheme (i.e., just using ten correspondence candidates without any learning).

The learnt correspondence candidate finder has two main advantages over the previous naive method, which is to find the top ten features with most similar luminance and texture descriptors as correspondence candidates. Experiments showed that the recognition rates increased by 6% on average. More importantly, the overall recognition time is dramatically reduced. It takes only 1/6 computation time of the previous scheme on average when matching a pair of images. This is because the candidate finder can largely reduce the search space of the next deliberate spectral matching algorithm.

Comparison with PCA-SIFT

In order to see if linear features are effective in our problem setup, we compared the performance of our line feature based approach with a PCA-SIFT based

method [13]. The same matching algorithm is used for both cases, only difference
is feature types.

Fig.10 shows the recognition rates of two methods for all 145 test images. In
the most object classes, the line feature based approach outperforms the PCA-
SIFT based method. In only the *wall clock* object class, the PCA-SIFT feature
is a little superior to the line feature. The main reason is that a *wall clock* has
enough texture variation on it, for example, numbers and long/short hands in
circular *wall clocks*, to be captured by the PCA-SIFT algorithm as shown in
Fig.11. Another reason is that general indoor scenes have many rectangles made
by walls, doors, or floors (see Fig. 11(b)). This situation could often be confused
with our rectangular objects like *refrigerators*, *monitors*, and *microwave ovens*.
It is one of most common false positive cases in our system, which may be an
inherent limitation of a line based approach.

As shown in Fig.10, the performance of the monitor object class detection
was quite poor. Fig.12 shows what causes the failures of detection of *monitors*.
Apparently, *monitors* are quite similar to other objects which have rectangular
shapes, especially *microwave ovens*. This kind of confusions occurs with other

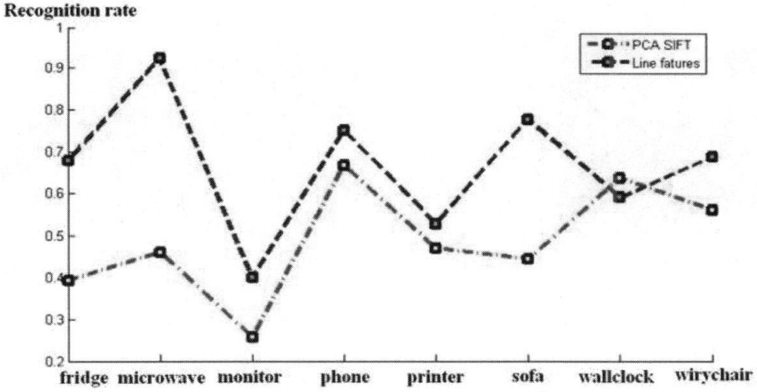

Fig. 10. Comparison of recognition rate between PCA-SIFT [13] and line features

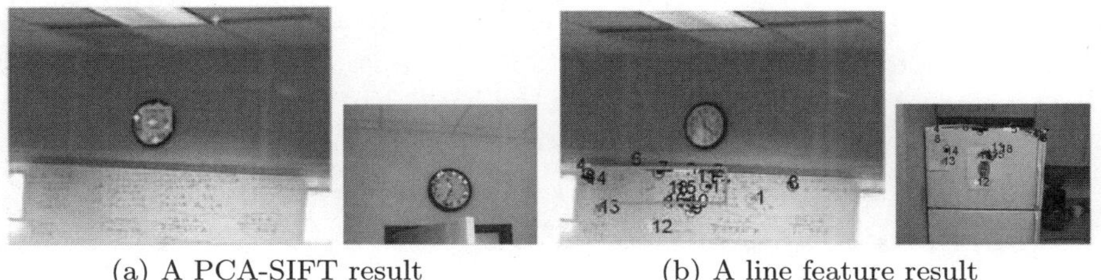

(a) A PCA-SIFT result   (b) A line feature result

Fig. 11. *Wall clock* examples. The left pictures are test images (i.e., novel images),
and the right small pictures are best matched training images.

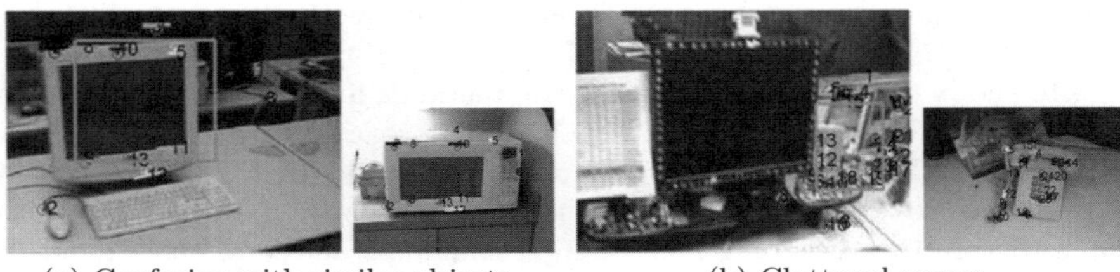

(a) Confusion with similar objects   (b) Cluttered scenes

Fig. 12. *Monitor* examples

(a) Detection of multiple objects (1st best matched: a phone, 2nd best matched: a monitor).

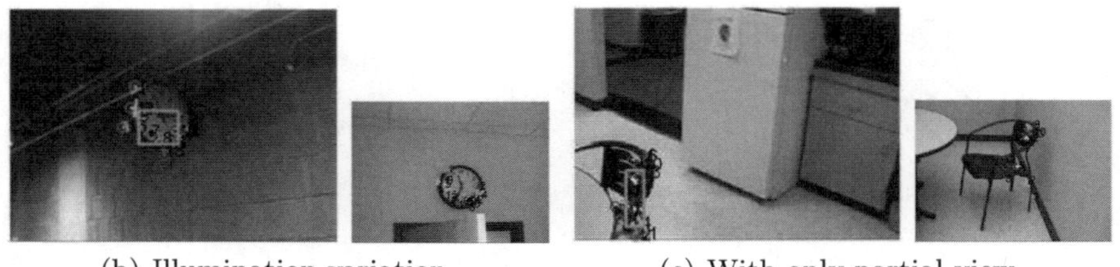

(b) Illumination variation   (c) With only partial view

Fig. 13. Examples of correct matching

pairs of objects such as *sofas/wiry chairs*, and *fridge/printers*. Another explanation is that *monitors* are generally located on the table and other objects on the desktop could distract the system. Fig. 12(b) is a good example of this.

Fig. 13(a) shows an example of multiple object detection. There are a *monitor* and a *phone* in the test scene, and they are successfully detected as the first and second best matches. Also, our recognition system is robust in the variation of object instances and illumination to some extent (13(b)) It is generally successful for scenes with clutter and with only partial view of the target object (13(c)).

6   Conclusions and Future Work

This paper presents preliminary results of a line feature-based object recognition system for textureless indoor objects. We started with very simple linear features to represent object boundaries and we augmented by incorporating geometric consistency such as pairwise relationships between features and a local structure with neighbors.

Our matching approach consists of two steps; 1) find candidate correspondences, and 2) use spectral matching algorithm to find the subset of correspondences which is the most consistent. By integrating them, we can achieve better recognition performance and reduction of recognition time.

We evaluated the developed recognition system with our own database, which is composed of eight indoor object classes. Experiments showed that our recognition system is feasible to the detection of textureless objects.

This preliminary approach is still limited. In particular, we mainly used pairwise relationship and local structures with several neighbors for geometrical consistency. We may infer more valuable information from global structures of objects, for example, in the form of large, graph-structured data. Furthermore, the current system is a complete image based approach. In other words, we do not have a unified or clustered category models but each image becomes an example to be matched. To reduce computation complexity or better performance, we may require a more sophisticated representation method.

Acknowledgement

This research was performed for the Intelligent Robotics Development Program, a 21st Century Frontier R&D Programs funded by the Ministry of Commerce, Industry, and Energy of Korea.

References

1. Lowe, D.G.: Distinctive image features from scale-invariant keypoints. International Journal of Computer Vision 60(2), 91–110 (2006)
2. Vidal-Naquet, M., Ullman, S.: Random subwindows for robust image classification. In: International Conference on Computer Vision (2003)
3. Mikolajczyk, K., Zisserman, A., Schmid, C.: Shape recognition with edge-based features. In: British Machine Vision Conference (2003)
4. Ferrari, V., Tuytelaars, T., Gool, L.V.: Object Detection by Contour Segment Networks. In: European Conference on Computer Vision (2006)
5. Opelt, A., Pinz, A., Zisserman, A.: A Boundary-Fragment-Model for Object Detection. In: European Conference on Computer Vision (2006)
6. Leordeanu, M., Hebert, M., Sukthankar, R.: Beyond Local Appearance: Category Recognition from Pairwise Interactions of Simple Features. In: Conference on Computer Vision and Pattern Recognition (2007)
7. Leordeanu, M., Hebert, M.: Efficient MAP approximation for dense energy functions. In: International Conference on Machine Learning (2006)
8. Berg, A.C., Berg, T.L., Malik, J.: Shape matching and object recognition using low distortion correspondences. In: Conference on Computer Vision and Pattern Recognition (2005)

9. Carmichael, O., Hebert, M.: Shape-based recognition of wiry objects. IEEE Transaction on Pattern Analysis and Machine Intelligence 26(12), 1537–1552 (2004)
10. Canny, J.A.: Computational Approach to Edge Detection. IEEE Transaction on Pattern Analysis and Machine Intelligence 8, 679–714 (1986)
11. Leung, T., Malik, J.: Representing and recognizing the visual appearance of materials using three-dimensional textons. International Journal of Computer Vision 43(1), 29–44 (2001)
12. Shi, J., Malik, J.: Normalized Cuts and Image Segmentation. IEEE Transaction on Pattern Analysis and Machine Intelligence 22(8), 888–905 (2000)
13. Ke, Y., Sukthankar, R.: PCA-SIFT: A More Distinctive Representation for Local Image Descriptors. In: International Conference on Computer Vision (2004)

Modelling of Second Order Polynomial Surface Contacts for Programming by Human Demonstration

Peter Slaets, Wim Meeussen, Herman Bruyninckx, and Joris De Schutter

Department of Mechanical Engineering, Katholieke Universiteit Leuven, Celestijnenlaan 300B, B-3001 Leuven, Belgium
peter.slaets@mech.kuleuven.be

Summary. This paper presents a contribution to *automatic model building* in *quadratic polynomial* environments, in the context of programming by human demonstration. A human operator moves a demonstration tool equipped with a probe, in contact with an unknown environment. The motion of the demonstration tool is sensed with a 3D camera, and the interaction with the environment is sensed with a force/torque sensor. Both measurements are uncertain, and do not give direct information about the different objects in the environment (such as cylinders, spheres, planes, ...) and their geometric parameters. This paper uses a Bayesian Sequential Monte Carlo method or *particle filter*, to recognize the different discrete objects that form the environment, and simultaneously estimate the continuous geometric parameters of these different quadratic polynomial objects. The result is a complete geometric model of an environment, with different quadratic polynomial objects at its building blocks. The approach has been verified using real world experimental data, in which it is able to recognize three different unknown quadratic polynomial objects, and estimate their geometric parameters.

Keywords: model building, compliant motion, Bayesian estimation, particle filter, human demonstration, task segmentation, polynomial surfaces.

1 Introduction

Compliant motion refers to robot tasks where an object is manipulated in contact with its environment, such as for example an assembly task [1]. One of the challenges in compliant motion is the robust specification of a task, because compliant motion tasks often include many simultaneous contacts and many contact transitions [2, 3, 4]. Therefore, many task specification methods in compliant motion require a geometric model of the environment where the task will be executed [5]. This paper presents a method to automatically build a geometric model of an environment. While previous work was limited to polyhedral environments in terms of vertices, edges and faces [6], this paper extends the possible building blocks of the environment to quadratic polynomial objects such as spheres, cylinders, ellipsoids, ... While this previous approach was based on multiple Kalman filters (one filter for each possible contact formation), this

S. Lee, I.H. Suh, and M.S. Kim (Eds.): Recent Progress in Robotics, LNCIS 370, pp. 269–282, 2008.

papers uses a single particle filter. A particle filter can deal with a hybrid (partly discrete and partly continuous) state, and therefore is able to simultaneously distinguish between different discrete quadratic polynomial objects, and estimate the continuous geometric parameters of these polynomials, using only a single filter.

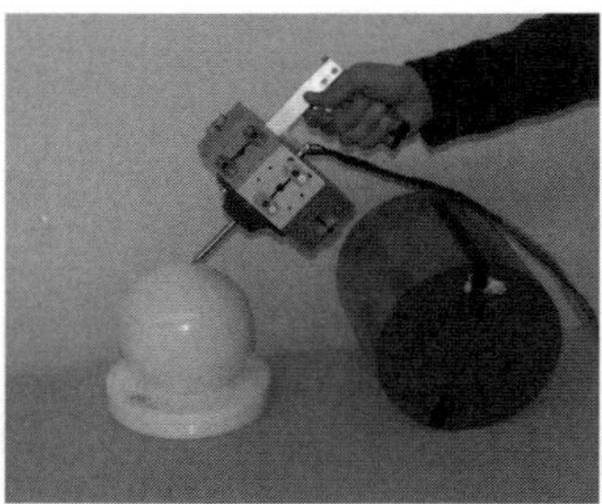

Fig. 1. Using a demonstration tool equipped with pose and force/torque measurements, a probe makes contact with different quadratic polynomial objects in the environment

The paper is organized as follows. First Section 2 presents the definitions of the reference frames together with the demonstration tool which is used to collect sensor data during human demonstration in compliant motion. Section 3 describes the concept of contact formations with polyhedral objects and the extension to contact formations of a probe with quadratic polynomial objects. The Bayesian interpretation of the sensor data obtained from the demonstration tool is covered in Section 4, where the different environmental objects identified and estimated. Section 5 describes the real world experiment that validates the presented approach. Finally, Section 6 contains conclusions and future work.

2   Definitions

In programming by human demonstration for compliant motion tasks, a human uses a demonstration tool to demonstrate a task in which an object moves in contact with its environment. This demonstration phase however, often requires a geometric environmental model. This paper presents an approach to use the demonstration tool to build a geometric environmental model prior to the task specification phase, based on the sensor measurements of the demonstration tool. These sensor measurements are defined with respect to specific reference frames as discussed in Subsection 2.1. Subsection 2.2 describes the data collection of the wrench and pose sensors mounted on the demonstration tool, and Subsection 2.3 discusses the unknown geometric parameters and the parameters describing the different discrete objects that can be distinguished in the environment.

2.1   Frames

The following reference frames (see Figure 2) are considered: $\{w\}$ attached to the world, $\{t\}$ attached to the demonstration tool, $\{c\}$ attached to the contact point and $\{e\}$ attached to the environment object. The position and orientation

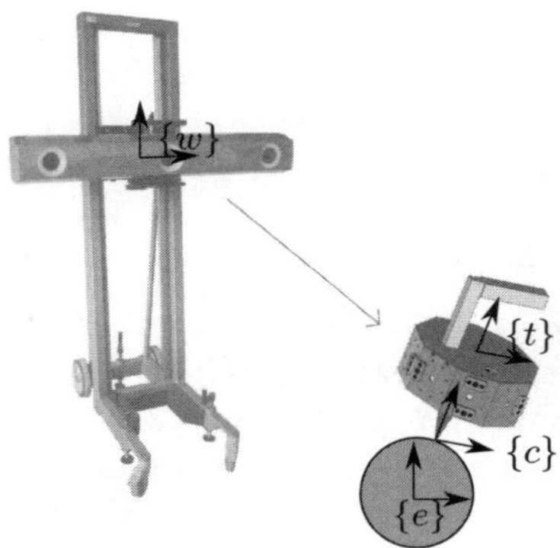

Fig. 2. The reference frame on the camera system $\{w\}$, demonstration tool $\{t\}$, contact point $\{c\}$ and the environmental object $\{e\}$

of a reference frame, relative to another reference frame, is called a pose. In this paper a pose is described by a homogeneous transformation matrix (T_a^b), which represents the position and orientation of a reverence frame a relative to a reference frame b:

$$P_a^b = \begin{bmatrix} R_a^b & T_a^b \\ 0 & 1 \end{bmatrix}, \tag{1}$$

where T_a^b represents the position vector from a to b, and R_a^b represents the orientation matrix between a and b.

2.2   Demonstration Tool

Figure 1 shows the demonstration tool during the real world experiment. A handle on top of the demonstration tool provides an easy grasp for the human demonstrator to manipulate the demonstration tool and the object attached to it. The demonstration is a hollow cylinder-like shape, consisting of nine faces in 40 [*degrees*] increments. On each of the faces, up to four LED markers can be mounted. Inside the demonstration tool, a *JR3* wrench sensor is mounted between the demonstration tool and the manipulated object, in order to measure the wrench $\boldsymbol{w}_m$ applied by the human demonstrator to the manipulated object. The wrench measurements are expressed with respect to the $\{t\}$ frame:

$$\boldsymbol{w}_m = \begin{bmatrix} F_x & F_y & F_z & \tau_x & \tau_y & \tau_z \end{bmatrix}^T . \tag{2}$$

where F denotes a linear force, and τ is a moment.

The pose ($\boldsymbol{P}_m$) of the demonstration tool, expressing the position and orientation of the $\{t\}$ frame with respect to the $\{w\}$ frame, is measured indirectly by the Krypton K600 6D optical system (see Figure 2). The K600 system measures the spatial positions of LEDs attached to the demonstration tool at 100 $[Hz]$, with a volumetric accuracy of 90 $[\mu m]$. Rutgeerts e.a. [7] combined the knowledge of the positions of the visible LED markers, to calculate the pose of the demonstration tool, by solving an over-constraint linear system using a least square technique.

2.3   Unknown Geometric Parameters

Three types of uncertain geometric parameters are considered in this paper:

- the geometric parameters describing the pose $\boldsymbol{P}_w^e$ of the reference frame $\{e\}$ with respect to the reference frame $\{w\}$
- the geometric parameters describing the polynomial geometry $\boldsymbol{G}_p^e$ of the environmental object with respect to the reference frame $\{e\}$, and
- the geometric parameters describing the position $\boldsymbol{G}_c$ of the contact point with respect to the reference frame $\{t\}$.

All these parameters are time-invariant with respect to their corresponding reference frame. They are called geometric parameters denoted by $\boldsymbol{\Theta}$:

$$\boldsymbol{\Theta} = \left[\, (\boldsymbol{P}_w^e)^T \; (\boldsymbol{G}_p^e)^T \; (\boldsymbol{G}_c)^T \,\right]^T . \tag{3}$$

3   Modelling Polyhedral and Quadratic Polynomial Contact Situations

A general contact model, connecting the measurements of the demonstration tool ($\boldsymbol{P}_m, \boldsymbol{w}_m$) with the unknown geometric parameters ($\boldsymbol{\Theta}$), consists of two types of equations. The first equation relates the manipulated measured object pose ($\boldsymbol{P}_m$) to the geometric parameters ($\boldsymbol{\Theta}$), and is called the *Closure equation*:

$$\boldsymbol{r}_t = h_d(\boldsymbol{\Theta}, \boldsymbol{P}_m) \tag{4}$$

The second equation, called the *consistency based wrench residue equation*, expresses the part of the measured wrench ($\boldsymbol{w}_m$) that is not explained by ideal frictionless contact, as a function of the unknown geometric parameters ($\boldsymbol{\Theta}$) and the measured object pose ($\boldsymbol{P}_m$):

$$\boldsymbol{r}_{6x1} = \left[\, \boldsymbol{W}(\boldsymbol{\Theta}, \boldsymbol{P}_m)\phi - \boldsymbol{w}_m \,\right]^T . \tag{5}$$

The residue should vanish when the wrench measurements and the frictionless contact model are consistent. For a given $\boldsymbol{P}_m$ and discrete object, the first order kinematics are represented by a wrench space $\boldsymbol{W}(\boldsymbol{\Theta}, \boldsymbol{P}_m)$. This wrench space contains all possible wrenches that can be applied between the contacting object at the current pose, and is spanned by the column vectors of $\boldsymbol{W}$. Once a spanning set is selected, every wrench can be represented by a coordinate vector ϕ describing the dependency of the measurement on the spanning set. This coordinate vector is derived using a weighted least square:

$$\phi = \left[(W(\Theta, P_m))^{\dagger \kappa} w_m \right]^T . \tag{6}$$

The operator $^{\dagger \kappa}$ represents the Moore Penroose weighted pseudo-inverse [8] of a matrix using a weighting matrix K,

3.1 Polyhedral Contact Situations

Contact Description

The notion of *principal contacts* (PCs) was introduced in [9] to describe a contact primitive between two surface elements of two polyhedral objects in contact, where a surface element can be a face, an edge or a vertex. Figure 3 shows the six non-degenerate PCs that can be formed between two *polyhedral* objects. Each non-degenerate polyhedral PC is associated with a *contact plane*, defined by a contacting face or the two contacting edges at an edge-edge PC. A general

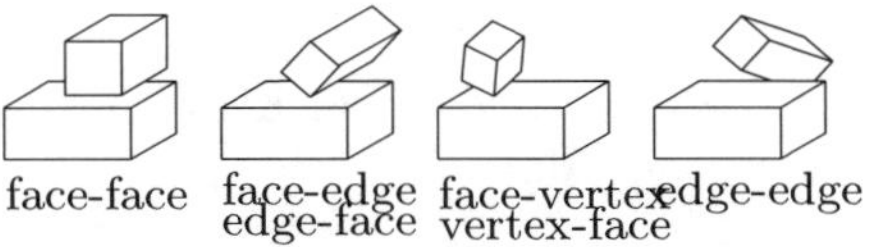

Fig. 3. The six possible non-degenerate principal contacts (PCs) between two polyhedral objects

contact state between two objects can be characterized topologically by the set of PCs formed, called a *contact formation* (CF). A PC can be decomposed into one

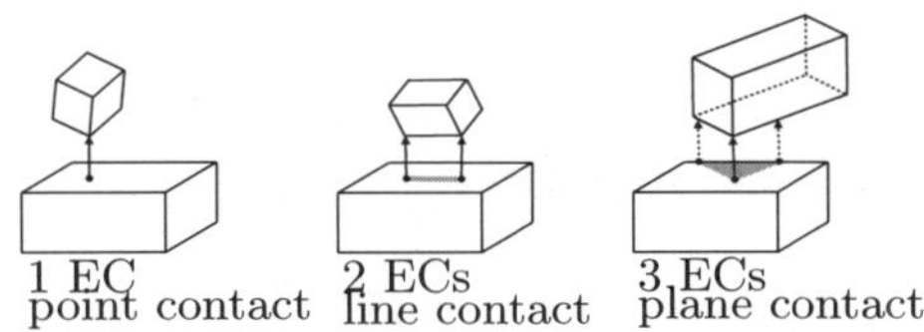

Fig. 4. A principal contact (PC) can be decomposed into one or more elementary contacts (ECs), which are associated with a contact point and a contact normal. The dotted arrows indicate the edge-edge ECs, and the full arrows indicate the vertex-face or face-vertex ECs.

or more *Elementary Contacts* (ECs), providing a lower level description of the CF, as shown in Figure 4. The three types of ECs (face-vertex, vertex-face and edge-edge) are shown in the two examples at the right of Figure 3. The building blocks for polyhedral ECs are a plane, a vertex and an edge. In subsection 3.2 the ECs are extended to a group of quadratic contacts,leading to additional building blocks: an ellipsoid, a hyperboloid, a cone and a cylinder.

Contact Modelling

Polyhedral contact models for the different ECs (face-vertex, vertex-face, edge-edge) where derived by Lefebvre [10]. Lefebvre describes each EC contact model by:

- a frame $\{c\}$ on the contact point,
- a wrench spanning set $\boldsymbol{W}$ expressed in this frame, and
- a signed distance function $(\boldsymbol{r}_t)$ between the two contacting polyhedral primitives.

A closure equation (4) and wrench residue equation (5) can be derived for every separate polyhedral EC. For example, a vertex-face contact consists of the parameters of the vertex $(t_p^{t,c})$ expressed in the reference frame $\{t\}$ and the parameters of the face (a normal n_e and a height d) expressed with respect to the reference frame $\{e\}$. To calculate the residue equation, all parameters need to be expressed in the same reference frame. Therefore a transformation between the reference frame $\{t\}$ and the reference frame $\{e\}$ is necessary.

$$_e p^{e,c} = T_e^t(\boldsymbol{P})_t p^{t,c} \tag{7}$$

This transformed vector $(_e p^{e,c})$ is used to calculate the signed distance function between the vertex and the face, defined by:

$$\boldsymbol{r}_t = \frac{(n_e)_e^T p^{e,c} - d}{\sqrt{n_e^T n_e}} \tag{8}$$

Similar reasoning is applicable to the wrench spanning set, where a screw transformation [11] is needed to transform $\boldsymbol{W}$ from the $\{e\}$ frame to the $\{t\}$ frame.

$$\boldsymbol{W}^e(\boldsymbol{\Theta}) = \begin{bmatrix} n_e \\ 0_{3x1} \end{bmatrix}. \tag{9}$$

3.2   Quadratic Contact Situations

Contact Description

In previous research the ECs where described using only three polyhedral building blocks: a vertex, face and an edge. An extension of these building blocks by second order contact surfaces, allows the description of ECs between a vertex and a quadratic polynomial surface [12]. The closed form analytical expression for quadratic polynomial surfaces expressed in a reference frame $\{w\}$, corresponds to a quadratic polynomial equation in three variables:

$$f(\boldsymbol{X}) = \boldsymbol{X}^T \Delta \boldsymbol{X} + 2C^T \boldsymbol{X} + a_{44} = 0 \tag{10}$$

where Δ is a symmetric 3×3 parametric matrix , $\boldsymbol{X}$ represents the contact point expressed in a $\{w\}$ reference frame, a_{44} is a scalar and C is a vector of parameters.

Equation (10) is reduced to a standard minimal form by proper choice of a new reference frame $\{e\}$. A standard minimal description leads to an easier classification and parameterization of the different surfaces. The transformation of the reference frame consist of a translational part $(\boldsymbol{T}_e^w)$ and a rotational part $(\boldsymbol{R}_e^w)$. The translation of the $\{w\}$ frame to the center of the polynomial results in the disappearance of the linear term C in equation (10). The rotation of the reference frame results in diagonal matrix Δ only if the rotation matrix $(\boldsymbol{R}_e^w)$ satisfies following constraint:

$$\Lambda = (\boldsymbol{R}_e^w)^{-1}\Delta(\boldsymbol{R}_e^w) = (\boldsymbol{R}_e^w)^T\Delta(\boldsymbol{R}_e^w). \tag{11}$$

The resulting transformation $(\boldsymbol{P}_e^w)$ reduces the polynomial equation (10) to a simplified quadratic form:

$$\boldsymbol{X}' = {}_e^w T\boldsymbol{X} \tag{12}$$

$$f(\boldsymbol{X}') = \boldsymbol{X}'^T\Lambda\boldsymbol{X}' + \delta = 0 \tag{13}$$

with α, β, γ the diagonal elements of Λ. The geometric parameters $(\boldsymbol{G}_p^e)$ describing a vertex-polynomial contact are:

- six parameters $(\boldsymbol{P}_e^w)$ representing the transformation of the reference frame from $\{w\}$ to $\{e\}$;
- four parameters $([\alpha\ \beta\ \gamma\ \delta])$ representing the polynomial object ;
- three parameters representing the contact point $(\boldsymbol{X}')$.

To reduce the complexity of this model some assumptions where made: (i) only quadratic polynomial having a *finite center* are considered, therefore the elliptic paraboloid and the hyperbolic paraboloid are omitted; (ii) Only *real surfaces* are considered, therefore all imaginary surfaces are omitted. Taking these assumptions into account equation (12) can be used to represent the six different quadratic surfaces shown in Figure 5.

Contact Modelling

Like polyhedral contacts models , quadratic polynomial ECs are described by:

- a reference frame $\{t\}$ on the demonstration tool,
- a wrench spanning set $\boldsymbol{W}$ expressed in this frame, and
- a signed distance function($\boldsymbol{r}_t$) between the contact point and the environmental object.

The distance from a point p to a quadratic surface defined by equation (12) is equivalent to the distance from the closest point p_{cl} on the surface to p, where $p - p_{cl}$ is the normal to the surface. Since the surface gradient $\nabla f(p_{cl})$ is also normal to the surface, a first algebraic condition for the closest point is

$$p - p_{cl} = \lambda(\nabla f(p_{cl})) = \lambda(2\Lambda p_{cl}) \tag{14}$$

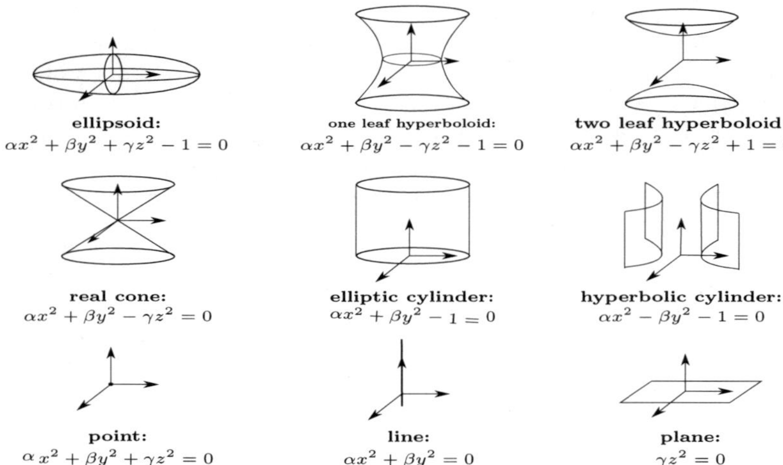

Fig. 5. An overview of the Euclidean classification of real quadratic surfaces having a finite center. The parameters α,β,γ are real and positive, δ equals -1, 0 or 1.

$$p_{cl} = (I + 2\lambda\Lambda)^{-1}p \tag{15}$$

for some scalar λ and I the identity matrix. A second algebraic condition states that the closest point p_{cl} needs to be a point on the quadratic polynomial surfaces. Substitution of $\boldsymbol{X}'$ by p_{cl} (14) in equation (12) leads to a sixth order polynomial constraint in λ:

$$f_c(\lambda) = ((I + 2\lambda\Lambda^{-1})p)^T \Lambda (I + 2\lambda\Lambda)^{-1}p + d = 0 \tag{16}$$

The smallest root ($\lambda = \lambda_{min}$) that complies with this constraint (16) is computed using the root-finding bisection algorithm [13]. This algorithm converges only linearly, but guarantees convergence. Super-linear methods like Newton-Raphson do not guarantee convergence. The algorithm we apply works by repeatedly dividing an interval in half and selecting the subinterval in which the root exists. A subinterval is determined by $\lambda = 0$ and an outer limit λ_{out}. This outer limit is determined by approximating $f_c(\lambda)$ around $\lambda = 0$ by its most dominant term:

$$f_c(\lambda) \approx \tag{17}$$
$$p_{dom}^T (1 + 2\lambda\Lambda_{dom})^{-1} \Lambda_{dom} (1 + 2\lambda\Lambda_{dom})^{-1} p_{dom} + d$$

with scalar Λ_{dom} an element of the diagonal of Λ and scalar p_{dom} is its corresponding direction coordinate. This second order equation (17) is bounded by

$$\lambda_{out1} = -\frac{1}{2\Lambda_{dom}} \quad \text{and} \quad \lambda_{out2} = \frac{\sqrt{(\frac{-\lambda_{dom}p_{dom}^2}{d}) - 1}}{2\lambda_{dom}}. \tag{18}$$

The resulting bounding box is $[0 \ \lambda_{out1}]$ when $f_c(\lambda = 0)$ and λ_{out1} have a different sign, otherwise the bounding box is $[0 \ \lambda_{out2}]$.

The Euclidean distance from a point to a polynomial is therefore calculated as follows[1]:

$$r_t(p, p_{cl}) = \lambda_{min} \ \| \nabla f(p_{cl}) \| \tag{19}$$

The wrench spanning set for a quadratic polynomial contact is defined by the direction of the normal in the contact point (p_{cl}).

$$G^c = \begin{bmatrix} n = \nabla f(p_{cl}) \\ 0_{3x1} \end{bmatrix} \tag{20}$$

4 Simultaneous Recognition of Discrete Objects and Estimation of Continuous Geometric Parameters

The identification of the surrounding environment is performed by gathering information $(\boldsymbol{P}_m, \boldsymbol{w}_m)$ of a human exploring the environment using a demonstration tool. This demonstration can be segmented into a contact sequence with different objects. At each different contact, the same analytical contact constraint applies (equation 16), only the values of the uncertain geometric parameters of the objects involved in the compliant motion task differ. To include non contacting discrete situation, the contact model needs to be extended (see Section 4.1). This results in a parameter space consisting of a discrete parameter representing the different discrete objects $(\mathcal{O})$, including the non contacting state, and a continuous parameter space $(\boldsymbol{\Theta})$ representing the union of subspaces (Ω) corresponding with the geometry of the discrete objects. The estimation problem consists of two connected sub-problems: the recognition of the (discrete) object and the estimation of (continuous) geometric parameters. This simultaneous recognition is performed using a hybrid *Probability Density Function* (PDF), shown in Figure 23. In this paper particle filters [14] are used to implement this hybrid estimation problem due to their ability to cope with multi-modal PDFs. The particle filter algorithm (Section 4.2) updates the hybrid PDF, in a two step procedure. First the system model makes a prediction of the geometric parameters that are related to the different objects. In a second step the measurement model corrects this prediction based on sensor data.

4.1 Extended Contact Model

The contact model of equations (4) and (5) is extended to detect loss of contact during a demonstration. In equation (4) the wrench measurements are represented by a coordinate vector ϕ describing the dependency of the measurement on the spanning set $\boldsymbol{W}$. If $\boldsymbol{W}$ is orthonormal, then the coordinate vector ϕ is

---

[1] When $d = 0$, λ becomes ∞, substitution into equation (19) and evaluating this limit results in $r_{v-p}(p, p_{cl}) = \| p \|$

proportional to the applied contact force. Therefore ϕ indicates when a contact is established:

$$P(\mathcal{O}_k = \text{no contact}) = \text{N}(\phi; 0, \text{R}) \tag{21}$$

where $N(\phi; 0, R)$ represents a normal PDF in ϕ with mean 0 and where the covariance R corresponds to the noise on the wrench sensor measurements. The extended contact model (21) together with the previous contact models (4),(5) define the 'likelihood' of a measurement z_k, given that the time-invariant continues geometric parameters $\boldsymbol{\theta}$ belong to the subspace Ω_k corresponding with the one-dimensional discrete object $\mathcal{O}_k = j$ at timestep k :

$$P\left(z_k \mid \boldsymbol{\theta} \in \Omega_j, \mathcal{O}_k = j\right). \tag{22}$$

The result of the estimation process is a hybrid PDF expressing the believe that the demonstrator is in contact with the environment (or not) ($\mathcal{O}$) is quantified together with the estimation of the geometric parameters ($\boldsymbol{\theta}$).

$$P\left(\boldsymbol{\Theta} = \boldsymbol{\theta}, \mathcal{O}_k = j \mid \boldsymbol{Z}_{1...k} = z_{1...k}\right).^2 \tag{23}$$

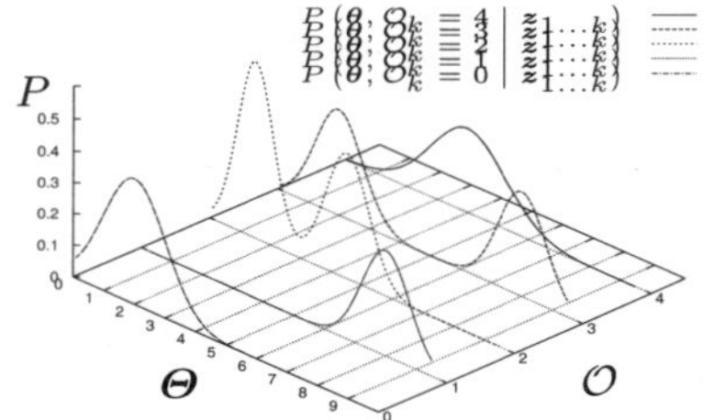

Fig. 6. An example of a probability density function (PDF) of a hybrid joint density, with a one-dimensional continuous geometric parameter $\boldsymbol{\Theta}$ and a one-dimensional discrete state $\mathcal{O}$

4.2   Particle Filter

The particle filters estimation process consists of a two step procedure: a prediction step followed by a correction step.

System Model –Prediction

A prediction step uses a system model to make a prediction for the hybrid joint density at time step k, given the hybrid joint density at time step $k - 1$:

---

² In the rest of the paper, the notation $\boldsymbol{A} = \boldsymbol{a}$ is shortened into $\boldsymbol{a}$, wherever the distinction between a stochastic variable and an actual value is unambiguous.

$$P\left(\boldsymbol{\theta}, \mathcal{O}_k = j \mid \boldsymbol{z}_{1\ldots k-1}\right) =$$
$$\mu P\left(\boldsymbol{\theta}, \mathcal{O}_{k-1} = j \mid \boldsymbol{z}_{1\ldots k-1}\right) + (1-\mu)P_{init}\left(\boldsymbol{\theta}, \mathcal{O}_{k-1} = j\right). \tag{24}$$

This prediction step states that if there is consistency between the measurements $(\boldsymbol{z}_{1\ldots k-1})$ and the current parameter space $(\boldsymbol{\theta})$, the hybrid PDF is unchanged $(\mu = 1)$. Otherwise the estimated PDF is reset to its initial distribution (P_{init}). The consistency indicating parameter is derived by integrating the likelihood of equation (22) over the parameter space $\boldsymbol{\Theta}$ and the state space $\mathcal{O}_k$.

$$\mu \;=\; 1 \text{ when } \sum_{j=1}^{N} \int P\left(\boldsymbol{z}_{\mathrm{k}} \mid \boldsymbol{\theta}, \mathcal{O}_{\mathrm{k}} = \mathrm{j}\right) \mathrm{d}\boldsymbol{\theta} > \mathrm{T}_2$$
$$\text{else } \mu = 0 \tag{25}$$

Where T_2 is a threshold dependent on the dimension of the model equations.

Measurement Model – Correction

Every correction step uses the likelihood model to calculate the hybrid joint posterior density over the state and parameter vector at time step k, given the prediction based on the measurements until timestep $k-1$:

$$P\left(\boldsymbol{\theta}, \mathcal{O}_k = j \mid \boldsymbol{z}_{1\ldots k}\right) =$$
$$P\left(\boldsymbol{z}_k \mid \boldsymbol{\theta}, \mathcal{O}_k = j\right) P\left(\boldsymbol{z}_k \mid \boldsymbol{z}_{1\ldots k-1}\right) P\left(\boldsymbol{\theta}, \mathcal{O}_k = j \mid \boldsymbol{z}_{1\ldots k-1}\right). \tag{26}$$

5   Experiments

This section reports on the real world experiment to validate the presented approach. In the experiment, a human demonstrator manipulates a pin through

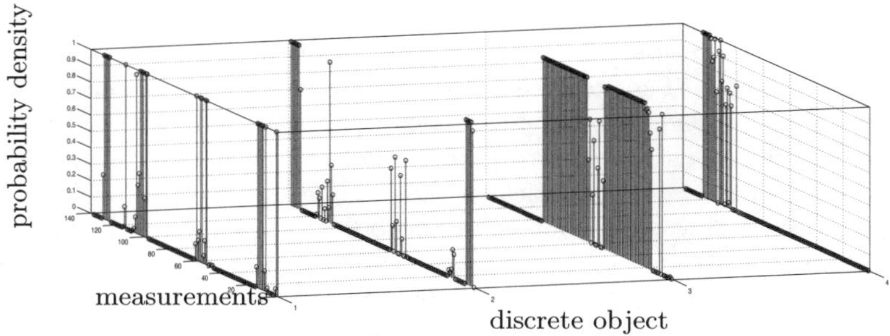

Fig. 7. This figure shows the evolution of the estimated probability on each of the four different discrete objects. The shown probabilities are obtained by integrating over the continuous parameters. The time evolution of the different discrete objects is shown in table 8.

a sequence of contacts in an environment consisting of different discrete objects: a plane table, a cylinder and a sphere. Figure 1 shows the experimental setup and table 8 show the chosen uniform uncertainty on the elements of the 13-dimensional continuous geometric parameters (Θ) and the discrete parameter ($\mathcal{O}$).

parameter	no contact	plane	cylinder	sphere
$(p_x)_w^e$	-	0[mm]	200[mm]	160[mm]
$(p_y)_w^e$	-	0[mm]	0[mm]	160[mm]
$(p_z)_w^e$	-	160[mm]	200[mm]	160[mm]
$(\theta_x)_w^e$	-	0.2[rad]	0.2[rad]	0[rad]
$(\theta_y)_w^e$	-	0.2[rad]	0[rad]	0[rad]
$(\theta_z)_w^e$	-	0[rad]	0.2[rad]	0[rad]
$r = \sqrt{d}$	-	0[mm]	100[mm]	100[mm]
$(p_z)^c$	-	100[mm]	100[mm]	100[mm]
$\mathcal{O}$	1	2	3	4

Fig. 8. The sequence of contacts of the pin at the demonstration tool with different discrete objects

The Gaussian PDF on the wrench residue has a sigma boundary of 3.0 $[N]$ for the forces and 0.33 $[Nm]$ for the torques, while the The Gaussian PDF on the distance at an EC has a sigma boundary of 0.0025 $[m]$. The particle filter uses 80.000 particles, and the joint posterior PDF is dynamically re-sampled using importance sampling, once the effective number of particles drops below 40.000.

5.1   Sequence of Contact Formations

Figure 7 shows the evolution of the estimated probability on each of the four different discrete objects (no contact, plane contact, cylinder contact and sphere contact). Each discrete object is represented by the value of the discrete state as indicated in table 8, and its probability is obtained by integrating over the continuous parameter. The demonstrator subsequently makes contact with a plane, a cylinder, a sphere and then returns in contact with the initial plane. These contact situations are interrupted by non contacting situations. During most of the experiment, the particle filters assign the highest probability to the discrete object that corresponds to the objects that is touched with the demonstration tool. At some points during the experiment, a (false) contact loss is detected because the contact force became too low.

6   Conclusions

This paper present an extension of the polyhedral contact modelling to second order curvatures represented by a polynomial equation. Because the extended model is non-implicit and non-linear stochastical linear filtering techniques like Kalman filters, e.a. are the wrong tools. A particle filter, which uses a discrete sampled representation of a PDF, overcomes these problems and further makes it possible to compare the probability of different discrete objects. Future work will focus on the extension of the particle to cope with even larger uncertainties by taking 100 times more samples at a reasonable timing cost. This acceleration can be achieved by developing more specific hardware to speed up mathematical calculation by manually developing a parallel pipelined hardware implementation on a field-programmable gate array (FPGA).

Acknowledgment

All authors gratefully acknowledge the financial support by K.U.Leuven's Concerted Research Action GOA/05/10.

References

1. De Schutter, J., Van Brussel, H.: Compliant Motion I, II. Int. J. Robotics Research 7(4), 3–33 (1988)
2. Mason, M.T.: Compliance and force control for computer controlled manipulators. IEEE Trans. on Systems, Man, and Cybernetics SMC-11(6), 418–432 (1981)
3. Kröger, T., Finkemeyer, B., Heuck, M., Wahl, F.M.: Compliant motion programming: The task frame formalism revisited. J. of Rob. and Mech. 3, 1029–1034 (2004)
4. De Schutter, J., De Laet, T., Rutgeerts, J., Decré, W., Smits, R., Aertbeliën, E., Claes, K., Bruyninckx, H.: Constraint-based task specification and estimation for sensor-based robot systems in the presence of geometric uncertainty. Int. J. Robotics Research 26(5), 433–455 (2007)
5. Meeussen, W., Rutgeerts, J., Gadeyne, K., Bruyninckx, H., De Schutter, J.: Contact state segmentation using particle filters for programming by human demonstration in compliant motion tasks. IEEE Trans. Rob. 23(2), 218–231 (2006)
6. Slaets, P., Lefebvre, T., Rutgeerts, J., Bruyninckx, H., De Schutter, J.: Incremental building of a polyhedral feature model for programming by human demonstration of force controlled tasks. IEEE Trans. Rob. 23(1), 20–33 (2007)
7. Rutgeerts, J., Slaets, P., Schillebeeckx, F., Meeussen, W., Stallaert, B., Princen, P., Lefebvre, T., Bruyninckx, H., De Schutter, J.: A demonstration tool with Kalman Filter data processing for robot programming by human demonstration. In: Proc. IEEE/RSJ Int. conf. Int. Robots and Systems, Edmonton, Canada, pp. 3918–3923 (2005)
8. Doty, K.L., Melchiorri, C., Bonivento, C.: A theory of generalized inverses applied to robotics. Int. J. Robotics Research 12(1), 1–19 (1993)
9. Xiao, J.: Automatic determination of topological contacts in the presence of sensing uncertainty. In: Int. Conf. Robotics and Automation, Atlanta, GA, pp. 65–70 (1993)

10. Lefebvre, T.: Contact modelling, parameter identification and task planning for autonomous compliant motion using elementary contacts, Ph.D. dissertation, Dept. Mech. Eng., Katholieke Univ. Leuven, Belgium (May 2003)
11. Parkin, J.A.: Co-ordinate transformations of screws with applications to screw systems and finite twists. Mechanism and Machine Theory 25(6), 689–699 (1990)
12. Tang, P., Xiao, J.: Generation of point-contact state space between strictly curved objects. In: Proc. of Rob. Sci. Sys., Cambridge, USA, pp. 31–38 (June 2006)
13. Brent, R.P.: Algorithms for minimization without derivatives. Prentice-Hall, Englewood Cliffs, NJ (1973)
14. Doucet, A., Gordon, N.J., Krishnamurthy, V.: Particle Filters for State Estimation of Jump Markov Linear Systems. IEEE Trans. Signal Processsing 49(3), 613–624 (2001)

Robot Self-modeling of Rotational Symmetric 3D Objects Based on Generic Description of Object Categories

Joon-Young Park, Kyeong-Keun Baek, Yeon-Chool Park, and Sukhan Lee

Intelligent Systems Research Center, School of Information and Communication Engineering, Sungkyunkwan University, Suwon, Korea
jypark7811@gmail.com, {parkpd,Lsh}@ece.skku.ac.kr

Summary. The next generation of service robots capable of such sophisticated services as errands, table/room settings, etc. should have in itself a database of a large number of object instances human use daily. However, it may be impractical, if not impossible, that human should construct such a database for robots and update it each time new objects are introduced. To reduce the level of human involvement to a minimum, we propose a method for a robot to self-model the objects that are referred to or pointed out by human. The approach starts with the generic description of object categories assumed available a-priori. The generic description represents a category of objects as a geometric as well as functional integration of parts the 3D shape of which can be depicted as generalized cones or cylinders representing geometrical primitives or Geons. Given the 3D point clouds from the scene, 3D edge detection segments out the target object referred by human by removing out the background and/or neighboring objects. At the same time, it decomposes the target object into its parts for modeling. Here, we show that a rotational symmetric part or object, a special case of generalized cylinder, can be modeled precisely with only partial 3D point cloud data available. The proposed approach based on the generic description of object categories allows the self-modeling to infer a modeling procedure, estimate a full model from partial data, and inherit functional descriptions associated with parts. Experimental results demonstrate the effectiveness of proposed approach.

1 Introduction

It would be interesting and inspiring if robots are able to automatically model the objects shown to them by human. In fact, this kind of robot ability may not simply be a matter of inspiration but of practical significance too. This is because, a higher level of robotic services such as errand, table/room setting, cooking, etc. that the next generation of service robots are to provide for human require the recognition, handling and manipulation of various objects. Therefore, the next generation of service robots need to maintain a database of a large number of object instances often found in our daily life. And, it is impractical, if not infeasible, for human to construct such a database for robot and update it whenever new objects are introduced.

Modeling a 3D object from its surface data such as 3D point clouds has been a subject of intense investigation in the past. There have been several approaches

S. Lee, I.H. Suh, and M.S. Kim (Eds.): Recent Progress in Robotics, LNCIS 370, pp. 283–298, 2008.
springerlink.com

available for object modeling from the 3D point clouds captured as object surface data. For instance, researchers started to investigate parametric modeling of a class of geometrically well defined objects such as Superquadric objects [1], Convex/concave and planar surfaces [2], Gaussian and mean curvatures-based patches [3], Spin-Images [4], Tensor Voting [5], and mesh oriented [6]. And Octree based modeling approach was proposed to generate a model using volumetric representation [7]. Leibe and Shiele also suggested appearance-based modeling approach which models an object based on appearance and outlines from a set of samples in limited object categories [8].

It is interesting to refer to the generative approach proposed by Ramamoorthi and Arvo [9], where an object instance is generated from the generative model of object categories. The generative model describes an object category by the way object instances can be generated from a simple 3D geometry called a root model.

This paper deals with a problem of robot self-modeling, not simply modeling of an object as seen in the literature. Robot self-modeling of an object shown or referred to by human has some distinctive issues. First, it requires a robot to be able to segment out the target object from the background and neighboring objects. Second, the 3D point clouds captured as object surface data may be only partial, not covering the entire object surface. Third, robot self-modeling need to take into consideration how much the prior geometric knowledge be used for modeling. In the case of human modeling of an object, the human process of object modeling seems not merely a bottom-up process of geometrically describing the 3D point clouds obtained from vision. But, it seems a process of interaction between the top-down geometric knowledge and the bottom-up geometric data: data invoke relevant knowledge while relevant knowledge is refined by data.

In this paper, we assume that a generic description of object categories is available as a priori knowledge. In the generic description we propose, a category of object is described as the geometric as well as functional integration of parts, where the 3D geometry of a part can be depicted as a generalized cone or cylinder [9]. This assumption on object categories has similarity to the concept of Geons proposed by Biederman, where an object is segmented into an arrangement of simple geometric shapes such as cylinders and hexahedrons [10]. The generative approach proposed by Ramamoorthi has also similarity where the generic description is used. However, it only stores algebraic model description and simple object category, and controls points of the model's parametric curves, not deals with geometric and functional relationship among the parts of the model.

The advantages of using generic description of object are: 1) The method or procedure for self-modeling of the object can be either embedded in generic description or inferred from it, For instance, as for the cup example, we can infer that finding the rotational axis may be a way to model the container after segmenting out a handle or any surrounding objects and separating the inner from outer surface. 2) Inherit semantic information such as functions associated with parts described by generic model. For instance, the semantic meaning of container can be directly attached to the geometric model of the container once modeled. 3) Partial data may be sufficient for completing the model since the generic description allows predicting the whole from the partial information.

In this paper, we consider that parts or objects are of rotational symmetric, a special case of generalized cylinder, for the sake of proof of concept.

Through the modeling process guided by the generic description, it is possible to produce intuitive and simple geometric model that can be easily controlled and deformed. This modeling method uses less data to represent an object than mesh or volumetric expression so that the robot can manipulate the object faster. Also, the modeling method robustly estimates and recovers a complete model even under incomplete visual input data (partial or occluded).

In the following Sections we present the architecture in a detailed manner, also providing experimental results aimed at illustrating the functioning of the various components. Specifically, Section 2 delineates the design of the architecture based on generic descriptions; Section 3 describes the segmentation process to separate the object from surrounding objects, as well as for separation of parts of the object, while Sections 4 describes model generation and refinement process. Finally, Section 5 describes the employed experimental setup and the obtained results, and Section 6 presents some concluding remarks and hints on future works.

2 Overview of the Algorithm for Rotational Symmetric Object Modeling

In real world, there are infinite types of objects ranging from simple geometric primitives to articulated objects or deformable objects. However, there is no universal modeling method that can model all those various objects. Only there are just specific modeling methods designated for specific types of objects. In our research, among the various types and shapes of real world objects, the rotational symmetric object is selected as the initial target to be modeled. Fig. 1 illustrates a whole processes of the proposed algorithm for the self-modeling.

2.1 Self-modeling Algorithm

The following is the brief description of each process.

Step 1) 3D point cloud acquisition: Typically, sensors used for acquiring 3D point cloud can be divided into three types: Stereo camera utilizes disparity of two images. Structured light camera uses active light, and laser scanner. For the experiment, we use the developed our own structured light camera. Structured light camera can be the most suitable sensor for object self-modeling. While slower than stereo camera, nevertheless it provides more accurate and rich point cloud. When compared to laser sensor, it is faster and generates 2D image as well.

Step 2) Generic model: From the generic model, the modeling procedures or hints such as segmentation and geometric representation of the objects(ex: cup, can, etc.) are determined.

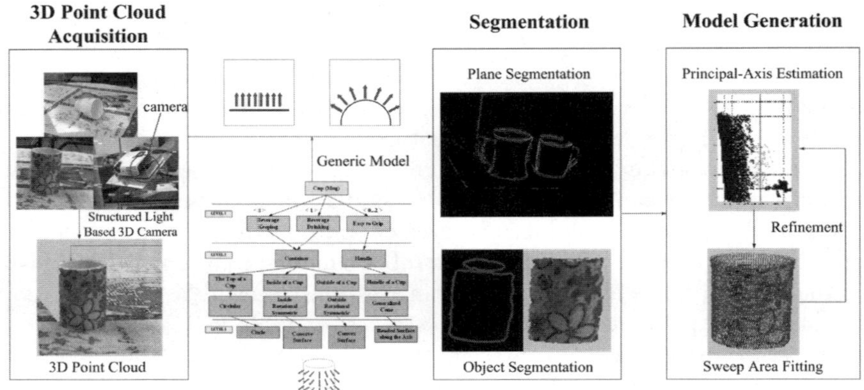

Fig. 1. Algorithm for Rotational Symmetric Object Modeling

Step 3) Segmentation: The target object must be extracted from background. This process can be divided into three parts. One is to extract the objects from the plane, the Second is to extract the target object from surrounding objects and the last is to separate parts of the object. These three segmentation steps are done by using normal vector map and 3D edge operator.

Step 4) Principal-axis estimation: Principal axis is a line that passes through the center of the rotational symmetric object, so that the object can be rotated based on the axis. It is estimated by SVD and normal vectors from the surface of an object.

Step 5) Shape fitting To recover the shape of the object, divide the main axis at uniform interval, estimate disc's dimension in each sector along the principal axis, and integrate those to complete the model.

2.2   Configuration of the Generic Model

As mentioned above, we assume that robot has generic description of the categories of objects usually found in our home. The generic description represents a category of objects as a geometric as well as functional integration of parts the 3D shape of which can be depicted as generalized cones or cylinders. For instance, a cup is described generically as an object consisting of a container or a container with a handle, where the diameter of the container is small enough for human to grasp, when no handle is applied. In order to represent the geometry of the most of the cups often we use, we may describe the container as an object of rotational symmetry with the concave volume inside to hold liquid(Fig. 2).

The structure of generic model we designed is illustrated in Fig.3, and consists of three layers: Functional information is contained in Level 1, Geometric information goes to Level 2, and Evidence/Feature information, which will be matched with sensory data, is included in Level 3. An example of the cups generic model that is based on the structure is demonstrated in Fig. 4. It also illustrates how the three layers can relate and associate from one to another.

Fig. 2. Semantic description of a cup

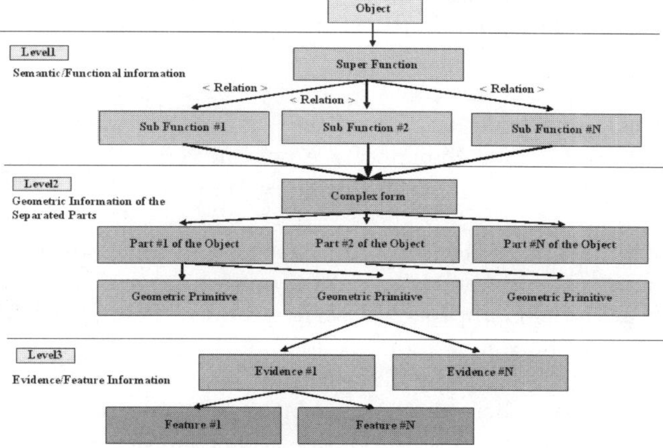

Fig. 3. Relationship between functions and geometric features of an object

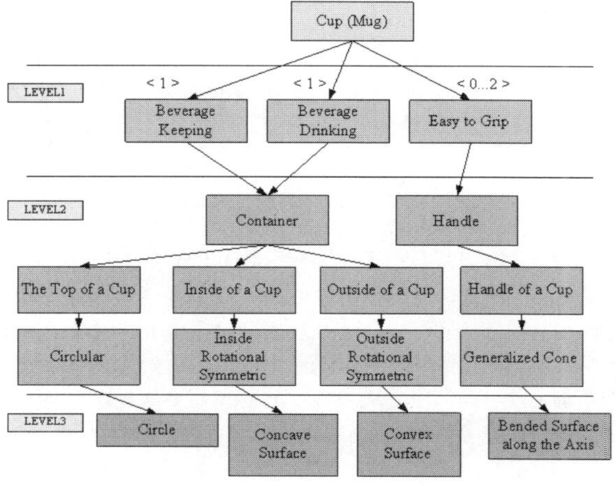

Fig. 4. An example of generic description of a cup

Given the generic information of a cup illustrated in Fig. 4, the modeling procedure is determined or inferred like as: 1) A cup can be segmented and modeled into convex/curvature surface with rotational symmetric characteristics, a handle with generalized cone characteristics, and the top with circle. 2) The parts have their own representation method.

3   Segmentation

What we want to segment are the geometric primitives of the rotational symmetric objects. There has been a lot of researches on classifying curvature through extraction of principle curvature from range data and calculation of mean curvature and Gaussian curvature [11][13][14].

However, what we really need is not the calculation of the curvature values but the extraction of the characteristics of the geometric primitives (Fig. 5).

As for the 3D plane and curved surface, geometric characteristics can be understood very well by the change of the normal vector. The adjacent normal vectors on a plane have no change, while those on a curvature or ellipsoid show the changes (Fig 5).

When considering that range data from a sensor, it might be a partial surface of an object, thus this approach is useful to connect range data to the geometric primitives. To utilize this approach described above, points in 2.5D should be represented as point normal vectors. A method to extract the normal vectors from neighboring points is briefly described as follows.

Assume 3 by 3 mask where the point located in the center has eight adjacent points. Directional vectors can be obtained by connecting central point and adjacent points. Then, through calculating cross-product of those directional vectors, we can yield the normal vectors of a plane that is generated by central point and other adjacent two points.

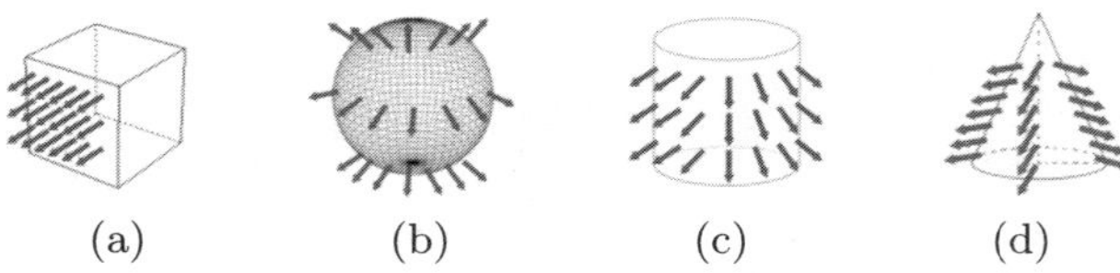

(a)   (b)   (c)   (d)

Fig. 5. Characteristics of geometric primitives: (a)box, (b)sphere, (c)cylinder, (d)cone

From the neighboring points, we obtained 8 plane vectors $p(n_i = 1, 2, \ldots, 8)$, as shown in Fig. 6. And the point normal of the center points represents the average of the 8 plane vectors. Now, we can get the whole point normal vectors from the input image (640 by 480).

To remove out the 3D plane in an image, and to separate the object from neighbors and parts of an object, we borrowed Wanis 3D edge-region based segmentation method [15]. The method can generate 3 types of 3D edges from 3D range data. Those are the Fold Edge, Boundary Edge, and Semistep Edge. The

$$\underline{x}_i = \left(x_i, y_i, z_i \right)$$

$$\underline{n} = \left(n_x, n_y, n_z \right)$$

$$\underline{n} = \frac{\sum_{i=1}^{N} \underline{n}_i}{\left| \sum_{i=1}^{N} \underline{n}_i \right|}$$

Fig. 6. The point normal from the neighboring vectors

Fig. 7. 3D Edges; Fold Edge(red), Boundary Edge(green), Step Edge(green)

FE is the edges corresponding to pixels where surface normal vector discontinuity occurs, and BE are formed by pixels that have at least one immediate neighbor pixel belong to the background, and the SE occur where one object occludes another, or when a part of an object occludes itself. Fig. 7 shows the 3D edge extraction result. Therefore, we can segment the scene to a few object and planar regions based on the extracted 3D edges.

From the generic model of a cup as mentioned in section 2.2, the segmentation procedure can be inferred. There are three parts of a cup in the third layer of the generic model: convex/curvature surface, a hole, and an ellipsoid. Thus, we can infer that a cup is possible to be segmented into the three parts based on the geometric representations such as inside rotational symmetric(inner surface), outside rotational symmetric(outer surface), generalized cylinder(bended handle surface), and circle(the top of a cup). Based on these sequences, we segmented a cup like as Fig. 8.

4   Model Generation

4.1   Principal Axis Estimation

3-Dimentional cylinders are symmetrical in a round shape in each direction of height. Therefore, when calculating the cross product for point normals in equal

Fig. 8. Three types of parts segmented from a range image; Parts to be segmented are decided by the geometric model

heights, the direction of the vector of the cross product tends to form in the direction of the axis or similar directions.

Using this tendency, the vector product in a certain domain was calculated and voted about the direction of the vector of the product. The cross product between the point normal and the directional vector that received the most points in the voting is calculated to determine perpendicularity. If it surpasses a certain number of point normals, that directional vector which includes those point normals becomes the axis of the cylinder(Fig. 9).

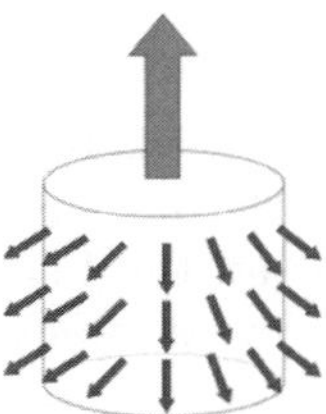

Fig. 9. The principal axis of cylinder and dot product results of the point normals

Also, from this converging point, the distance to the other points becomes equal. From these facts, we can come up with the following expression.

$$\mathbf{N_1} = \begin{pmatrix} \dfrac{x_{p_1} - x_c}{r_1} \\ \dfrac{y_{p_1} - y_c}{r_1} \\ \dfrac{z_{p_1} - z_c}{r_1} \end{pmatrix} \qquad \mathbf{N_2} = \begin{pmatrix} \dfrac{x_{p_2} - x_c}{r_2} \\ \dfrac{y_{p_2} - y_c}{r_2} \\ \dfrac{z_{p_2} - z_c}{r_2} \end{pmatrix}$$

Here, (x_{p1}, x_{p2}) are points on the same height as the selected point, (N_1, N_2) are normal vectors of each point, and x_c is the central point, the converging point on the principal axis. Using the above method, we can obtain the survived, the central points that meet the both equations, and we can estimate the principal axis by applying Covariance Matrix and SVD to these central points.

4.2 Shape Fitting

The following factors have to be found for the matching of the point cloud and cups primitive, the cylinder.

Cylinder := {radius, height, the location of the axis in world coordinates}

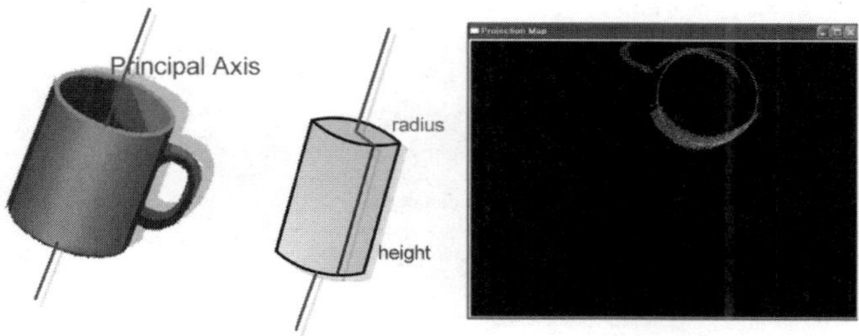

Fig. 10. Matching factors of a cup(left), Circle fitting using RANSAC in 2D space(right)

To find out the above information, the directional vector of the cup, i.e., the principal axis needs to be estimated by the methods described in section 4.1. For calculating the cylinders height, the principal axis needs to be rotated perpendicularly on the x-y plane in the world coordinates using the calculated rotational matrix. And then, the rotated point clouds are divided at an equal length in the direction of the z-axis and the points belong to the same cross-section are projected onto the x-y plane. Finally, using the RANSAC, the circles center points and radius can be estimated from the surface points. This process is applied to all cross-sections along the principal axis. After that, we can estimate the whole center points on the cylindrical surface.

4.3 Principal Axis Refinement

The extracted principal axis might have some errors because of the noisy and incomplete data caused by ill-conditioned environment such as reflection, occlusion and illumination change(Fig. 11). Therefore it needs to be refined.

The algorithm of the refining process we have used is as follows.

As confirmed in Fig.13, the central points found above are distributed so widely that they are not all on the same line. Therefore, we cannot say that this principal axis is accurate. However, we can still estimate a rough pose of the cylinder. To find the accurate principal axis, we locate the extracted principal axis be parallel to the z-axis and recalculate the dimension of a disc partition(cross-section), and project the cross-section on the xy-plane (Fig. 14 (left)).

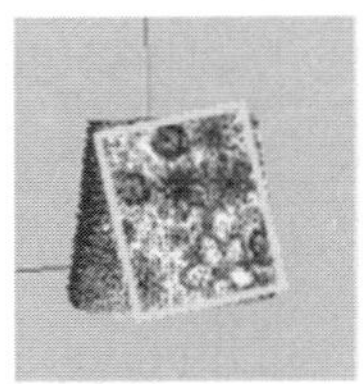

Fig. 11. Poorly estimated object model(red points) and the original pose of the object(rectangle)

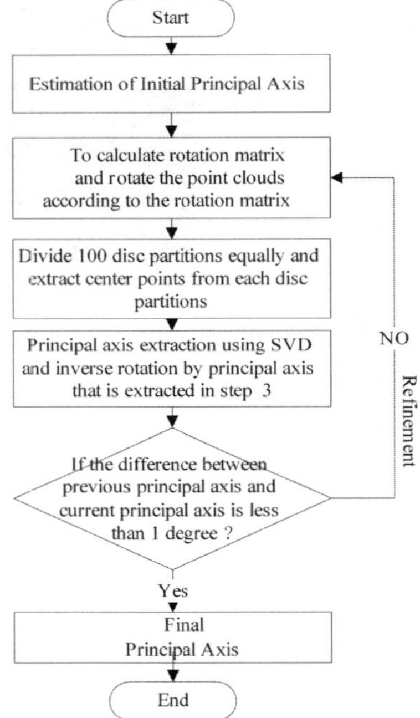

Fig. 12. Algorithm of Principal Axis Refinement

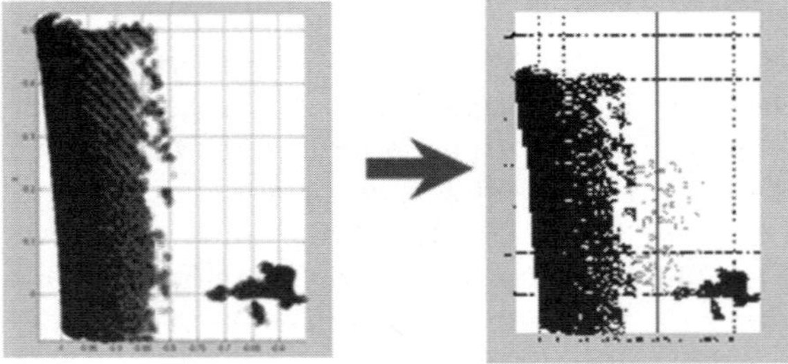

Fig. 13. Principal axis estimation using center points

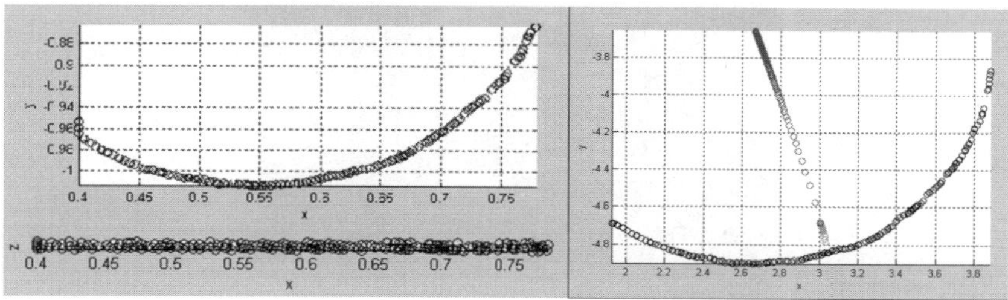

Fig. 14. Cross-section of a disc(left), Estimation of a center point(right)

Therefore, assuming that the cross-sections are arcs and fitting them as a circle, then each cross-section can be expressed as a circle and using the least square, we can find the circles central point(Fig. 14 (right)). The following is an expression to find the circles central point.

P_c is the final center point of a cross section. If a certain center point at time k is equal to previous center point at time k-1, it concludes that the final center point is same as the point at time k.

$$P_{c_k} = E(P') - E(D) \times E(\vec{V})$$

$$E(P') = \frac{1}{n} \sum_{i=1}^{n} (P_i')$$

$$E(D) = \frac{1}{n} \sum_{i=1}^{n} (P_{c_{k-i}} - P_i'), \text{ where n is the number of 3D points}$$

$$E(\vec{V}) = \frac{1}{n} \sum_{i=1}^{n} (\vec{V}_{P_{c_{k-1}}, P_i'})$$

P_c is the center point at time k, $E(P')$ is the average of the surface points according to the principal axis, $E(D)$ is the average distance between previous center point and all the surface points, and $E(\vec{V})$ is the average of the vectors from previous center point and all the surface points. Then, assuming that the central point of a circle must be a point on the principal axis, we can get a more accurate principle axis iteratively using the methods used to find the initial principal axis. This process is stopped when the difference between the old one and the newly calculated axis is less than 1 degree.

5   Experimental Result

The experimental environment is set like as follows. The sensor used in the experiment was a structure light camera, which obtains both 2-dimensional images and 3-dimentional range data at once. The resolution of the structured light camera

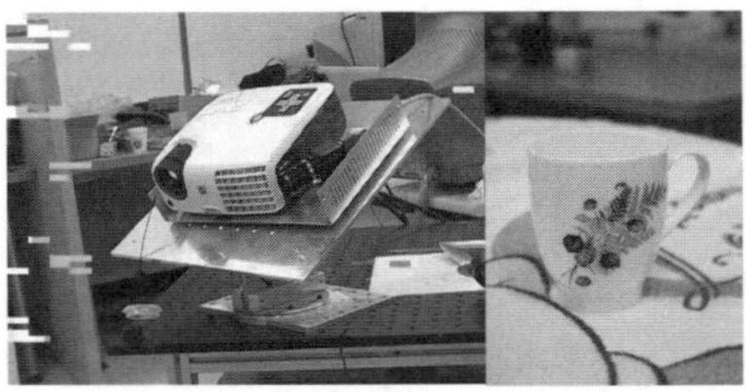

Fig. 15. 3D structured light camera(left) and Target object(right)

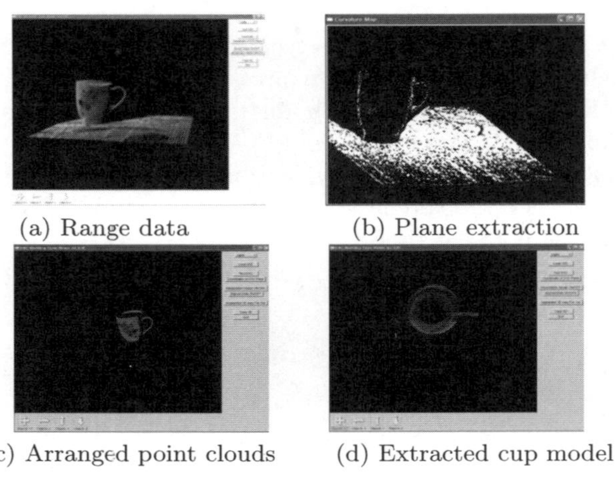

(a) Range data   (b) Plane extraction

(c) Arranged point clouds   (d) Extracted cup model

Fig. 16. Model generation example of a cup

is 640*480 and the accuracy is 5mm. The system used in the experiment is Pentium 4 1.8Ghz, with 1G of RAM. Object used in the experiment were basically rotational symmetric objects such as a cone, ball (sphere), cylinder, cup, and solid geometric primitives. The distance between the object and the camera is 1m, and the illumination of the experiment environment is 800lux.

5.1   Evaluation on Accuracy of Principal Axis

To measure the accuracy of the extracted main axis, the angle between the normal vector of the plane and the principal axis is calculated.

Before we measure the accuracy of the axis, we have to first find the normal vector of the plane and measure the precision of these values. Therefore, to evaluate the precision of the normal vector of the plain, we select three points to find a plane and repeat 500 times estimation and select the best plane normal vector that is valid(RANSAC), Also, an average of these normal vectors $\overline{X}_{plane}$ and a variance of these vectors s_p^2lane are calculated to measure the precision.

Table 1. Estimation error of the initial principal axis

Statistic	x	y	z
Average	0.159339	0.30606	0.392502
Stddev	0.197665	0.392299	0.64012
Variance	0.039071	0.153898	0.409753

Then, using the average of the valid planes, the error between the normal vector and the directional vector are calculated as the accuracy factors.

In Table 1, the range of the deviation of the old one is larger than the refined one. On the other hand, when refined axis was used, the average error was 0.02, 0.04, 0.04, i.e., which is significantly lower than previous one, and standard deviation is also less than 0.05 which is significantly lower than previous one as well.

Table 2. Estimation error of the refined principal axis

Statistic	x	y	z
Average	0.022697	0.048964	0.045191
Stddev	0.012556	0.038955	0.020851
Variance	0.000158	0.001517	0.00043

The result shows more than 15 times in the x-axis, 10 times in the y-axis, and more than 30 times in the z-axis in performance enhancement between the previous and the refined one.

5.2 Evaluation on Curve Fitting Accuracy

To evaluate the accuracy of the shape reconstruction, we find the average and the standard deviation of the difference between the radius of the estimated disc and the radius of the real point clouds.

By slicing cylindrical objects(cups and solid models) into 90-100 pieces, which has the same diameter throughout, we can have 90-100 cross sections as well. And then, the average and the standard deviation of the errors are calculated. There are 485 total measured data, the result shows more than 8 times of performance enhancement.

Table 3. Initial fitting accuracy(left) and refined fitting accuracy(right)

Initial estimation		After refinement	
Average	1.90182	Average	1.07894
Stddev	12.0377	Stddev	1.44583
Variance	144.905	Variance	2.09042

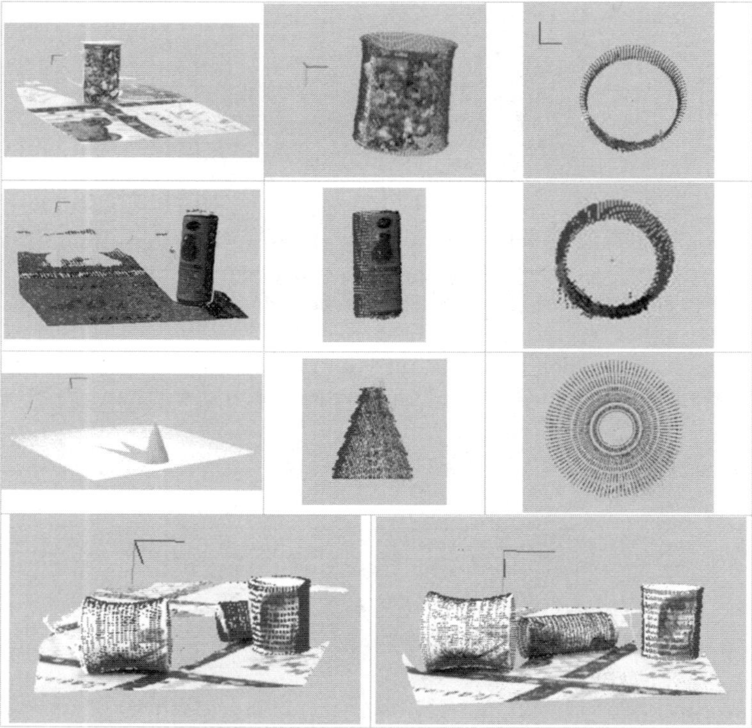

Fig. 17. Self-modeling examples

6   Conclusions and Future Work

The number of specific models of which should be stored in the robot database would be potentially very large and the update of the database is required. From these reasons, it is not practical, if not impossible, to encode every specific instances of the objects in the working domain, i.e., a household. To tackle these problems, we use the generic description that includes both symbolic and geometric representations. And we proved the concept by applying the generic descriptions to the rotational symmetrical object.

Through the modeling process guided by the generic model, it is possible to segment parts of the object and produce intuitive and well fitted geometric model, even under incomplete visual input data (partial or occluded), and semantic information at the same time.

The experimental results show that the proposed method can reconstruct a specific 3D object model automatically even though it has only partial 2.5D information or occlusion. As shown in the experimental results, we proved that the principal axis and shape of the object can be estimated fast and accurately by using the introduced approach.

Since an object (e.g. a cup) in everyday life has countless shape variations, our approach will help to solve memory and time consuming problem for knowledge storing.

The future works are to integrate it to the robot and find a good way to separate the handle and the inner from the outer surface, and to generate more complex model such as curved, articulated, and deformable objects.

References

1. Pentland, A.P.: Perceptual organization and the representation of natural form. Artificial Intelligence 28, 293–331 (1986)
2. Jain, A.K., Hoffman, R.L.: Evidence-based recognition of 3-D objects. IEEE Transactions on Pattern Analysis and Machine Intelligence 10, 783–802 (1988)
3. Besl, P.J.: Surfaces in Range Image Understanding. Springer Series in Perception Engineering. Springer, Heidelberg (1988)
4. Johnson, A.: Spin-Images: A Representation for 3-D Surface Matching, doctoral dissertation, tech. report CMU-RI-TR-97-47, Robotics Institute, Carnegie Mellon University (1997)
5. Tang, C.K., Medioni, G., Lee, M.S.: N-Dimensional Tensor Voting, Application to Epipolar Geometry Estimation. IEEE Transaction on Pattern Analysis and Machine Intelligence 23(8), 829–844 (2001)
6. Hoppe, H., DeRose, T., Duchamp, T., McDonald, J., Stuetzle, W.: Mesh optimization. In: SIGGRAPH 1993, pp. 19–26 (1993)
7. Ballard, D.H., Brown, C.M.: Computer Vision. Prentice Hall, Englewood Cliffs (1982)
8. Leibe, B., Schiele, B.: Analyzing Appearance and Contour Based Methods for Object Categorization. In: Proc. IEEE Conf. Computer Vision and Pattern Recognition (2003)
9. Ramamoorthi, R., Arvo, J.: Creating Generative Models from Range Images. In: SIGGRAPH 1999, pp. 195–204 (1999)
10. Biederman, I.: Human Image Understanding: Recent Research and a Theory. Computer Vision, Graphics, and Image Processing 32, 29–73 (1985)
11. Trucco, E., Verri, A.: Introductory Techniques for 3-D Computer Vision. Prentice Hall, Englewood Cliffs (1998)
12. Liu, Y., Emery, R., Chakrabarti, D., Burgard, W., Thrun, S.: Using EM to learn 3D models of indoor environments with mobile robots. In: Proceedings of the 18th Conference on Machine Learning, Williams College (2001)
13. Lee, S., Jang, D., Kim, E., Hong, S.: Stereo Vision Based Real-Time Workspace Modeling for Robotic Manipulation. In: IROS 2005. Proc. IEEE Conf. International Conference on Intelligent Robots and Systems (2005)
14. Csakany, P., Wallace, A.M.: Representation and Classification of 3-D Objects. IEEE Trans. Systems, Man, and Cybernetics-part B: Cybernetics 33(4) (2003)
15. Wani, M.A., Batchelor, B.G.: Edge-Region-Based Segmentation of Range Images. IEEE Transactions on Pattern Analysis and Machine Intelligence 16(3) (1994)

16. Dickinson, S., Pentland, A., Rosenfeld, A.: From Volumes to Views: An Approach to 3-D Object Recognition. Computer Graphics, Vision, and Image Processing: Image Understanding 55(2), 130–154 (1992)
17. Froimovich, G., Rivlin, E., Shimshoni, I.: Object Classification by Functional Parts. In: IEEE Proceeding of 3D Data Processing Visualization and Transmission (2002)
18. Besl, P., McKay, N.: A method for registration of 3D shapes. IEEE Transactions on Pattern Analysis and Machine Intelligent 14, 239–256 (1992)
19. Hoffman, R., Jain, A.K.: Segmentation and classification of range images. IEEE Transactions on Pattern Analysis and Machine Intelligence 9(5) (1987)

Chapter III

Human-Robot Interaction and Intelligence

Summary of Human-Robot Interaction and Intelligence

Il Hong Suh

Part 1: Human-Robot Interaction

As one of indispensable capacities for viable robotic service to human, there will be included natural interaction with human by means of speech, gesture, haptic display, and even cooperative manipulation. Several human-robot interface and/or interaction technologies have been proposed for development of such interaction capacity in ICAR 07.

Koch et al propose that a robot has to serve as interaction partner to give information and/or control various devices in ambient environment over a unified speech interface. For this, they integrate a speech engine into mobile service robots for lab tour guide and food information service, where the speech engine includes several functional components such as registration of dialogs, parser for dialog files, speech synthesis and recognition, dialog processing engine, and callback.

Wang and Wang apply Adaboost learning algorithm to SIFT features to implement a reliable hand posture recognition system. They show that their recognition system is better than systems involving Viola-Jones detector in the sense of in-plane rotation invariance as well as insensitivity to background noise.

Brell and Hein show that vibrotactile display is valuable for performance enhancement of classical computer navigated surgery, where one or two of vibrators mounted on hand of a surgeon is selected to give directional information. And then, both amplitude and frequency of vibration signal for the selected vibrator are modulated to code the distance between current and desired location of tool tip.

Edsinger and Kemp present a robot manipulation scheme supporting three design themes: "cooperative manipulation", "task relevant features", and "let the body do thinking," which enable a humanoid robot to help a person place everyday objects on a shelf.

Part 2: Intelligence

It is necessary for mobile robots to share and reuse their knowledge regardless of the form of knowledge representation to improve their services to human in versatile environments. In general, knowledge is often represented by employing logics, software component, visual programming framework, or successful task trajectories. There are reported various frameworks to program and process such knowledge in ICAR 07.

Braun, Wettach and Berns propose a highly flexible framework for visualization and sensor simulation in three-dimensional environments which allows programmable elements to be freely inserted for online scene modification. They demonstrate that two complex robotic applications can be effectively simulated by using their

S. Lee, I.H. Suh, and M.S. Kim (Eds.): Recent Progress in Robotics, LNCIS 370, pp. 301–302, 2008.
springerlink.com

framework, where two complex robotic applications require both a high quality simulation of cameras and lasers scanners and an intuitive 3D visualization.

Aleotti and Caselli describe a technique to program walking paths for humanoid robots based on imitation, where NURBS (Non Uniform Rational BSpline) is employed to synthesize demonstrated paths and the synthesized paths are modified by the robot based on local and dynamic information. They also report simple experimental results to support their imitation-based programming by utilizing a Robosapien V2 low-cost humanoid toy robot.

Kim et al describe a software framework for component-based programming of intelligent robotic tasks which is developed by Center for Intelligent Robotics (CIR), KIST, Korea. The framework employs a popular 3-layered control architecture consisting of layers for deliberate, sequencing and reactive behaviors. They show that their component-based programming framework is useful for the rapid development of versatile robot systems under distributed computing environments, different operating systems, and various programming languages. The framework has been practically applied to robot platforms such as T-Rot, Kibo and Easy Interaction Room, where T-Rot and Kibo have been exhibited at the 2005 Asia-Pacific Economic Cooperation (APEC) in Korea.

Choi, et al propose a method to construct semantic contextual knowledge for mobile robots to recognize objects by means of logical inference, where OWL has been used for representing object ontology and contexts. They propose a four-layered context ontology schema to represent perception, model, context, and activity for intelligent robots. And, axiomatic rules have been used to generate semantic contexts using ontology. They show that an invisible object is found by applying their ontology-based semantic context model in a simple real world environment.

Dynamic Speech Interaction for Robotic Agents

Jan Koch[1], Holger Jung[1], Jens Wettach[1],
Geza Nemeth[2], and Karsten Berns[1]

[1] University of Kaiserslautern, Robotics Research Lab
Gottlieb-Daimler-Straße, 67653 Kaiserslautern, Germany
`{koch, h_jung, wettach, berns}@informatik.uni-kl.de`
[2] Budapest University of Technology and Economics
1117. Budapest, Magyar tudosok krt.2., Hungary
`nemeth@tmit.bme.hu`

Summary. Research in mobile service robotics aims on development of intuitive speech interfaces for human-robot interaction. We see a service robot as a part of an intelligent environment and want to step forward discussing a concept where a robot does not only offer its own features via natural speech interaction but also becomes a transactive agent featuring other services' interfaces. The provided framework makes provisions for the dynamic registration of speech interfaces to allow a loosely-coupled flexible and scalable environment. An intelligent environment can evolve out of multimedia devices, home automation, communication, security, and emergency technology. These appliances offer typical wireless or stationary control interfaces. The number of different control paradigms and differently lay-outed control devices gives a certain border in usability. As speech interfaces offer a more natural way to interact intuitively with technology we propose to centralize a general speech engine on a robotic unit. This has two reasons: The acceptance to talk to a mobile unit is estimated to be higher rather than to talk to an ambient system where no communication partner is visible. Additionally the devices or functionalities to be controlled in most cases do not provide a speech interface but offer only proprietary access.

Keywords: HCI, mobile robots, transactive systems, ambient intelligence.

1 Introduction

To access different types of hardware devices and information technology, people have to learn to deal with a lot of proprietary interfaces. In most times this is related to remembering which buttons to press on the devices. We envision highly ambient technology in home environments with lots of information and functionality to offer. Human speech especially for elderly people seems to be a better choice to interact with technology. Though generally the quality of speech recognition is not perfect yet, the border of usability is lower as with pressing buttons because a lot of elderly people are scared to come in touch with technology. We propose a generalized concept for speech integration on a mobile robotic unit. The robot serves as interaction partner as it seems more convenient to talk to a certain visible autonomous device, like a pet, than to a light switch for example.

S. Lee, I.H. Suh, and M.S. Kim (Eds.): Recent Progress in Robotics, LNCIS 370, pp. 303–315, 2008.
springerlink.com   © Springer-Verlag Berlin Heidelberg 2008

Generally mobile robots offer great opportunities in the area of human-machine interaction [1]. When integrated into the home environment the robot becomes a transactive agent. It processes control and information transfer between human users and ambient technology. Controllable and information sharing devices register their speech interface dynamically at a dialog engine which runs on the robot. Users this way are able to use the robot for interaction with systems that by default do not have the capability to offer speech interfaces. The robot this way does not only become interactive, it becomes transactive as it is relaying the control information to other components and responds in their role. The mobile robot additionally offers several features regarding human recognition to improve the audio-related quality of speech recognition in particular.

The speech engine, the mobile unit's characteristics as well as the implemented appliances are presented within this work. In the next chapter we start with an overview on development and examples of speech interaction on mobile systems and in intelligent environments. The robot prototypes that currently have our speech engine running are introduced in 3. Our approach to implement a dynamic dialog system is explained in 4. Furthermore in chapter 5 we give some examples on applications that have been facilitated using the speech engine on the robots and discuss conclusions in 6.

2 Speech Interaction with Service Robots and Intelligent Environments

There are several important reasons to use man-machine dialog systems based on spoken language. Input by speech can be faster than with haptive hardware while hands and eyes are free to focus on other activities. Especially for older people or people who lack experience in computer usage, it is the simplest way to interact with technology by speech. Another advantage is the independence from the distance between user and machine when using room microphones. It is even possible to access the system remotely via telephone or internet. Today, there are many areas where speech systems already are used. The most important areas are:

- home automation (lighting, entertainment electronic, answerphone),
- industrial use (control of machines, e.g. for quality checks or dispatch),
- office use (database queries, dictation, organisation of documents),
- banking systems (bank assignments, stockjobbing, credit cards),
- public transport and other information services (time table informations, booking, weather forecast, event notes),
- medical appliances (diagnoses systems, interphone systems for patients, surgeon aids),
- elderly and disabled people (emergency systems),

There are currently several projects from industry, institutes and universities where dialog systems are developed. Many projects are developed for simple database queries. A big research project is Smartkom [2], a multimodal task-oriented dialog system for two main domains, information seeking in databases and device control. To improve the acceptance and to make control easier, dialog systems have also been implemented on many robotic systems. Usually, dialog systems are used here to offer controls for easy tasks, as to use the robot as information systems or to get information about its technical status. Examples are CARL from the University of Aachen [3] or HERMES from the University of Munich [4]. Both can perform several easy jobs when the user asks for it. Armar from Karlsruhe [5] and another robot called CARL from Aveiro, Portugal [6] can fulfil kitchen tasks. The Japanese robots Jijo-2 [7] and HRP-2 [8] can be asked for people, for their room numbers or for the path to their location within an office building. At the University of Bielefeld, the robot BIRON [9] is developed which can learn the name of artifacts if the user shows them. There is also a current research at the University of Saarbruecken [10], where a system is developed which tries to recognize the intention the user has.

All these projects are application-driven and the robot is optimized to perform certain tasks while providing a suitable specific speech interface. Our approach goes into the opposite direction where the robot only has rudimentary functionality but behaves as a transactive system, offering control and information of various other parts of the ambient environmet over a unified speech interface engine. It has to be mentioned that in this work the improvement of the quality of speech recognition or synthesis was not a subject. The idea is to keep the dialog system independent from the implementation of recognizer and synthesizer. Nevertheless the libraries we selected to perform these task in our examples are described in section 4.

Multiple research groups address the field of intelligent indoor environments. The intelligent room project at MIT uses architectural ideas of the area of mobile robotics [11] and focuses on supporting humans by interpretation of their actions, gestures, and speech. The Amigo-project in Germany [12], the Aware Home in Atlanta or the I-Living project are also examples for intelligent environments which shall serve humans in their daily life.

3 Robot Platform

In this work the mobile prototypes MARVIN[1] and ARTOS carry speech capabilities in terms of our speech engine. MARVIN is a experimental platform for basic research on indoor exploration and human detection. ARTOS is has been optimized in size and power consumption for usage in home environments.

3.1 MARVIN

MARVIN (Mobile Autonomous Robot Vehicle for Indoor Navigation, see fig.1) is a test platform for autonomous exploration of indoor environments, 3D mapping and environmental modelling. The basic vehicle concept consists of a differential drive and two planar laser range scanners for obstacle detection and SLAM. The

---

[1] http://www.agrosy.informatik.uni-kl.de

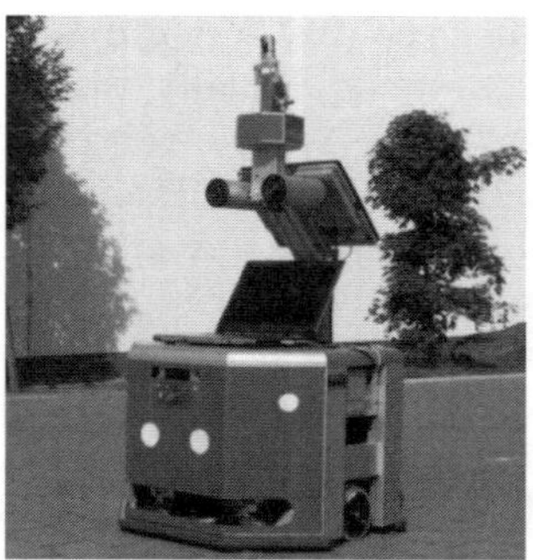

Fig. 1. MARVIN

Fig. 2. ARTOS

robot control system is implemented as behaviour-based approach on several levels of abstraction beginning on basic collision avoidance and trajectory control, up to exploration and mapping. In this context a behavioral network has been developed that enables the robot to explore a structured, but a-priori unknown indoor environment without any user intervention and to create a topological map of all visited rooms and their connections (see [13]).

3.2 ARTOS

ARTOS (Autonomous Robot for Transport and Service) is under development for the BelAmI project[2]. Within that project, an assisted living environment for elderly care is assembled. ARTOS shall serves three purposes:

- Transport aid for items of the daily life to help disabled people
- Teleoperated or autonomous emergency recognition
- Multimedia and system interaction agent

Table 1. Some Facts about ARTOS

Weight	20kg
Height, Width, Length	25cm, 30cm, 50cm
Max Speed	60cm per sec
Sensors	Laser, Infrared, Ultrasonic, Bumper
Multimedia	Speakers, Micro, CCD Camera
Navigation	Odometry, RFID Landmarks
Operation time	6h at 40W
Control Framework	MCA, behavior-based collision avoidance

ARTOS uses an Hokuyo laser range finder, ultrasonic and infrared sensors for collision avoidance. The Modular Controller Architecture (MCA), see section 4.5, is used as control framework. For the navigational subsystems a behavior-based

---

[2] http://www.belami-project.de

control approach has been chosen as we use it on all of our robots [14]. Artos is controlled by a low-power PC so that the overall power consumption with all sensors and multimedia activated at maximum drive speed does not exceed 40W. An overview of the technical details is given in Table 1.

As the interaction aspect is the main topic of this paper we will continue with a description of the speech engine and return to the robot and the implemented scenarios in section 5.

4   The Dynamic Speech System

As motivated in the introduction a mobile robot should be enabled to interact and transact providing its own functionalities and those of other services to human inhabitants of intelligent environments. A coarse overview of the system is given in Fig. 3. From the technical point of view the service-interaction is negotiated as follows: An application or the robot itself hands over the dialog description to the speech engine. The dialog is registered and set active. While traversing the dialog the engine calls back the registering application, passing state variables that might have been changed by the human during interaction. Finally dialogs may terminate or be unregistered. The dialog engine therefore consists of several parts:

- Registration and deregistration of dialogs
- Parser for dialog files
- Traversing and processing dialog representations
- Callback to registering application

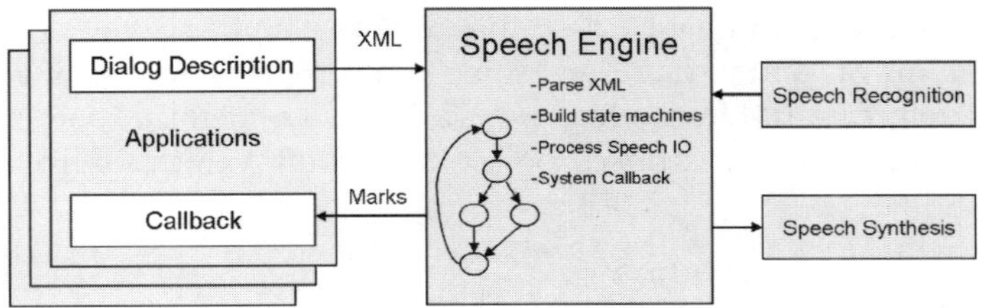

Fig. 3. Architecture of Dynamic Speech System

4.1   Dialog Specification

Dialogs are specified in XML using a syntax that has been developed in [15]. The scheme includes elements for describing speech inputs and outputs, to specify speech parameters, and to define condition and value variables for controlling and adapting the dialog flow. Upon dialog registration the given XML is parsed and mapped to state machines.

The settings part of a dialog specification contains parameters for influencing the behavior of the speech synthesis library. Special actions can be globally defined which are engaged if the user inputs some prescribed words or makes no input for a certain time.

In the main dialog section user inputs which have to be recognized by the speech recognition system and machine outputs which have to be performed by the speech synthesis are specified. Other elements define conditions which have to be fulfilled, e.g. using variables and corresponding values or by comparing two variables. therefore variables can be declared with a certain type and their value can be set initially and be changed during runtime by the user or the machine. Also an element to stop the control flow for a certain time is defined; jumps in the dialog flow are also possible. A very important element is defined for the interaction between the dialog engine and the application that registered the respective dialog.

The example presented in Fig. 4.1 is a simple dialog where a user can ask the robot ARTOS about foods comprised in an intelligent refrigerator. In the element `waiting_for_userinput` beginning in line 1, it is possible to design some `user_input` elements. Each of these elements begins with the element `sentence`. The sentence defined here has to be recognized by the speech recognition system, then elements inside of that special `user_input` element are processed. With these `user_input` elements inside of a `waiting_for_userinput` element, it is possible to define branches in the dialog flow depending of a users speech input. At the end of the `waiting_for_userinput` element, the branches are merged together again. Here in the first example, there is only one possibility to make a speech input. If the sentence "ARTOS, which food items are in my fridge?" is asked by user, the element `prompt` is affected. Here it is possible to define a sentence to be spoken by the speech synthesis system. The sign "%" marks a variable, the system substitutes automatically with it is values. This will be shown in more detail in the second example.

The second example in fig.4.1 describes a more complex dialog. Here the system triggers an output if there are some food items in the fridge which are expired. therefore at the beginning, two variables are defined, one, `itemCount`, concludes the number of expired foods. Its default value is zero. The other

```
0 <dialog dialog_id="main">
1   <waiting_for_userinput>
2     <user_input>
3       <sentence>ARTOS, which food items are
                  in my fridge?
       </sentence>
4       <prompt>You have %items in your fridge.
       </prompt>
5     </user_input>
6   </waiting_for_userinput>
7 </dialog>
```

Fig. 4. Example 1 of a dialog definition

variable, `itemNames`, contains the food items' names. Beginning in line four, a condition depending on the value of `itemCount` is defined. If its value is zero, there are no expired foods and nothing is to do. Otherwise if the value is higher than zero, it is necessary distinguish again between singular and plural for building a grammatically correct sentence. Depending on these inputs, a sentence is generated by the speech synthesis library. As already mentioned, the variables marked with "%" are substituted by their values. If the system sets the food variables, one of the two conditions in line four and five becomes true and the corresponding sentence is synthesized. In general it is not necessary to define sentences here, any other XML dialog element could also be used inside of the condition elements.

After handling the conditions, the variables are set back to their default values. Then, there is a jump back in the dialog flow from line nine to three and the dialog system waits again if variables are externally set and one of the two conditions becomes true.

```
 0 <dialog dialog_id="main">
 1    <var type="int" name="itemCount" value="0"/>
 2    <var type="string" name="items" value=""/>
 3    <mark name="begin"/>
 4    <if var="itemCount" value="1" relation="==">
        Consider that %itemNames has expired.
 5       <elseif var="numberFoods" value="1"
         relation=">">
           Consider that %itemNames have expired.
         </elseif>
 6    </if>
 7    <set_variable name="itemNames" value=""/>
 8    <set_variable name="itemCount" value="0"/>
 9    <goto mark="begin"/>
10 </dialog>
```

Fig. 5. Example 2 of a dialog definition

4.2 Synthesis

We are using the speech synthesis engine *Profivox* developed at the Speech Technology Laboratory at Budapest University of Technology and Economics [16], [17]. Profivox is under development since 1997. Its goals are intelligible human-like voice, robust software technology for continuous running, automatic conversion of declarative sentences and questions, the possibility of tuning for application-oriented special demands and real-time parallel running on minimum 30 channels. Profivox is built up modular to make the system multilingual and easier to develop by simply changing single modules. A control module manages every step necessary to produce synthesized speech.

4.3   Recognition

The speech recognition engine *AmiASR* comes to use. It has also been developed at Budapest University of Technology. This software is based on the *Hidden Markov Model Toolkit* (HTK)[3] [18]. Input for the recognition system are a dictionary where all valid words are listed and a grammar file with the words connected to sentences to be detected. Generally spoken, a speech recognizer works better if the set of words to be recognized is limited to the necessary amount. In our system we therefore generate dictionary and grammar on the fly for each point in the dialog.

4.4   Dialog Engine

The speech engine becomes active when dialogs are registered. These are modelled in XML and are converted by the speech engine to states and transitions of state machines. Each dialog is represented by a state machine. Within states, speech output can be triggered. Transitions are traversed when condition variables are valid or the user has given matching speech input. All dialogs share a common start state. Three possibilities exist to walk into a certain dialog from here:

- Matching speech input
- Valid condition variable
- Manually triggered start

When reaching a state the engine at first generates and triggers the speech synthesis if corresponding output is specified for that state. Furthermore the dictionary and grammar files are produced containing the words and sentences whose match might be the condition for transitions emerging from this state. Afterwards the recognizer is activated if speech input is defined. Finally the transitions that require the fulfillment of conditions regarding dialog variables are checked.

The engine continues then to periodically check whether either a matching user input has been given or whether a condition has been fulfilled. Within the dialog specification also a timeout may be given that leads to a special transition. Also transitions for help or termination may be specified that are connected to each state of the dialog. If a dialog reaches its final state, the engine returns to the starting state.

During runtime, new files with dialogs can be added or files can be removed. Then new states and transitions are added to the finite state machine or states and transitions are removed. When an application removes a dialog the design decision has been to complete the currently running dialog and then remove it from the set of state machines.

4.5   System Integration and Callback

For communication between speech engine and application the Modular Controller Architecture (MCA) is used. MCA has been developed for mobile robotics

---

[3] http://htk.eng.cam.ac.uk

research, initially at the FZI in Karlsruhe[4] [19] and is under continuous development[5]. MCA enables primitive and object-oriented communication, re-usability, component-based development, hardware access and application distribution. From the robot side the speech engine is accessed by those particular MCA modules that are designed to share data and control with a human user. When an ambient service wishes to register its interface it has to be written either in MCA, too, and this way perform a native connection or use one of the other middleware solutions that have been developed within the BelAmI project, see [20].

System call-backs are triggered by asynchronous messages. States of variables are communicated via the shared-memory concept of MCA. The XML representation for interaction dialogs is stored within the shared data structure by the registering application. A notification of the application triggers parsing and translation of the internal state machine representation of the dialog by the speech engine. If dialogs have been registered that require activation by the application another notification is necessary.

Applications provide marker variables within their XML dialog description as described in 4.1. The value of markers is changed when reaching certain states. While traversing the state machine the speech engine sets the markers to the according values. These may be boolean variables for simply reaching the dedicated states or string variables that are filled with default values or even dynamic string values that have been derived from human speech input. The dialog registering application is then notified and may access the state of its marker variable within the shared memory. In the same way the application may have specified variables that are subject to change during runtime. These variables are stored in a different shared memory.

We continue by showing how these mechanisms have been utilized to facilitate application scenarios.

5   Application Scenarios

The topological map, described as one feature of MARVIN in section 3.1 is used for a 'tourist guidance tour' through the RRLAB as demonstration application for speech-based human-machine interaction. This functionality in concept does not differ much from the well-known robotic museum tour guides that have been in use for some time now [21]. We improved the human detection however to enhance the quality of interaction fusing information from different sensors. That is, detection of sound sources with a stereo microphone system [22], of pairs of legs with the laser scanners and consequently of faces with a webcam mounted on a pan-tilt unit helps to perform the speech interaction more naturally. Hence the appearance of a human-being hypothesis in the robot's vicinity triggers the start of the dialog. The lab guidance tour is accomplished by approaching different outstanding positions in the robots map where the robot introduces itself, presents historical information about the research group and explains several

---

[4] http://www.fzi.de/ids
[5] http://sourceforge.net/projects/mca2

Fig. 6. MARVIN giving a tour through the RRLAB

Fig. 7. ARTOS in the Assisted Living Laboratory

actual and future research topics as well as its colleagues, see Fig. 6. To achieve this functionality a new behaviour component has been added to the higher level control system which interacts which the dialog control system to trigger speech synthesis, recognition and robot motion as required.

The robot ARTOS uses speech for giving reminder and system messages within the Assisted Living Lab at Fraunhofer IESE [23] [24]. The Assisted Living Lab is a fully equipped living area, see Fig. 7. Several dialogs may be registered at runtime as described in the following section. They are distinguished as human-triggered and system-triggered. Some examples are given as follows: We have an intelligent refrigerator that checks the expiry of food items using RFID. The dialog description, used as example in section 4.1 represents this funcionality. Expiry warnings are presented to the human by the robot ARTOS. The refrigerator therefore has to register its dialogs at the speech engine of ARTOS. It consists of warning messages containing the food descriptions that are provided by the refrigerator as variable strings. As shown in Fig. 4.1 the dialog differs if one ore more items have expired. Also the robot may be asked by the human about the refrigerator content.

Furthermore dialogs can be dynamically registered by a generic warning and reminder service that collects and distributes information in our intelligent environment. Here the robot serves as true agent for the system, while all other examples are more or less related to its own functionality.

A lot of more applications however are planned to be integrated. A service for reminding elderly people to consume enough liquids and nutrition is already included in the assisted living area. Also a home automation system that controls all door sensors, light and shading switches is available. Two systems to gather the inhabitants location have been integrated. These systems will be added to the speech interface repository in the near future. For example we aspire simple transport functionalities for the robot as well as communication relaying. The human therefore may call the mobile unit to approach his location. Our indoor localisation system will register this interaction dialog at the robot.

6 Conclusion and Outlook

We presented an engine that dynamically processes speech dialogs for human-machine interaction with transactive mobile robotic units in intelligent home environments. The dialogue machine is installed on the mobile robots ARTOS and MARVIN and is continuously used and tested within several applications and scenarios. Especially the possibility to dynamically add and remove dialogs turned out to be helpful for fast prototyping and testing as well as for the ambient character of our test environment.

Apart from the existing appliances a number of additional services will be included within the next months. Further work will be done regarding the speech recognition and synthesis quality. Problems with dialogs starting with the same condition variable or speech input triggers have to be addressed. Also the middleware connectivity that is currently built in MCA shall be migrated to a more open standard.

Acknowledgements

We gratefully acknowledge the funding of the BelAmI project by the State Ministry of Science, Education, Research and Culture of Rhineland-Palatinate, by the Fraunhofer-Society and by the German Federal Ministry of Education and Research.

References

1. Tapus, A., Matarić, M., Scassellati, B.: The grand challenges in socially assistive robotics. IEEE Robotics and Automation Magazine 14(1) (March 2007)
2. Alexandersson, J., Becker, T.: Overlay as the basic operation for discourse processing in a multimodal dialogue system. In: 2nd IJCAI Workshop on Knowledge and Reasoning in Practical Dialogue Systems, Seattle (2001)
3. Dylla, F., Lakemeyer, G.: A speech interface for a mobile robot controlled by golog. In: CogRob-2000 (2000)
4. Bischoff, R., Graefe, V.: Dependable multimodal communication and interaction with robotic assistants. In: 11th IEEE International Workshop on Robot and Human Interactive Communication (2002)
5. Stiefelhagen, R., Fugen, C., Gieselmann, P., Holzapfel, H., Nickel, K., Waibel, A.: Natural human-robot interaction using speech, head pose and gestures. In: IEEE/RSJ International Conference on Intelligent Robots and Systems (2004)

6. Lopes, L.S.: Carl: from situated activity to language level interaction and learning. In: Proceedings of the 2002 IEEE/RSJ Int. Conference on Intelligent Robots and Systems (2002)
7. Matsui, T., Asoh, H., Fry, J., Motomura, Y., Asano, F., Kurita, T., Hara, I., Otsu, N.: Integrated natural spoken dialogue system of jijo-2 mobile robot for office services. In: 16th National Conference on Artificial Intelligence and the Eleventh Innovative Applications of Artificial Intelligence Conference (1999)
8. Ido, J., Matsumoto, Y., Ogasawara, T., Nisimura, R.: Humanoid with interaction ability using vision and speech information. In: Proceedings of IEEE/RSJ International Conference on Intelligent Robots and Systems, pp. 1316–1321 (2006)
9. Toptsis, I., Haasch, A., Hüwel, S., Fritsch, J., Fink, G.: Modality integration and dialog management for a robotic assistant. In: European Conference on Speech Communication and Technology (2005)
10. Wilske, S., Kruijff, G.J.: Service robots dealing with indirect speech acts. In: IROS. IEEE/RSJ International Conference on Intelligent Robots and Systems, Beijing, China (2006)
11. Brooks, R.: The intelligent room project. In: CT. 2nd International Cognitive Technology Conference, Aizu, Japan (1997)
12. Magerkurth, C., Etter, R., Janse, M., Kela, J., Kocsis, O., Ramparany, F.: An intelligent user service architecture or networked home environments. In: 2nd International Conference on Intelligent Environments, Athen, Greece, July 5-6 2006, pp. 361–370 (2006)
13. Schmidt, D., Luksch, T., Wettach, J., Berns, K.: Autonomous behavior-based exploration of office environments. In: 3rd International Conference on Informatics in Control, Automation and Robotics - ICINCO, Setubal, Portugal, August 1-5, 2006, pp. 235–240 (2006)
14. Proetzsch, M., Luksch, T., Berns, K.: Behaviour-based motion control for offroad navigation. In: HUDEM. International Workshop on Humanitarian Demining, Brussels, Belgium (June 16-18, 2004)
15. Hauck, C.: Conception and integration of speech services into an ambient intelligence platform (2006)
16. Nemeth, G., Olaszy, G., Olaszi, P., Kiss, G., Zainko, C., Gordos, G.: Profivox - a hungarian text-to-speech system for telecommunications applications. International Journal of Speech Technology 3, 201–215 (2000)
17. Koutny, I., Olaszy, G., Olaszi, P.: Prosody prediction from text in hungarian and it's realization in tts conversion. International Journal of Speech Technology 3 (2000)
18. Young, S., Evermann, G., Hain, T., Kershaw, D., Moore, G., Odell, J., Ollason, D., Povey, J., Valtchev, V., Woodland, P.: The HTK Book. Microsoft Corporated (2000)
19. Scholl, K.U., Albiez, J., Gassmann, G.: Mca- an expandable modular controller architecture. In: 3rd Real-Time Linux Workshop, Milano, Italy (2001)
20. Anastasopoulos, M., Klus, H., Koch, J., Niebuhr, D., Werkman, E.: Doami - a middleware platform facilitating (re-)configuration in ubiquitous systems. In: International Workshop on System Support for Ubiquitous Computing (UbiSys) at the 8th International Conference of Ubiquitous Computing (Ubicomp), New Port Beach, California, USA (2006)

21. Burgard, W., Cremers, A., Fox, D., Hähnel, D., Lakemeyer, G., Schulz, D., Steiner, W., Thrun, S.: The interactive museum tour-guide robot. In: AAAI-98. Proceedings of the Fifteenth National Conference on Artificial Intelligence (July 26-30, 1998)
22. Lantto, J.: Sound source detection system for control of an autonomous mobile robot, a behaviour-based approach. In: ROMANSY. 16-th CISM-IFToMM Symposium on Robot Design, Dynamics, and Control (2006)
23. Nehmer, J., Karshmer, A., Becker, M., Lamm, R.: Living assistance systems - an ambient intelligence approach. In: ICSE. Proceedings of the 28th International Conference on Software Engineering, Shanghai, China (May 20-28, 2006)
24. Becker, M., Ras, E., Koch, J.: Engineering tele-health in the ambient assisted living lab solutions in the ambient assisted living lab. In: AINA. 21st International Conference on Advanced Information Networking and Applications, Niagara Falls, Canada (2007)

Hand Posture Recognition Using Adaboost with SIFT for Human Robot Interaction

Chieh-Chih Wang[1] and Ko-Chih Wang[2]

[1] Department of Computer Science and Information Engineering and Graduate Institute of Networking and Multimedia, National Taiwan University, Taipei, Taiwan
bobwang@ntu.edu.tw
[2] Graduate Institute of Networking and Multimedia, National Taiwan University, Taipei, Taiwan
casey@robotics.csie.ntu.edu.tw

Summary. Hand posture understanding is essential to human robot interaction. The existing hand detection approaches using a Viola-Jones detector have two fundamental issues, the degraded performance due to background noise in training images and the in-plane rotation variant detection. In this paper, a hand posture recognition system using the discrete Adaboost learning algorithm with Lowe's scale invariant feature transform (SIFT) features is proposed to tackle these issues simultaneously. In addition, we apply a sharing feature concept to increase the accuracy of multi-class hand posture recognition. The experimental results demonstrate that the proposed approach successfully recognizes three hand posture classes and can deal with the background noise issues. Our detector is in-plane rotation invariant, and achieves satisfactory multi-view hand detection.

1 Introduction

When robots are moved out of factories and introduced into our daily lives, they have to face many challenges such as cooperating with humans in complex and uncertain environments or maintaining long-term human-robot relationships. Communication between human and robots instinctively and directly is still a challenging task. As using hand postures/gestures is natural and intuitive for human-to-human interaction and communication, hand detection and hand posture recognition could be essential to human-robot interaction. Figure 1 illustrates an example of human robot interaction through hand posture in which our NTU PAL1 robot and an image from an onboard camera are shown. In this paper, the issues of hand detection and posture recognition are addressed and the corresponding solutions are proposed and verified.

As the Viola-Jones face detector based on an Adaboost learning algorithm and Harr-like features [11] has been successfully demonstrated to accomplish face detection in real time, these approaches are also applied to detect other objects. Unfortunately, it failed to accomplish the hand detection task because of its limited representability on articulated and non-rigid hands [6]. In addition, hand detection with the Viola-Jones detector can be accomplished with about 15° in-plane rotations compared to 30° on faces [5]. Although rotation invariant hand detection can be accomplished using the same Adaboost framework in a way of treating the problem as a multi-class classification

S. Lee, I.H. Suh, and M.S. Kim (Eds.): Recent Progress in Robotics, LNCIS 370, pp. 317–329, 2008.
springerlink.com

 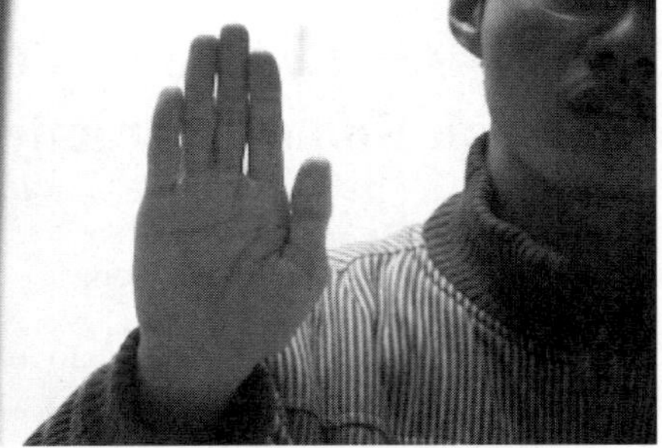

(a) A person interacts with the NTU PAL1 robot via hand posture.

(b) An image from the onboard camera.

Fig. 1. Hand posture based human robot interaction

problem, the training process needs much more training images and more computational power is needed for both training and testing. In this paper, a discrete Adaboost learning algorithm with Lowe's SIFT features [8] is proposed and applied to achieve in-plane rotation invariant hand detection. Multi-view hand detection is also accomplished straightforwardly with the proposed approach.

It is well understood that background noise of training images degrades detection accuracy significantly in the Adaboost learning algorithm. In the face detection applications, the training images seldom contain background noise. However, it is unlikely to show an articulated hand without any background information. Generating more training data with randomly augmented backgrounds can solve this background noise issue with a highly computational cost [1]. With the use of the SIFT features, the effects of background noise in the training stage are reduced significantly and the experimental results will demonstrate that the proposed approach performs with high accuracy.

Given that hand detection is successfully accomplished, hand posture recognition can be done in a way that one classifier/detector is trained for each hand posture class [4]. An *one versus all* strategy is often used where the results from all classifiers are computed and the class with the highest score is labeled as the class of the test image. The computational cost of these sequential binary detectors increases linearly with the number of the classes. The one versus all strategy do not always generate correct recognition. In this paper, we apply a sharing feature concept proposed by Torralba *et al.* [10] to separate sharing and non-sharing features between different hand posture classes. Sharing features of different hand posture classes are used for detecting hand robustly. As non-sharing features represent the discrimination among classes, these non-sharing features are used to increase recognition accuracy and to speed up the recognition process.

The remainder of this paper is organized as follows. Related work is reviewed in Section 2. The details of hand detection using the Adaboost learning algorithm with SIFT features are described in Section 3. Hand posture recognition based the sharing feature concept is described in Section 3.5. Ample experimental results and comparisons are demonstrated in Section 4. Conclusions and future work are in Section 5.

2   Related Work

The Adaboost learning algorithms are currently one of the fastest and most accurate approaches for object classification. Kölsch and Turk [6] exploited the limitations of hand detection using the Viola-Jones detector. A new rectangle feature type was proposed to have more feature combinations than the basic Haar-like features proposed by Viola and Jones. As the feature pool for learning contains about 10^7 features, a highly computational cost is needed for training. Ong and Bowden [9] applied the Viola-Jones detector to localize/detect human hands, and then exploited shape context to classify differences between hand posture classes. Anton-Canalis and Sanshez-Nielsen [1] proposed to collect more training images for reducing the background noise effects. Their approach is to collect images under several controlled illuminations and to randomly augment the training images with various backgrounds to increase the robustness of the detectors. Just *et al.* [4] integrate a variant of Adaboost with a modified censes transform to accomplish illumination invariant hand posture recognition.

In addition to the Adaboost-based approaches, Athitsos and Scalaroff [2] formulated the hand posture recognition problem as an image database index problem. A database contains 26 hand shape prototypes, and each prototype has 86 difference viewpoint images. A probabilistic line matching algorithm was applied to measure the similarity between the test image and the database for recognizing hand posture class and estimating hand pose.

In this paper, the discrete Adaboot learning algorithm is integrated with SIFT features for accomplishing in-plane rotation invariant, scale invariant and multi-view hand detection. Hand posture recognition is accomplished with the sharing feature concept to speed up the testing process and increase the recognition accuracy.

3   Hand Detection and Posture Recognition

In this section, the SIFT keypoint detector and the Adaboost learning algorithm are briefly reviewed. The modification to integrated Adaboost with SIFT and the sharing feature concept are described in detail.

3.1   SIFT

The Scale Invariant Feature Transform (SIFT) feature introduced by Lowe [7] consists of a histogram representing gradient orientation and magnitude information within a small image patch. SIFT is a rotation and scale invariant feature and is robust to some variations of illuminations, viewpoints and noise. Figure 2 shows the extracted SIFT features from five hand images. Lowe also provided a matching algorithm for recognize the same object in different images. However, this approach is not able to recognize *a category of the objects*. Figure 3 shows some examples of hand detection using the SIFT matching algorithm in which most of the pair images only contain less than five SIFT keypoint matches.

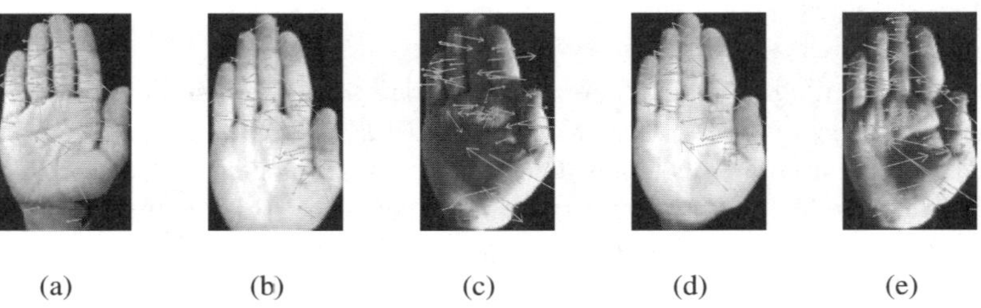

(a)   (b)   (c)   (d)   (e)

Fig. 2. The SIFT features are extracted and shown. From left to right, 67, 57, 63, 70 and 85 SIFT features are extracted from the hand images respectively.

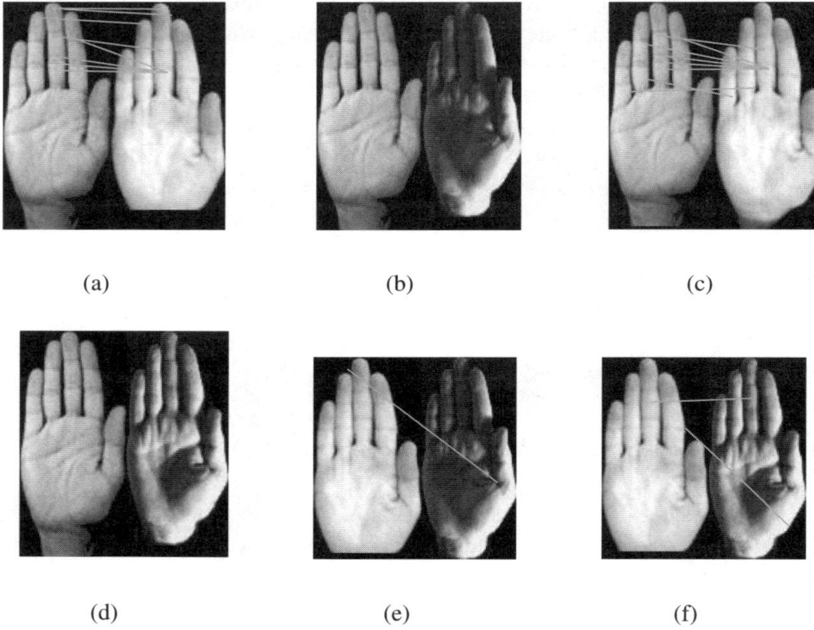

(a)   (b)   (c)

(d)   (e)   (f)

Fig. 3. Hand Detection using the SIFT matching algorithm. Most of the pair images only contain less than five SIFT keypoint matches.

3.2   The Adaboost Learning Algorithm

The Adaboost learning algorithms provide an excellent way to integrate the information of a category of objects. As a single weak classifier can not provide a satisfactory result, Adaboost combines many weak classifiers to form a strong classifier in which a weak classifier can be slightly better than randomly guess to separate two classes. Given a set of positive and negative images, the Adaboost learning algorithm chooses the best weak classifier from the large pool. After choosing the best weak classifier, Adaboost adjusts the weights of the training images. The weights of misclassified training images of this round are increased and the weight of correct ones are decreased. In the next round, the Adaboost will focus more on the misclassified images and try to correctly classify the misclassified images in this round. The whole procedures are iterated until a predefined performance requirement is satisfied.

3.3   Adaboost with SIFT

Our hand detection approach applies Adaboost with SIFT features. Compared to the existing Adaboost-based hand detection approaches, the proposed approach has three advantages. First, thanks to the scale-invariant characteristic of SIFT, it is unnecessary to scale the training images to a fixed resolution size to adapt the characteristic of the Harr-like features in the original Viola-Jones approach. Second, rotation-invariant and multi-view detection is straightforwardly accomplished because of the rotation-invariant characteristic of SIFT features. Finally, the background noise issue is taken care easily. During the training data collection stage, the background of positive training images is set to a single color without any texture. Therefore, the extracted SIFT features from the positive training images exist only in the hand areas of the images. The classification performance is achieved without increasing the number of training samples.

Here the modifications to integrate the discrete Adaboost with SIFT features are described in detail. Let $\{I_i, i = 1, \ldots, N\}$ be the training image set where N is the number of the training images. Every image is associated with a label, $\{l_i, i = 1, \ldots, N\}$ and $l_i = 1$ if the image contains a hand, otherwise $l_i = 0$. Each image is represented by a set of SIFT features $\{f_{i,j}, j = 1, \ldots, n_i\}$, where n_i is the number of SIFT features in image I_i.

The weights, $\frac{1}{2N_p}$, $\frac{1}{2N_n}$, are initially set to positive training samples and negative training samples respectively where N_p is number of positive sample and N_n is number of negative sample. Each weak classifier, h_m, consists of a SIFT keypoint (f), a threshold (t) and a polarity (p).

A weak classifier, h_m, is defined as:

$$h_m(I_i) = \begin{cases} 1, \text{ if } \ p * f_m(I_i) < p * t \\ 0, \text{ } otherwise \end{cases} \tag{1}$$

The next step is to choose m weak classifiers and combine them into a strong one. Our detector uses the function $F(f, I_i) = min_{1 \leq j \leq n_i} D(f, f_{i,j})$, where D is Euclidean distance, to define the distance between an image and a feature. Algorithm 1 shows the details of classification using the discrete Adaboost learning algorithm with SIFT features.

3.4   Hand Detection

With the learned SIFT features, hand detection is accomplished as follows. The SIFT features are firstly extracted from the test image. For each weak classifier, the distance between its associated SIFT feature and the extracted SIFT features from the test image

---

Algorithm 1. Training using the Discrete Adaboost Algorithm with SIFT features

---

Require: Given training images $(I_1, l_1) \ldots (I_n, l_n)$ where $l_i = 0, 1$, for negative and positive examples respectively.

1: Initialize weights $w_{1,i} = \frac{1}{2N_p}, \frac{1}{2N_n}$ for $l_i = 0, 1$ respectively, where N_p and N_n are the number of negatives and positives respectively.

2: **for** $m = 1, \ldots, T$ **do**

3:     Normalize weight of all training samples such that $\sum_{i=1}^{N} w_{m,i} = 1$;

4:     Choose a SIFT keypoint feature (f_m), a threshold (t_m) and a polarity (p_m) to form a weak classifier such that the error is minimize. We define the error is;

$$e_m = \sum_{i=1}^{N} w_{m,i} |h_m(I_i) - l_i| \qquad (2)$$

5:     Define $h_t(x) = h(x, f_m, p_m, t_m)$ where f_m, p_m, t_m are the minimizers of e_m

6:     Update the weights:

$$w_{m+1,i} = w_{m,i} \beta_m^{1-e_i} \qquad (3)$$

   where $e_i = 0$ if example I_i is classified correctly, $e_i = 1$ otherwise, and $\beta_m = \frac{e_m}{1-e_m}$

7: **end for**

8: The final strong classifier is:

$$H = \begin{cases} 1, & \text{if} \quad \sum_{t=1}^{m} \alpha_t * h_t > T \\ 0, & otherwise \end{cases} \qquad (4)$$

   where $\alpha_t = \log \frac{1}{\beta_t}$

9: **return** strong classifier $H(x)$

---

are computed. The best match with the shortest distance shorter than a threshold t_m is treated as a valid result from this weak classifier. Then the weight factors α_m of all valid weak classifiers are summed. If this summed value is greater than the threshold of the strong classifier, T, the test image is classified as a hand image. Algorithm 2 shows the details of the hand detection algorithm.

3.5   Multi-class Recognition

With the use of the proposed hand detection method, multi-class hand posture recognition is done using the sharing feature concept. As different object classes still have sharing and non-sharing features, our method use non-sharing features to speed up the recognition process with a higher accuracy than the one versus all approaches.

In the detection phase, the sharing feature set is used to detect hand robustly. If the test image does not exist any sharing feature, the image is discarded. In the posture recognition stage, only non-sharing features are used in the sequential classification process. The class with the highest score is labeled as the image class. It should be noted that the current system trains each classifier independently. All classifiers could be trained jointly to get a better performance.

Algorithm 2. Detection

Require: Given a strong classifier (T, W), T is the threshold of strong classifier. $W : (h_1, ..., h_m)$ is a set of weak classifiers. h_i consists of $(\alpha_i, f_i, t_i, p_i)$. α_i, f_i, t_i, p_i are the weight, SIFT feature, threshold and polarity of h_i, respectively. An image: I

1: Initialize $WeightSum = 0$
2: S = Extracting SIFT features from I
3: **for** $i = 1, \ldots, m$ **do**
4:     S_x = Find the nearest SIFT feature of f_i in S
5:     **if** EuclideanDistance(S_x, f_i)* $p_i < t_i$ * p_i **then**
6:         $WeightSum + \alpha_i$;
7:     **end if**
8: **end for**
9: **if** $WeightSum > T$ **then**
10:     return 1
11: **else**
12:     return 0
13: **end if**

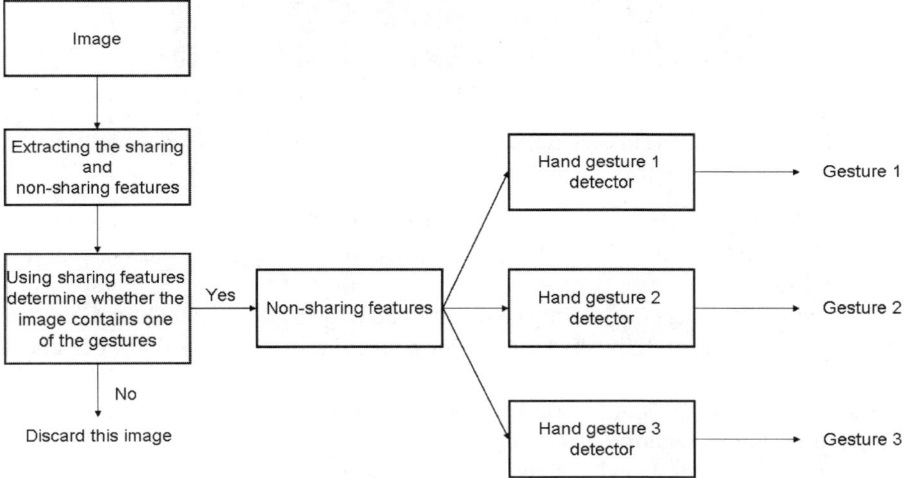

Fig. 4. Multi-class hand posture recognition using the sharing feature concept

4   Experimental Results

In this paper, three targeted hand posture classes, "palm", "fist" and "six", are trained and recognized. 642 images of the "palm" posture class from the Massey hand gesture database provided by Farhad Dadgostar *et al.*[1] are used as positive samples. As the Massey hand gesture database does not contain images of "fist" and "six", 450 "fist" images and 531 "six" images were collected by ourself under different lighting

---

[1] http://www.massey.ac.nz/fdadgost/xview.php?page=farhad

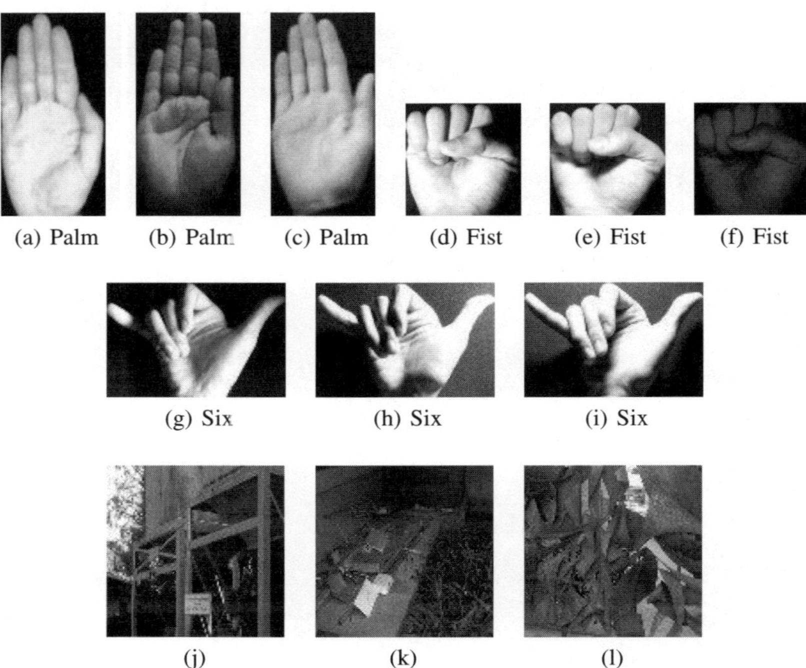

(a) Palm   (b) Palm   (c) Palm   (d) Fist   (e) Fist   (f) Fist

(g) Six   (h) Six   (i) Six

(j)   (k)   (l)

Fig. 5. The training images of the "palm", "fist" , "six" and the backgrounds

conditions. The negative/background images are consist of 830 images from the internet and 149 images collected in the building of our department. Figure 5 shows examples of the training data.

For testing, 275 images were collected using the onboard Logitech QuickCam Pro 5000 with a resolution of 320x240. Figure 6 shows the sharing and non-sharing features determined by our algorithm.

Figure 7 shows samples of correct hand detection and posture recognition using the proposed algorithms. Figure 8 shows some of correct hand detection but incorrect posture recognition. Tables 1 and 2 show the performances of multi-class hand posture recognition using our proposed detection algorithms without and with using the sharing feature concepts. The experimental results show that the approach using the sharing feature concept is superior.

We will show the quantitative results in terms of background noise, in-plane rotation variant recognition and multi-view recognition.

4.1   Background Noise

The performances of hand detection using the Viola-Jones detector and the proposed approach are compared.

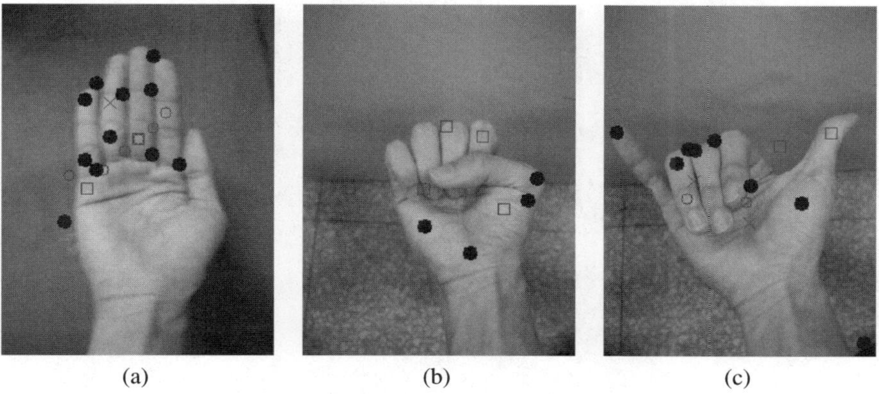

(a)   (b)   (c)

Fig. 6. Sharing and non-sharing features. Blue solid circles indicate sharing features. Red circles indicates non-sharing features detected by the "palm" detector. Red rectangles are non-sharing features detected by the "fist" detector. Red 'X' indicates a non-sharing feature detected by the six detector. Note that the weights of the detected features are different.

Table 1. Hand Posture Recognition without using the sharing feature concept

Truth	PALM	FIST	Result SIX	Total	Accuracy
PALM	80	3	8	91	87.9%
FIST	0	94	4	98	95.9%
SIX	3	7	76	86	88.3%
Total				275	90.9%

Table 2. Hand Posture Recognition using the sharing feature concept

Truth	PALM	FIST	Results SIX	Total	Accuracy
PALM	89	2	0	91	97.8%
FIST	1	97	0	98	98.9%
SIX	3	6	77	86	89.5%
Total				275	95.6%

The training results are shown by the ROC curve of the detector in Figure 9. Given the 90% detection rate, hand detection using the Viola-Jones detector generates 3 - 4% false positive rate. This result is unsatisfactory and is worse than the same approach applied on face detection. As the face training images almost do not contain any background. The experimental results show that background noise in the training images could degrade the performance of the Adaboost learning algorithm significantly. On the contrary, the results of hand detection using Adaboost with SIFT are satisfactory.

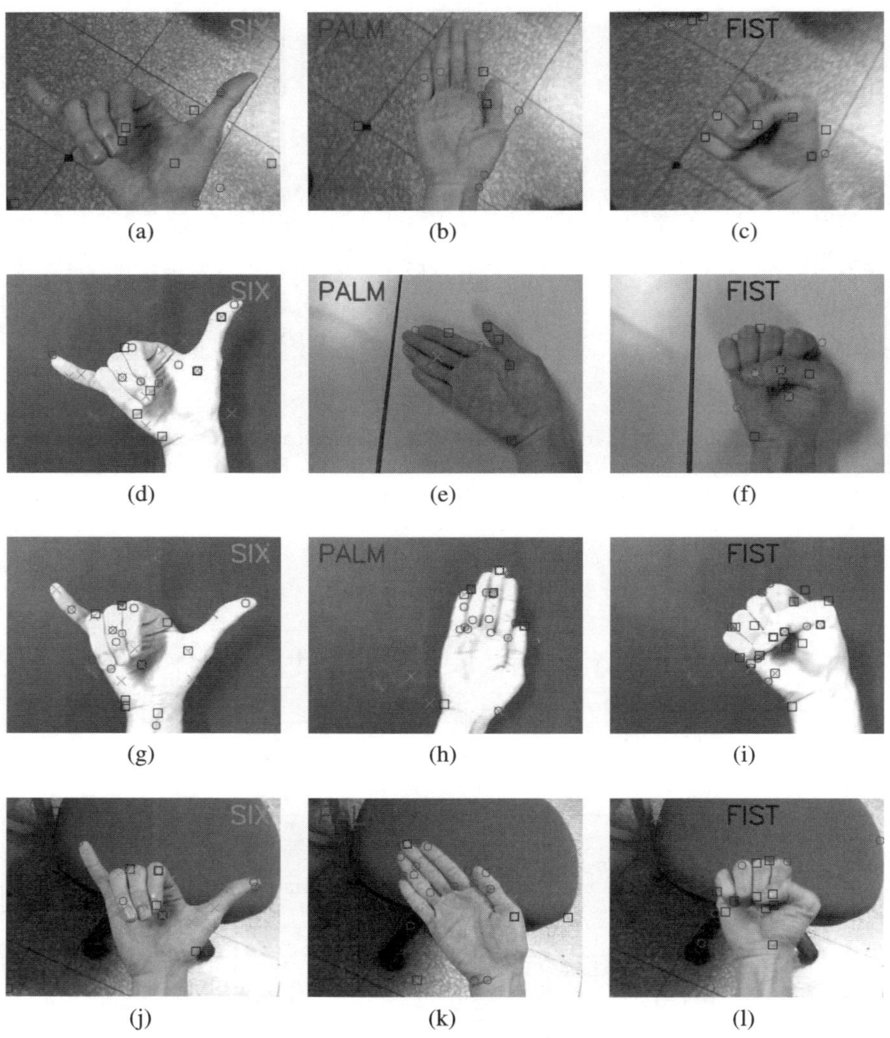

Fig. 7. Correct hand gestures recognition results

4.2   Rotation Invariant Recognition

Figure 10 shows the ROC curve in which the in-plane rotation invariant recognition is demonstrated. Three data sets were collected to test $0°$, $90°$ and $-90°$ of in-plane rotations. It is clearly shown that the performances between different in-plane rotations are very similar. The proposed approach use only one detector to accomplish in-plane rotation invariant hand detection.

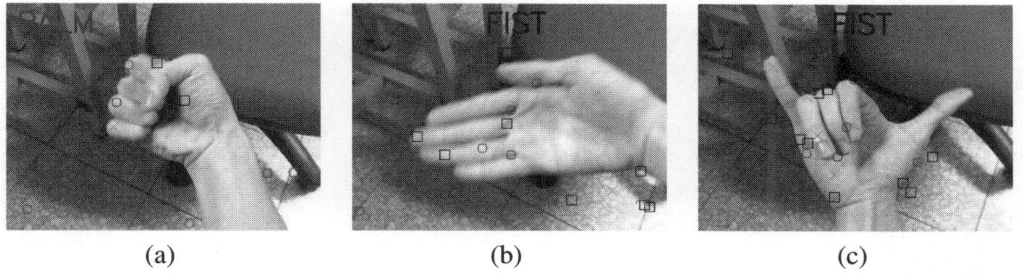

(a)   (b)   (c)

Fig. 8. Incorrect hand gestures recognition results. In (a), the palm detector may detect too many features between fingers and the background. In (b), the test image could be too blurred. In (c), both the six detector and the fist detector found many features.

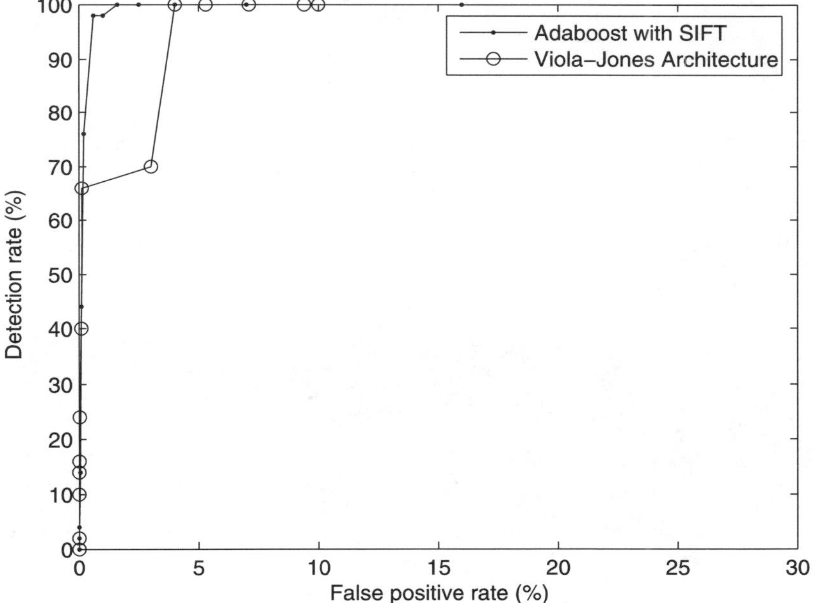

Fig. 9. The ROC curves of hand detection using the Viola-Jones detector and the proposed approach

4.3   Multi-view Recognition

Here we further verify if the proposed approach can achieve multi-view hand posture detection. Although more data from different viewpoints can be collected and trained to achieve multi-view hand detection, only data with a fixed viewpoint are used in this experiment. Figure 11 shows that the detector can still work in the situation of the 40 degree viewpoint .

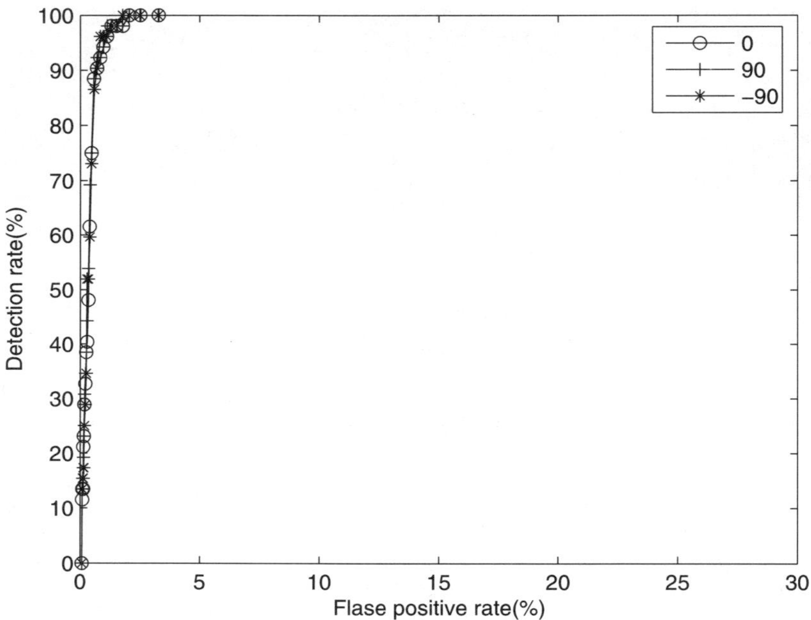

Fig. 10. The ROC curve shows that the proposed approach accomplishes in-plane rotation invariant recognition

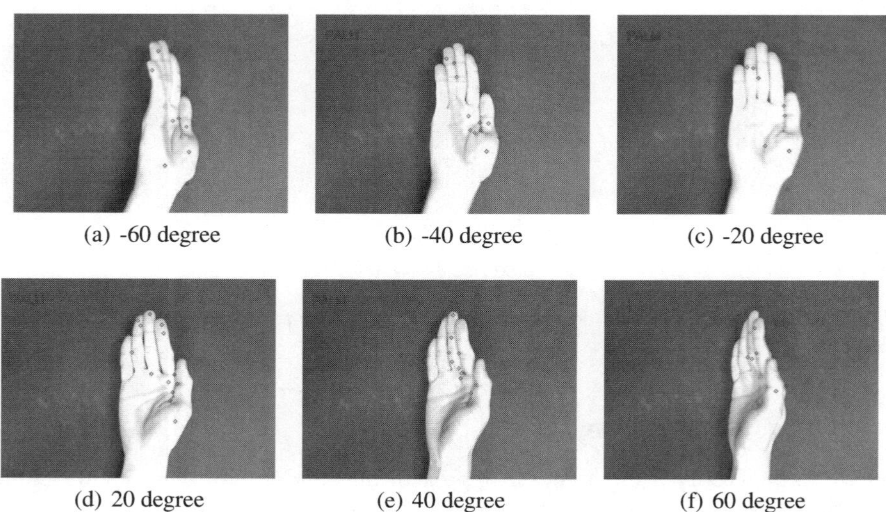

(a) -60 degree   (b) -40 degree   (c) -20 degree

(d) 20 degree   (e) 40 degree   (f) 60 degree

Fig. 11. Multi-view hand detection. The detector works until that the viewpoint is larger than 40 degree. The images with red "PALM" texts are the correct recognition results.

5   Conclusion and Future Work

In this paper, we presented a robust hand detection and posture recognition system using Adaboost with SIFT features. The accuracy of multi-class hand posture recognition is improved using the sharing feature concept. The experimental results demonstrated that the proposed hand detector can deal with the background noise issues. Our detector is in-plane rotation invariant, and achieves satisfactory multi-view hand detection.

The future work is to add more hand posture classes for analyzing the performances and limitations of the proposed approaches. Different features such as contrast context histogram [3] will be studied and applied to accomplish hand posture recognition in real time. The system will be integrated with the NTU PAL1 robot for performing human-robot interaction. It should be of interest to study the methodology of jointly training and testing multiple hand posture classes.

Acknowledgments

We acknowledge the helpful suggestions by an anonymous reviewer. This work was partially supported by grants from Taiwan NSC (#95-2218-E-002-039, #95-2221-E-002-433); Excellent Research Projects of National Taiwan University (#95R0062-AE00-05); Taiwan DOIT TDPA Program (#95-EC-17-A-04-S1-054); and Intel.

References

1. Anton-Canalis, L., Sanchez-Nielsen, E.: Hand posture dataset creation for gesture recognition. In: VISAPP 2006. International Conference on Computer Vision Theory and Applications, Setúbal, Portugal (February 2006)
2. Athitsos, V., Sclaroff, S.: Estimating 3d hand pose from a cluttered image. Computer Vision and Pattern Recognition (2003)
3. Huang, C.-R., Chen, C.-S., Chung, P.-C.: Contrast context histogram - a discriminating local descriptor for image matching. In: ICPR. International Conference of Pattern Recognition (2006)
4. Just, A., Rodriguez, Y., Marcel, S.: Hand posture classification and recognition using the modified census transform. In: AFGR. IEEE International Conference on Automatic Face and Gesture Recognition (2006)
5. Kölsch, M., Turk, M.: Analysis of rotational robustness of hand detection with a viola-jones detector. In: ICPR 2004. The 17th International Conference on Pattern Recognition (2004)
6. Kölsch, M., Turk, M.: Robust hand detection. In: IEEE International Conference on Automatic Face and Gesture Recognition (2004)
7. Lowe, D.G.: Object recognition from local scale-invariant features. In: International Conference on Computer Vision (1999)
8. Lowe, D.G.: Distinctive image features from scale-invariant keypoints. International Journal of Computer Vision 60(2), 91–110 (2004)
9. Ong, E.J., Bowden, R.: A boosted classifier tree for hand shape detection. In: AFGR. IEEE International Conference on Automatic Face and Gesture Recognition (2004)
10. Torralba, A., Murphy, K., Freeman, W.: Sharing features: efficient boosting procedures for multiclass object detection. In: CVPR. IEEE Conference on Computer Vision and Pattern Recognition (2004)
11. Viola, P., Jones, M.J.: Robust real-time face detection. International Journal of Computer Vision 57(2), 137–154 (2004)

Multimodal Navigation with a Vibrotactile Display in Computer Assisted Surgery

Melina Brell[1,2] and Andreas Hein[2]

[1] Division of Automation and Measurement Technology (AMT), KISUM /
Department of Computing Science - University of Oldenburg, Germany
`melina.brell@informatik.uni-oldenburg.de`
[2] International Graduate School for Neurosensory Science and Systems, Medical
Physics, Faculty V, University of Oldenburg, Germany
`andreas.hein@informatik.uni-oldenburg.de`

Summary. A new concept of a tactile human-machine interface for surgical interventions via multimodal computer navigated surgery is presented in this paper. In contrast to conventional computer navigated surgery, information about the surgeon are explicitly processed and used to augment the human machine interaction by a vibro-tactile display. By means of this display the system is extended by non visual data presentation through tactile information transmission. Different tactile activation schemes (mono linear, duo linear) in which different numbers of tactors are activated to transmit direction information, are tested and compared. The experimental evaluation adds up with a mean error of -0.35 mm and a standard deviation of 0.51 mm for the mono linear scheme and a mean error of -0.01 mm and a standard deviation of 0.62 mm for the duo linear scheme.

1 Introduction

The advanced development of imaging technologies like computer tomography (CT) or magnetic resonance imaging (MRI) is primarily responsible for the entrance of computer navigated surgery in clinical practice. Although the imaging technologies themselves offer the surgeon a lot of advantages like the possibility of a 3-d reconstruction image of the patient and the possibility of a preoperative planning of the intervention, they did not solve the problem of the correlation between this pre-operative planning data and the real circumstances in the operating room. During the intervention the patient never has the same position as he had when the image dataset was taken. Finding a certain, preplanned position at or in the patient or finding the intended, right adjustment of an instrument like a needle are other problems, such as high accuracy when handling fine structures. Even bodily fluids or a too small aditus during a minimally invasive surgical intervention can cause poor visibility. Since the design of novel stereo cameras as pose measurement systems computer aided surgery has become common practice in complex interventions as neuron- or spine surgery to support the surgeon and solve the above mentioned problems. With the aid of a planning component the surgeon can indicate fine structures, vessels or nerves as well as pre-planned incision lines, work areas or safety areas can be

S. Lee, I.H. Suh, and M.S. Kim (Eds.): Recent Progress in Robotics, LNCIS 370, pp. 331–343, 2008.
springerlink.com

defined. After an initial registration between the system and the image data set of the patient the system can show the relative pose between the instrument and the patient and provides visibility on a screen while visibility is poor in reality.

2   State of the Art

State of the art for supporting surgical interventions is computer-aided surgery (CAS). Systems of CAS can be divided into navigation systems and robot systems.

2.1   Navigation Systems

systems can be classified in two different ways. The first classification depends on the kind of the supporting scheme. The systems can provide image guided navigation that shows the relative position of the instrument at the patient. These image data sets of the patient can also be taken intraoperativ. These systems provide navigated imaging [1] supplemental. In the third variation the systems have an additional planning component to plan the intervention [2]. The paper will focus on the third variation of plan based navigation systems. Another way of classifying the systems depends on the measurement technology of the integrated pose measurement systems. Integrated pose measurement systems can either be electromechanical, electromagnetic or optical systems. Electromechanical systems measure the positions by a mechanical construction (Leksell Stereotactic system, Elekta Inc., Stockholm, Sweden; Mayfield-clamp, Integra Radionics). Electromagnetic measurement systems use the technology of electro-magnetic fields to sense special coils (extended version of the ARION [3]). Optical systems contain stereo cameras (CCD, infrared, etc.) and are based upon the principle of triangulation of active or passive markers (VectorVision, BrainLab). Because optical systems do not restrict the surgeon's freedom of movement, they are the most frequently used systems in computer navigated surgery. Navigation systems are a good accepted concept because they lead to higher accuracy due to the possibility of planning. Nevertheless the actual disadvantage of navigations systems is the limitation to only visual indication of the important navigation information via a monitor. Normally the surgeon is looking at the field of surgery during the intervention. He is constrained to avert the gaze from the field of surgery and the patient to look at the monitor and get the information which causes the concentration of the surgeon to be disturbed (see Fig. 1). The perspective of the scene displayed is not necessarily identical to the surgeon's perspective on the patient, so that the surgeon has to transform the displayed information. Mini displays [4] which also can be fixed directly at the instrument are a coming trend but actual not widely used. Head mounted displays [5] and projection systems are in development, but the acceptance by surgeons is not clear.

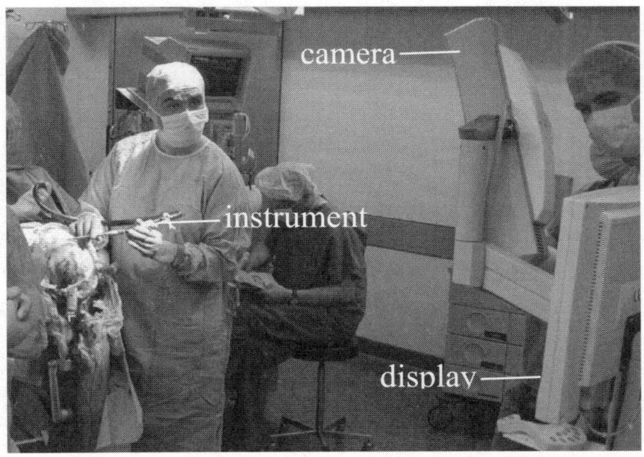

Fig. 1. Disadvantage of Computer aided surgery: The surgeon has to avert the gaze from the field of surgery (Charit Berlin, Germany, 2003)

2.2 Robot Systems

Robotic systems cover a wide field of surgical applications and can be distinguished by their control principles in automatic, telemanipulative and interactive systems as described in [6]. Interactive robotic systems can be termed as expansions of navigation systems. Beside the navigation component they are able to hold, guide, control and/or position surgical instruments in interaction with the surgeon. Aside to the advantages of the navigation this results in still higher accuracy, because forces can be scaled and safety areas can be prohibited. Next to the disadvantages the navigation component implicates other problems are the limited work area of most robot systems and the loss of the tactile feedback of the surgeon. Due to that robot systems are not as good accepted as navigation systems. To solve these problems possible alternatives to transmit navigation information have been surveyed which combine the advantages of navigation and robot systems. Acoustical transmission did not seem to be a suitable alternative, because navigation information is too complex to be transmitted by acoustical signals and can be disordered by background noise. Tactile signals are an alternative, because complex, spatial information can be coded by tactile signals [7] for navigation tasks, the tactile feedback of the surgeon is kept and the work are is not limited. The usage of tactile signals for information transmission is discussed in the next section.

2.3 Tactile Information Transmission

Tactile Displays in Medical Applications

Generally tactile information is provided by tactile displays in several ways. Tactile displays are containing specialized actors (called tactors) like vibration motors to generate a tactile sensation on the skin. To provide stable information transmission tactile displays can be integrated in garment like gloves or other textiles as flexible bands or belts equipped with special actors [8]. Instruments can also be expanded with tactors to provide a tactile feedback of the sensations occurring on the tip of the instrument [9, 10]. Nevertheless this principle is not yet established in computer aided surgery. It is rarely used during real interventions but rather for medical training applications or in robotics and other navigation applications as discussed in the next section. More often the use of tactile signals on the fingertip and palm for tactile feedback in telemanipulative tasks or augmented reality can be found. The perception of tactile stimuli is investigated extensively at these parts of the body so the perception can be easily used in telemanipulative robotic or medical applications. Even here tactile displays are used in clothing for example in gloves or in tool holders of telemanipulative kinematics systems [11]. They are used to feel the surface of virtual or remote objects and to facilitate their handling but not for positioning tasks.

Tactile Pattern for Guidance Tasks

Tactile signals are even used to transmit directional information to guide a person. The kind of signal coded with the tactile stimulation is determined by the spatial arrangement of the tactors and the arising stimuli. By means of these patterns directional information can be transmitted. Tactile signals for direction are actually often vibrations and only used for navigation of a person itself but not in combination with a navigation or robot system in computer aided surgery. For the navigation of a person the tactile displays are fixed via flexible bands directly or indirectly on the user's skin [7]. The displays can likewise be fixed in the clothing [12].

3   Concept

Recapitulating one can say that there is no tactile display in medical applications for the navigation of the surgeon or the surgical instruments. To fill this gap the navigation system conTACT [13] is developed. The system is used in computer aided surgery with a new kind of multimodal spatiotemporal information transmission via tactile signals. The concept of human-machine-interface is based on the assumption that the position of the surgeon's hand can be affected by the tactile signals. To achieve a constant signal transmission a vibrotactile human-machine-interface is constructed that contains vibration motors (cylindrical motors, part 4TL-0253B, JinLong Machinery, China) which are arranged on the back of the hand so that the surgeon's dexterity is not disturbed. First results of the system setup and tactile pattern perception on the back of the hand with different modes are discussed in [13]. Since the pose measurement system provides a real time (15Hz) high precisions information sampling, it is possible to process complex situations consisting of coordinate systems for the patient (*pat*), the instrument (*tool*) the surgeon's hand (*hand*) and the human-machine-interface (*tac*). So in addition to the position of the patient and the instrument also the position of surgeon's hand and the actors can be measured and directional information can be processed by the tactile display in correct positional arrangement. The respective items are described in detail in the next sections.

3.1 Navigation System Overview

The navigation system is to be classified in the group of plan based navigation systems with an integrated optical (MicronTracker, Claron Technology, Toronto, Canada) or electromagnetic pose measurement system (Aurora, NDI, Waterloo, Canada). For the communication with the pose measurement system a uniform interface has been implemented. Furthermore the system contains all typical elements of a navigation system like imaging components to load preoperatively taken image data sets and a planning component to define work or safety areas. To correlate the image data sets with the anatomy of the patient there are integral parts for the registration of the patient and the calibration of the surgical instrument. After the registration and calibration the actual positions of the patient and instrument are interpreted as spatial data in relation to plan data. The results of this analysis are sent to an embedded controller part which sends control values to the communication interfaces of the user (tactile human-machine-interface, screen). The paper will focus on the tactile human-machine-interface, its calibration and signal transmission. An Overview of the system is shown in Fig. 2.

The non visual data channel to communicate with the surgeon is the tactile signal transmission. The signal is transmitted by a vibrotactile human-machine-interface as mentioned above. The display arranges a certain number of tactors at quasi static positions on the back of the hand and the fingers of the surgeon (see Fig. 3). The remaining distance and direction to the target point or border is

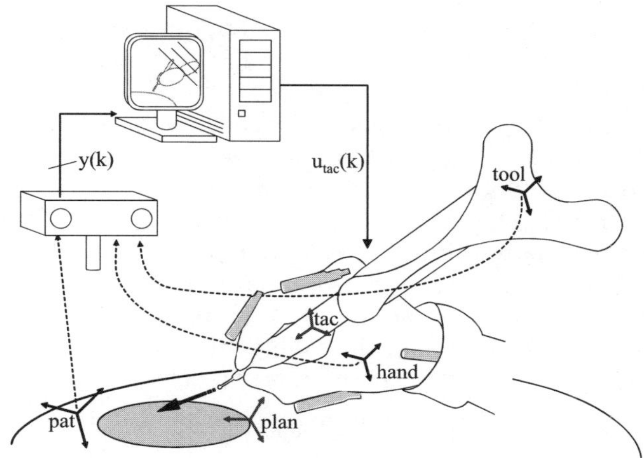

Fig. 2. Overview of the system including the information flow

transformed to the tactors coordinate system and afterwards coded by the signal form of the actors. Thereby the position of the tactor on the back of the hand determines the direction towards the target point or border. It is assumed that the surgeon's motor reaction is affected by the tactile signals in a way preferable for the navigation process.

Calibration and Registration

All real objects like the patient and the instrument are represented by virtual objects in the system. Each virtual object is containing an image data set. The patient's virtual object is composed of CT or MRI images and the instruments are compost of some STL-Data. After loading these image objects the registration between the real world situation and the image objects has to be built up. This step is not only needed to register the patient but also to calibrate the instruments and the human machine interface. Only if at least one instrument is calibrated, it can be used to register the patient and the human-machine interface To measure the position and orientation of the instruments, markers of the pose measurement system are fixed on each instrument (below called *tool*). During the calibration step of the instrument the transformation between the real object represented by the tool marker and the internal coordinate system of the STL object is generated. In addition a transformation matrix $^{tool}\mathbf{T}_{tcp}$ is built up which describes the transformation from the fixed marker to the tool center point (called *tcp*). By means of the rotation and position of the marker the position and rotation of the whole STL object (as the instrument) can be deduced and displayed in the 3-d scene of the system. If the transformation between the tool marker and the tcp is known, the global pose of the tcp and the instrument in the camera coordinate system *cam* can be calculated:

$$^{cam}\mathbf{T}_{tcp} = ^{cam}\mathbf{T}_{tool} \cdot ^{tool}\mathbf{T}_{tcp} \tag{1}$$

Now the instrument can be used to register the patient. There are several methods known to register the patient. The basics are described by [14]. By means of the registration a transformation matrix $^{pat}\mathbf{T}_{ima}$ between the patient marker *pat* and the internal coordinate system of the patient's image data set *ima* is calculated. As well as the instrument the tactile human-machine-interface must be calibrated to transform the navigation information in the coordinate system of the human-machine-interface. Because it is not realizable that each tactor has its own maker, there is one marker (below called *hand*) for the whole tactors' (human-machine-interface) coordinate system in which the positions of the tactors are defined. This coordinate system is built out of three defined position on the hand of the surgeon concerning the hand marker (Fig. 3). By touching the first position *p1* with the tip of the calibrated tool the origin of the tactors coordinate system is defined by the position part $\mathbf{p}$ of the transformation matrix:

$$^{hand}\mathbf{p}_{pi} = ^{hand}\mathbf{T}_{cam} \cdot ^{cam}\mathbf{p}_{tcp} \tag{2}$$

Together with this origin the second and third positions are used to define the x- and y-axes of the new coordinate system of the human-machine interface. By

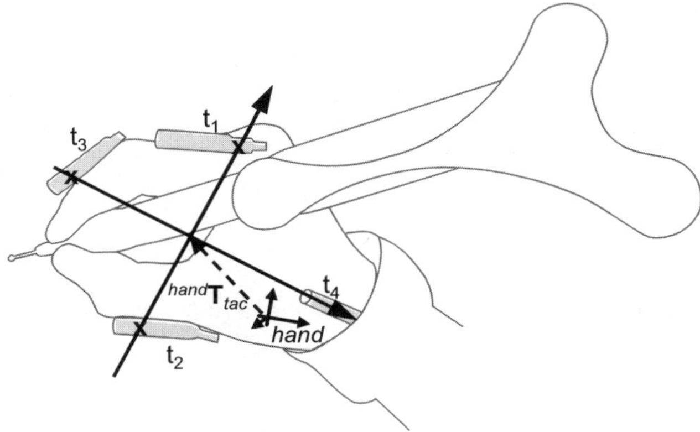

Fig. 3. Calibration of the tactile human-machine-interface

means of the cross product of the two axes the z-axis can be calculated. To assure orthogonality of the matrix, the y-axis is even redefined by the cross product. The result is a transformation matrix $^{hand}\mathbf{T}_{tac}$ from the measurable hand marker to the virtual tactor coordinate system tac of the human-machine-interface. This construction implies that the tool is grasped by the surgeon in the same pose during the calibration as during the usual handling. After calibrating the tool the tactors can be positioned free on the back on the hand. Their positions are saved in the tactors coordinate system even by touching them with the tip of the instrument. The way the navigation information is presented via the tactors in the tactors' coordinate system is described in section 3.2. Actual only the planar movement is determined. The 3-d STL objects are separated in 2-d outlines by the section of the STL object along the plane defined by the tactors at the tcp position (Fig. 4).

Plan Objects and Spatial Interpretation

To increase the accuracy and to minimize the risks of a surgical intervention, the intervention can be planned beforehand. The planning takes place in the preloaded image data set of the patient and thus directly in the coordinate system of the image data. Certain structures can be defined as working areas or safety areas (Fig. 4) during the planning. The defined areas are saved in a set of STL objects as described above. On account of the registration the planning data can even be visualized in the 3-d scene of the navigation system.

To protect sensible structures, vessels or veins the system will warn the surgeon during the intervention if the instrument leaves a working area or reaches a safety area (Fig. 5).

To check, whether the instrument comes closer the border of the defined area, the actual spatial arrangement of the instrument relative to the patient $y(k)$ must be analyzed and interpreted. Therefore the positions of the instrument and the patient at time k are measured in terms of transformation matrices

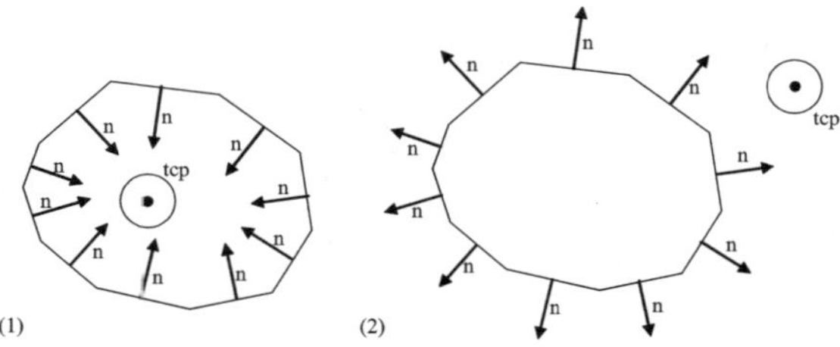

Fig. 4. 2-d planning data as working (1) or safety areas (2)

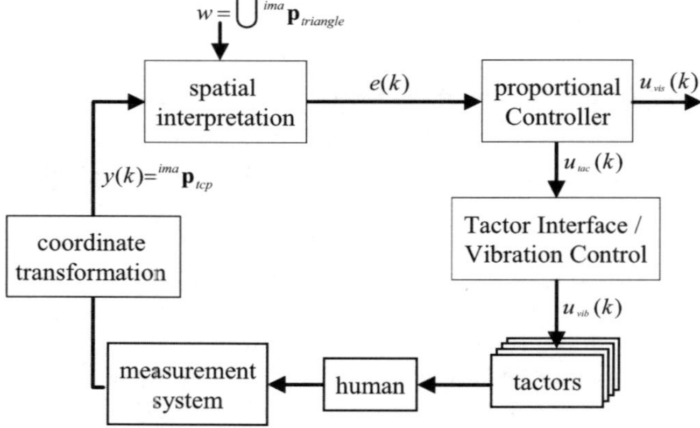

Fig. 5. Control cycle of the tactile navigation

between the pose measurement system and the markers (*pat, tool*) $^{cam}\mathbf{T}_{pat}$ and $^{cam}\mathbf{T}_{tool}$. Via matrix multiplication of the transformation calibration matrices and the actual pose information the actual tcp position of the instrument inside of the patient's image data $^{ima}\mathbf{P}_{tcp}$ is calculated:

$$\mathbf{y}(k) = {}^{ima}\mathbf{P}_{tcp} = {}^{ima}\mathbf{T}_{pat} \cdot {}^{pat}\mathbf{T}_{cam} \cdot {}^{cam}\mathbf{T}_{tool} \cdot {}^{tool}\mathbf{P}_{tcp} \tag{3}$$

The deviation vector $e(k)$ between the border of the planned data and the tcp is the distance of the tcp to the nearest facet of the plan data's STL object surface w. This distance is calculated by the difference of the tcp position $^{ima}\mathbf{P}_{tcp}$ and the projection of the tcp position $^{ima}\mathbf{P}_{tcpProj}$ via the normal n onto the area of the facet considering the radius r of the spheric milling head (4) as shown in Fig. 6 If the projected point does not meet the facet the nearest vertex or the intersection with straight line between the two nearest vertices is used.

$$\mathbf{e}(k) = \frac{{}^{ima}\mathbf{P}_{tcpProj} - {}^{ima}\mathbf{P}_{tcp}}{\left|{}^{ima}\mathbf{P}_{tcpProj} - {}^{ima}\mathbf{P}_{tcp}\right|} \cdot \left(\left|{}^{ima}\mathbf{P}_{tcpProj} - {}^{ima}\mathbf{P}_{tcp}\right| - r\right) \tag{4}$$

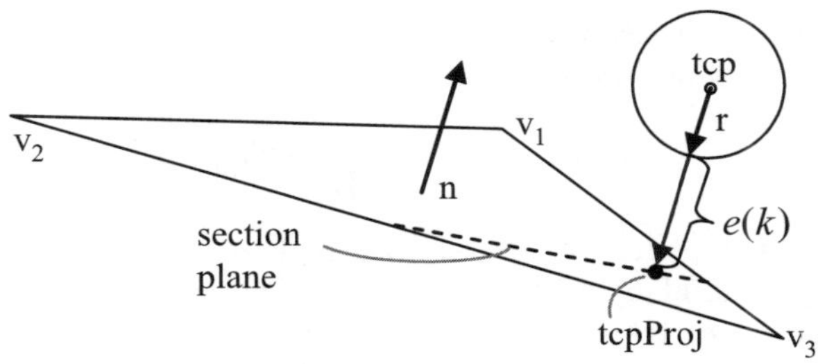

Fig. 6. Calculation of the deviation between one facet of the planning data's surface and the tcp position

Subsequently the controller part of the navigation system processes the deviation vector $e(k)$ and computes an adequate control vector $u(k)$ for the visualization and the tactor interface and so the output to the surgeon. The controller is a proportional controller with the proportional factor k_p and limitation of the control signal:

$$\mathbf{u}(k) = \begin{cases} u_{vis}(k) = k_{pvis} \cdot e(k) \\ u_{tac} \end{cases} \tag{5}$$

$$\mathbf{u}_{tac}(k) = \begin{cases} 0 & , |e(k)| > e_{max} \\ \frac{e(k)}{|e(k)|} \cdot (k_{ptac} \cdot |e(k)| + u_{max}) & , 0 \leq |e(k)| \leq e_{max} \\ \frac{e(k)}{|e(k)|} \cdot u_{max} & , |e(k)| < 0 \end{cases} \tag{6}$$

$$k_{ptac} = -\left(\frac{u_{max} - u_{min}}{e_{max}} \right) \tag{7}$$

3.2   Tactor Interface and Vibration Control

There are several possibilities to code a distance with a vibration signal: the amplitude, the frequency and a superposed pulse or frequency modulated signal. A modulation of the amplitude is hardly perceivable and most of the time sensed as a variation in frequency. Modulation of the frequency is overall perceived in a very low range and does not produce the desired effect. In contrast to the modulation of the amplitude or the frequency a variation in a superposed signal causes a well perceivable changing pulse effect. A similar problem of coding the distance in a vibration signal is discussed by [15] for waypoint navigation of a person. In the conTACT system a variation of the frequency of the superposed signal is used to code the distance. A tactor control interface is responsible for the mapping of the controller output vector $u(k)$ to the tactors which should be activated. The transferred signal u_{vib} from the vibration control to the tactors contains the frequency for each tactor that depends on a weighting factor w indicating the strength by which a tactor should vibrate (8). The weighting factor is defined as the positive vector product of a tactor's position $^{tac}\mathbf{p}_{tac(n)}$ and the controller output vector $\mathbf{u}_{tac}(k)$ in the tactors coordinate system. By doing so the signal is distributed to several tactors and directions not exactly matching one tactor can be presented.

$$w_n = max\left(0, \left(\left\|{}^{tac}\mathbf{p}_{tac(n)}\right\| \cdot {}^{tac}\mathbf{T}_{ima} \cdot u(k)\right)\right) \tag{8}$$

The actual system implements two different activation schemes. In these activation schemes the number of the active tactors is restricted to only one active tactor (mono linear) or two active tactors (duo linear). This means that in the mono linear activation scheme the weighting factor is simplified and could be therefore only 0 (no vibration) or 1 (vibration). Only the highest weighted tactor is activated with 100% of the calculated total frequency. All other tactors are inactive:

$$u_{vib}(k) = \begin{pmatrix} 0 \\ \vdots \\ freq(w_i) \\ \vdots \\ 0 \end{pmatrix} |w_i = max(w_1, \ldots, w_n) \tag{9}$$

In the duo linear activation scheme the weighting factor is calculated as described above but only the two tactors with the highest weighting factors were activated, each with a frequency calculated by their weighting factor:

$$u_{vib}(k) = \begin{pmatrix} 0 \\ \vdots \\ freq(w_i) \\ \vdots \\ freq(w_j) \\ \vdots \\ 0 \end{pmatrix} \begin{matrix} |w_i = max(w_1, \ldots, w_n) \\ |w_i = max(w_1, \ldots, w_{i-1}, w_{i+1}, \ldots, w_n) \end{matrix} \tag{10}$$

4   Experiments

In these experiments the intuitive understanding of the two tactile activation schemes of mono linear (i) and duo linear (ii) activation were compared in terms of accuracy, reaction time and the sensation of the subjects. The experiment was multimodal supported by a screen and the tactile human-machine-interface. They were performed by four male subjects. Each experiment (i) and (ii) was done three times by each subject.

4.1   Setup and Procedure

As mentioned in other works [2] the middle ear surgery is an interesting field of application for navigation systems. Therefor the experiments were done with a simplified setup as shown in Fig. 7. A plastic skull with an integrated, easy to change plaster model in place of the ear channel was representing the patient. A circle with a diameter of 2.0 cm in the plaster model was defined as working area representing the area to get access to the middle ear during interventions. Patient and instrument were measured by the integrated optical pose measurement system. The task of the subjects was to mill free within this working area but not to exceed the border. Therefore the circle and the actual position of the instrument's tcp within the enabled working area were visualized on the screen. Additional to the visualization the distance of the last 5 mm to the circle's border was indicated by the tactile display with an increasing pulsing signal of the tactors with a frequency of 1Hz to 5Hz depending on the distance and activation scheme. On reaching the border the display was vibrating with maximum frequency. To determine the accuracy the distance between the given border of the working area and the real milling is measured afterwards.

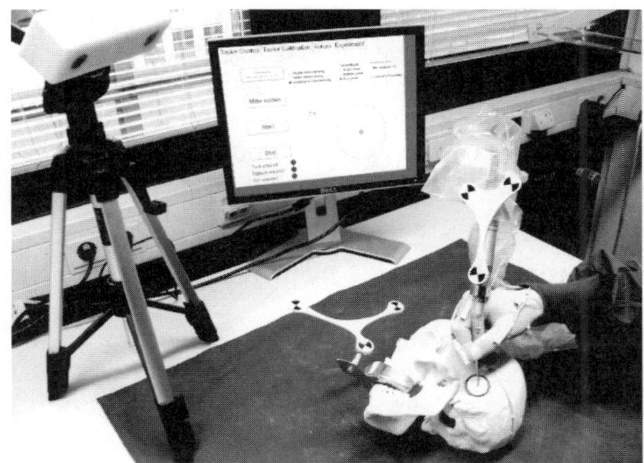

Fig. 7. Experimental setup with pose measurement system, instrument, tactile display and patient

4.2   Results and Discussion

The mono linear supporting scheme leads to a mean error of -0.35mm between the border of the circles geometry and the actually milled region and a standard deviation of 0.51mm. The mean time performing the experiments was 1:29 minutes. The duo linear supporting scheme leads to a mean error of -0.01mm. The low error is caused by the fact that there are to large circles as well as too small ones so that the mean value is almost the desired value. This is also reflected by the standard deviation of 0.62mm. The duo linear experiments were performed in an average time of 2:09 minutes. In both experiments there are high inter-individual differences between the single subjects concerning the accuracy and duration. But one can not say that a decreased duration leads to a decreased accuracy at the same time (Table 1).

Table 1. Results of the duo linear experiments

Subject	mean error	standard deviation	duration
P1	-0,79 mm	0,58 mm	01:25 min
P2	0,67 mm	0,80 mm	02:40 min
P3	0,34 mm	0,55 mm	02:13 min
P4	-0,25 mm	0,39 mm	02:59 min
	-0,01 mm	0,62 mm	

The supporting scheme of activating two tactors at the same time with different frequencies to transmit the direction between two tactors does not lead to significant better results than the supporting scheme with only one active tactor. The interviews with the subjects add up to the conclusion that a divided signal on two vibrating tactors with different frequency is not understood as weighting of these tactors to indicate a diagonal direction but is rather confusing. Even two tactors at the same finger are hardly perceivable as two separate signals. The high standard deviation and the simultaneous very low mean error could be caused by an inaccurate calibration of the plaster model, so that the transformation of the circle to mill was hardly an ellipse.

5   Conclusion and Outlook

The conTACT navigation system extends the group of navigation systems by a multimodal signal transmitting system. In addition to the standard visual support the systems also provides the navigation information directly in the hand's own coordinate system via a tactile human-machine-interface. Although the multimodal approach in itself seems to be more intuitive, the signal generated by more than one tactor vibrating with different, not synchronized frequencies is rather confusing but not transmitting direction information. Therefore the next step is a brief analysis of the signal generation to code direction information and the distance to a border at the same time.

References

1. Kirschstein, U., Hein, A.: Navigated Imaging for Neurosurgery. In: BioRob. Proc. of IEEE Intl. Conf. on Biomedical Robotics and Biomechatronics, Pisa, Italy (February 2006)
2. Hein, A., Lenzen, C., Brell, M.: Preliminary Evaluation of a Force-Sensing Human-Machine Interface for an Interactive Robotic System. In: IROS 2006. Int. Conf. on Intelligent Robots and Systems, Beijing, China, pp. 983–988 (October 2006)
3. Wegner, I., Vetter, M., Schoebinger, M., Wolf, I., Meinzer, H.P.: Development of a navigation system for endoluminal brachytherapy in human lungs. In: Proceedings of the International Society for Optical Engineering, vol. 6141 (March 2006)

4. Weber, S., Klein, M., Hein, A., Krueger, T., Lueth, T.C., Bier, J.: The navigated image viewer - Evaluation in Maxillofacial Surgery. In: Ellis, R.E., Peters, T.M. (eds.) MICCAI 2003. LNCS, vol. 2878, pp. 762–769. Springer, Heidelberg (2003)

5. Serefoglou, S., Lauer, W., Perneczky, A., Lutze, T., Radermacher, K.: Multimodal user interface for a semi-robotic visual assistance system for image guided neurosurgery. In: Lemke, H.U., Inamura, K., Doi, K., Vannier, M.W., Farman, A.G. (eds.) CARS 2005. International Congress Series: Computer Assisted Radiology and Surgery, Berlin, Germany, vol. 1281, pp. 624–629 (May 2005)

6. Lueth, Hein, Albrecht, Demirtas, Zachow, Heissler, Klein, Menneking, Hommel, Bier.: A Surgical Robot System for Maxillofacial Surgery. In: IECON 1998. Int. Conf. on Industrial Electronics, Control, and Instrumentation, Aachen, Germany, pp. 2470–2475 (August - Septmber, 1998)

7. Van Erp, J.B.F.: Presenting direction with a vibrotactile torso display. Ergonomics 48(3), 302–313 (2005)

8. Man, N.: Vibro-Monitor: A Vibrotactile display for Physiological Data Monitoring. In: Human Interface Technologies (HIT 2004) (2004)

9. Yao, Hayward, Ellis: A Tactile Enhancement Instrument for Minimally Invasive Surgery. Computer Aided Surgery 10(4), 233–239 (2005)

10. Rosen, Hannaford: Force Controlled and Teleoperated Endoscopic Grasper for Minimally Invasive Surgery - Experimental Performance Evaluation. IEEE Transaction on Biomedical Engineering 46(10) (October 1999)

11. Murray, A.M., Klatzky, R.L., Khosla, P.K.: Psychophysical Characterization and Testbed Validation of a Wearable Vibrotactile Glove for Telemanipulation. Presence: Teleoperators and Virtual Environments 12(2), 156–182 (2003)

12. Piateski, Jones: Vibrotactile pattern recognition on the arm and torso. In: Eurohaptics 2005. Conf. and Symp. on Haptic Interfaces for Virtual Environment and Teleoperator Systems, pp. 90–95 (2005)

13. Hein, A., Brell, M.: conTACT - A Vibrotactile Display for Computer Aided Surgery. In: WHC 2007. Second Joint Eurohaptics Conference and Symposium on haptic interfaces for virtual environment and teleoperator systems, Tsukuba, Japan, pp. 531–536 (March 2007)

14. Lavallee,: Registration for Computer-Integrated Surgery: Methodology, State of the Art. Computer-Integrated Surgery: Technology and Clinical Applications, 577–580 (September 1995)

15. Van Erp, Van Veen, Jansen, Dobbins: Waypoint navigation with a vibrotactile waist belt. ACM Transactions on Applied Perception 2(2), 106–117 (2005)

Two Arms Are Better Than One: A Behavior Based Control System for Assistive Bimanual Manipulation

Aaron Edsinger[1] and Charles C. Kemp[2]

[1] Computer Science and Artificial Intelligence Laboratory, Massachusetts Institute of Technology, Cambridge, MA, USA
`edsinger@csail.mit.edu`
[2] Health Systems Institute, Georgia Institute of Technology, Atlanta, GA, USA
`charlie.kemp@hsi.gatech.edu`

1 Introduction

Robots that work alongside people in their homes and workplaces could potentially extend the time an elderly person can live at home, provide physical assistance to a worker on an assembly line, or help with household chores. Human environments present special challenges for robot manipulation, since they are complex, dynamic, uncontrolled, and difficult to perceive reliably. For tasks that involve two handheld objects, the use of two arms can help overcome these challenges. With bimanual manipulation, a robot can simultaneously control two handheld objects in order to better perceive key features, control the objects with respect to one another, and interact with the user.

Addressing the challenges of manipulation in human environments is an active area of research. For example, the ARMAR project is investigating manipulation in human environments and has shown results including the bimanual opening of a jar [21]. Researchers working with the NASA Robonaut [1] have demonstrated a cooperative manipulation task where the robot employs a power drill to tighten lugnuts under human direction. Work at AIST has pursued fetch-and-carry tasks of everyday objects under partial teleoperation[18], while work at Stanford has recently investigated learning to grasp novel, everyday objects [16]. Many groups are also pursuing research on autonomous mobile manipulation in human environments [11, 19].

For most of these projects, the robots do not physically interact with people. They also tend to use detailed models of the world that are difficult to generalize and neglect opportunities for physical interactions with the world that can simplify perception and control. In contrast, our approach to robot manipulation emphasizes three design themes: *cooperative manipulation, task relevant features*, and *let the body do the thinking*. We have previously illustrated these themes with a behavior-based control system that enables a humanoid robot to help a person place everyday objects on a shelf [5]. Within this paper we extend this control system to enable a robot to perform tasks bimanually with everyday

S. Lee, I.H. Suh, and M.S. Kim (Eds.): Recent Progress in Robotics, LNCIS 370, pp. 345–355, 2008.
springerlink.com

handheld objects. The success of this extended system suggests that our approach to robot manipulation can support a broad array of useful applications, and demonstrates several distinct advantages of using two arms.

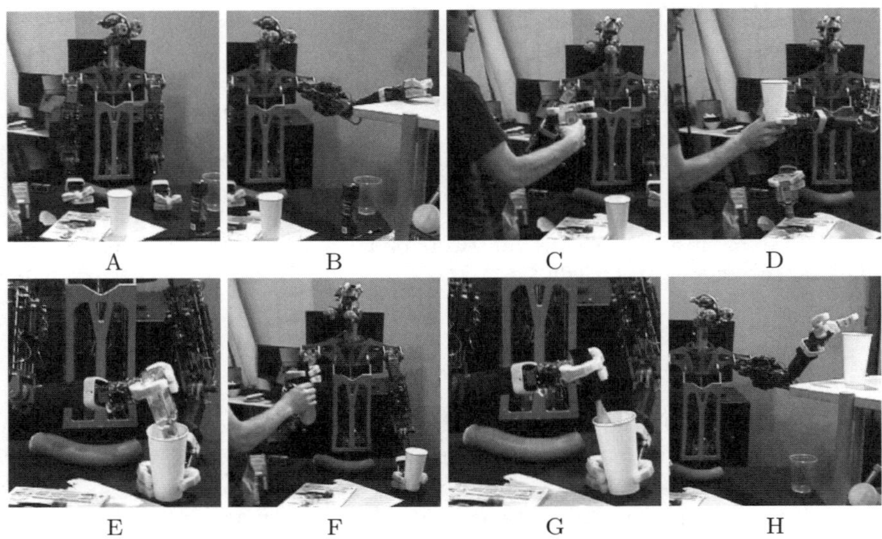

Fig. 1. The humanoid robot Domo assisting a collaborator in a task similar to making a drink. (A-B) Working at a cluttered table, Domo physically verifies the location of a shelf surface. (C-D) Upon request, Domo grasps a bottle and a cup handed to it by the collaborator. (E-F) Domo inserts the bottle into the cup, hands the bottle back to the collaborator, and then acquires a spoon from the collaborator. (G-H) Domo inserts the spoon into the cup, stirs it, and then puts the cup on the shelf.

Our work is implemented on the 29 degree-of-freedom humanoid robot, Domo, pictured in Figure 1. Domo is mechanically distinctive in that it incorporates passive compliance and force sensing throughout its body [7]. Its Series Elastic Actuators lower the mechanical impedance of its arms, allowing for safe physical interaction with a person [15, 20]. Working with unmodeled objects against a cluttered background, Domo is able to assist a person in a task akin to preparing a drink. As shown in Figure 1, Domo can socially cue a person to hand it a cup and a bottle, grasp the objects that have been handed to it, and conduct a visually guided insertion of the bottle into the cup. Domo can then repeat this process using a spoon to stir the interior of the cup, and place the cup on a shelf upon completion. This type of help might enable a person with serious physical limitations to maintain independence in everyday activities that would otherwise require human assistance. For a factory worker, this type of help could potentially offload physically demanding aspects of a task onto a robot.

2   Three Themes for Design

As previously described in [5], three themes characterize our approach to manipulation in human environments. We review these themes here. The first theme, *cooperative manipulation*, refers to the advantages that can be gained by having the robot work with a person to cooperatively perform manipulation tasks. The second theme, *task relevant features*, emphasizes the benefits of carefully selecting the aspects of the world that are to be perceived and acted upon during a manipulation task. The third theme, *let the body do the thinking*, encompasses several ways in which a robot can use its body to simplify manipulation tasks.

2.1   Cooperative Manipulation

For at least the near term, robots in human environments will be dependent on people. Fortunately, people tend to be present within human environments. As long as the robot's usefulness outweighs the efforts required to help it, full autonomy is unnecessary. With careful design robots can be made more intuitive to use, thereby reducing the effort required.

2.2   Task Relevant Features

Donald Norman's book *The Design of Everyday Things* [13], emphasizes that objects found within human environments have been designed to match our physical and cognitive abilities. These objects are likely to have common structural features that simplify their use. By developing controllers that are matched to these structural features, we can simplify robot manipulation tasks. Rather than attempting to reconstruct the world in its entirety, we focus the robot's sensory resources on elements of the world that are relevant to the current task.

2.3   Let the Body Do the Thinking

This theme bundles together design strategies that make use of the robot's body to simplify manipulation in three ways.

First, human environments, interactions, and tasks are well matched to the human body. For example, Domo's eye gaze, arm gesture, and open hand are similar in appearance to a human requesting an object, and are able to intuitively cue uninstructed, non-specialists [6].

Second, we can mitigate the consequences of uncertainty by trading off perception and control for physical design. This tradeoff is central to Pfeifer's notion of morphological computation [14]. For example, Domo uses passive compliance when inserting one object into another.

Third, a physically embodied agent can use its body to test a perceptual hypothesis, gain a better view on an item of interest, or increase the salience of a sensory signal. For example, in this work Domo simultaneously controls two grasped objects in order to better perceive their distal tips.

3   Behavior-Based Control

3.1   The Behavior System

Domo performs tasks through the coordination of its perceptual and motor behaviors over time. These behaviors (denoted in italics) are composed hierarchically, and run in a distributed, real-time architecture at $15 - 100hz$ on a 12 node Linux cluster. We have adopted a layered architecture similar to that of Brooks[2] and Connell[3]. We couple constant perceptual feedback to many simple behaviors in order to increase the task robustness and responsiveness to dynamics in the environment. For example, if a person removes the object from the robot's grasp at anytime during task execution, the active behavior will become inhibited and a secondary behavior will attempt to reacquire the object or to smoothly bring the arm to a relaxed posture.

3.2   Behaviors

A collaborator coordinates the robot's manual skills to accomplish a task. For example, the task of Figure 1 is accomplished using four manual skills: *ShelfPlace,*

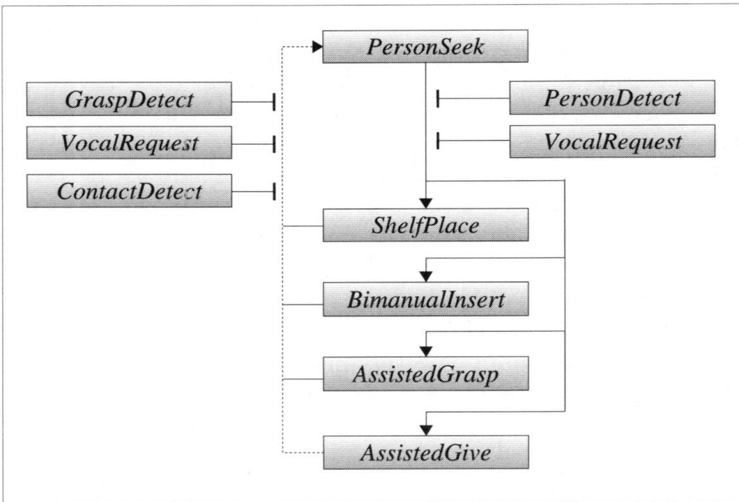

Fig. 2. A collaborator can compose a task using four manipulation behaviors: *Shelf-Place, BimanualInsert, AssistedGrasp,* and *AssistedGive.* Transitions (arrows) occur contingent on perceptual feedback (bars). Exceptions from the expected feedback result in a reset transition (dashed line). The collaborator coordinates the task through voice cues (*VocalRequest)* while the robot tracks the person in the scene (*PersonSeek, PersonDetect*). The person can ask the robot to take an object (*AssistedGrasp*), give back an object (*AssistedGive*), insert one object into another (*BimanualInsert),* or place an object on a shelf (*ShelfPlace*). The robot can reattempt a manual skill if failure is signaled (*GraspDetect, VocalRequest, ContactDetect*).

BimanualInsert, AssistedGrasp, and *AssistedGive.* As shown in Figure 2, these behaviors run concurrently, allowing a person to vocally request them at any time. If the collaborator notices that Domo is failing at a task, they can provide vocal (*VocalRequest*) or contact (*ContactDetect*) feedback to alert the robot. If Domo accidentally drops an object (*GraspDetect*), the person can pick it up and ask the robot to grasp it again (*AssistedGrasp*). Alternatively, at anytime the person can ask Domo to hand him or her a grasped object (*AssistedGive*). In this way, the robot and the person work as a team. The person provides task-level planning and guides the robot's action selection using intuitive modes of interaction, such as handing objects to the robot and simple verbal commands. In return, the robot performs requested manipulation tasks for the person using the provided objects.

The *AssistedGrasp, AssistedGive,* and *ShelfPlace* behaviors are fully described in [4] and [5]. In the next section we describe the implementation of the *BimanualInsert* behavior in more detail.

4   The Bimanual Insertion Task

In the *BimanualInsert* behavior, Domo grasps a common object such as a stirring spoon or bottle in one hand and a container such as cup or coffee mug in the other hand. It inserts the object into the container and then optionally stirs the contents. The specific geometric properties and appearance of each object and container are unknown, and their pose in the grasp is uncertain. The robot relies on visual sensing and manipulator compliance to overcome this uncertainty.

This behavior is related to the classic peg-in-hole task often studied in model-based manipulation under uncertainty [12]. For this task a single manipulator controls a peg with the goal of inserting it into a hole. Bimanual insertion is less common.

Through bimanual manipulation a robot can simultaneously control two grasped objects independently. In doing so, the robot can actively control the objects in order to simplify perception and control. For example, Domo wiggles both objects so that it can more easily perceive them through visual motion. Likewise, Domo is able to stabilize the container on a flat surface where it can easily view its opening, hold it steady while inserting the other object, and physically confirm the poses of the objects. Domo is also able to move the objects into its dexterous workspace, where it can more easily perform the physical motions necessary for the task. Finally, by holding both objects at all times, Domo clearly and unambiguously communicates to the person which objects it intends to use for the current task. This is important for cooperative tasks.

The following sections describe the sequential phases of the task in order.

4.1   *AssistedGrasp*

AssistedGrasp enlists the person's help in order to secure a grasp on a utensil and a container. By handing Domo the objects, the person directly specifies the objects that Domo will manipulate. In the case of tasks that involve two handheld objects, Domo clearly and unambiguously indicates which objects are in use by holding the objects in its hands. This approach to coordination is both intuitive and effective. It avoids the need for the person to select objects through speech or gesture, and makes it easier for the person to interpret the state or intentions of the robot. By handing the objects to the robot, the system also avoids the challenging robotics problem of locating and autonomously grasping selected objects. Robotic grasping of objects is still an active area of research and an open problem [17, 16].

AssistedGrasp locates a person in the scene, extends its arm towards the person, and opens its hand. By reaching towards the person, the robot reduces the need for the person to move when handing over the object. In assistive applications for people with physical limitations, the robot could potentially adapt its reach to the person's capabilities and effectively extend the person's workspace and amplify his or her abilities.

In addition, the robot cues the person through eye contact, directed reaching, and hand opening. This lets him or her know that Domo is ready for an object and prepared to perform the task. The robot monitors contact forces at the hand. If it detects a significant change, it performs a power grasp in an attempt to acquire an object. If the detector *GraspDetect* indicates that an object has been successfully grasped, the robot attempts to acquire another object with its free hand in the same way. Once the robot has an object in each hand, it proceeds to the next phase of the task.

4.2   *ContainerPlace*

After *AssistedGrasp*, the orientation of the grasped object in the hand is uncertain. The *ContainerPlace* behavior reduces the orientation uncertainty of a grasped container. Using force control, the behavior lowers the container onto a table while keeping the impedance of the wrist low. This robot behavior is shown in Figure 3.

Since each of the container objects has a flat bottom that is parallel to its opening, this action aligns containers with the table, which results in a stable configuration that is favorable for insertion. This behavior takes advantage of common task relevant features of everyday containers, which have been designed to both accommodate the insertion of objects and stably rest on the flat surfaces that are often found in human environments. For example, people often rest a cup on a table before pouring a cup of coffee.

By using two arms, Domo is able to stably hold the container object against the table throughout the insertion operation. This is important, since compliant contact during the insertion that can generate significant forces and torques on the container. Moreover, throughout the insertion, Domo has the opportunity to physically detect whether or not the object is still being held against the table.

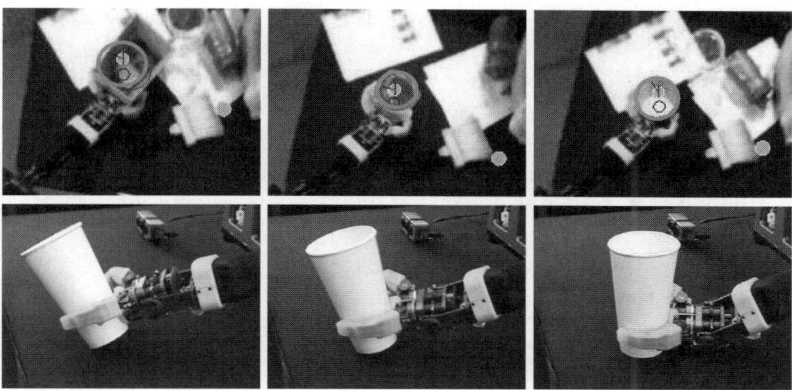

Fig. 3. Execution of the *ContainerPlace* behavior. (Top) The spatio-temporal interest point operator finds the roughly circular opening of a box, jar, and bowl. The detector is robust to cluttered backgrounds. (Bottom) The robot exploits active and passive compliance to align the container to the table.

4.3 *TipEstimate*

For a wide variety of tools and tasks, control of the tool's endpoint is sufficient for its use. For example, use of a screwdriver requires precise control of the tool blade relative to a screw head but depends little on the details of the tool handle and shaft.

The tip of an object is an important task relevant feature, and we have previously described a method to rapidly localize and control this feature [9, 10]. This method detects fast moving, convex shapes using a form of spatio-temporal interest point operator. As the robot rotates the object, it detects the most rapidly moving convex shape between pairs of consecutive images. Due to the tip's shape and distance from the center of rotation it will tend to produce the most rapidly moving, convex shapes in the image. The robot uses its kinematic model to estimate the 3D point in the hand's coordinate system that best explains these noisy 2D detections.

The *TipEstimate* behavior brings a grasped object into the field of view, rotates its hand, and then localizes the tip. The robot uses the same spatio-temporal interest point operator to detect the opening of the container as it is aligned to the table. As shown in Figure 3, using visual motion and the kinematic model enables the robot to robustly detect this opening on a cluttered table. This method works with a variety of containers such as drinking glasses, bowls, small boxes, and coffee mugs. The opening of the container serves as a form of object tip. Since the tip detector is edge-based, multi-scale, and sensitive to fast moving convex shapes, the edges of the container openings are readily detected.

4.4 *TipPose*

Once *TipEstimate* has localized the utensil tip within the hand's coordinate system, the *TipPose* behavior controls the feature by extending the robot's kinematic model by one link. This enables the robot to use traditional Cartesian space control. As the grasped object is moved, the spatio-temporal interest point operator provides visual detections of the tip. This enables the robot to visually servo the tip in the image [4].

Within the insertion task, the *TipPose* behavior visually servoes the object's tip to the container's opening. We adopt an approach similar to [8] where the object is aligned at a 45 degree angle to the table. This advantageous pose avoids visual obstruction of the tip by the hand and expands the range of acceptable misalignment when performing the insertion. During servoing, the tip is kept on the visual ray to the center of the container opening. The depth of the tip is then increased along the ray until the tip is just above the insertion location. This effectively compensates for errors in depth estimation.

Throughout this process, the use of two arms is important. The tip estimation is performed with respect to the hand's coordinate system. By continuing to rigidly grasp an object after estimating the location of its tip, the estimation continues to be relevant and useful. If the robot were to release one of the objects, the uncertainty of the tip's pose relative to the robot's body would be likely to increase, and additional perceptual mechanisms would be required to maintain the estimate, especially in the context of mobile manipulation.

4.5 *CompliantLower*

CompliantLower performs the insertion phase of the task by generating a constant downward force at the object's tip. The impedance of the manipulator wrist is also lowered in order to accommodate misalignment. Although the insertion forces are not used for control feedback, the sensed force between the object and the bottom of the container is used to signal task completion.

5 Results

Our three design strategies allow *BimanualInsert* to generalize across a variety of insertion objects and containers. In total, we have executed *BimanualInsert* in nearly one hundred informal trials with a variety of objects. To quantify its performance, we tested *BimanualInsert* in two experiments. In the first experiment, we tested the insertion of a mixing spoon, bottle, paint roller, and paint brush into a paper cup. In the second experiment, we tested the insertion of the mixing spoon into a paper cup, bowl, coffee mug, and jar. On these objects, the size of the container opening varies between 75-100mm and the size of the tool tip varies between 40-60mm. In each experiment, seven trials were conducted on each object pairing.

In a single experiment trial, the object was handed to the robot in an orientation that was deliberately varied between $\pm 20°$ along the axis of the hand's power grasp. The grasp location on the object was varied by approximately ± 50mm along its length. Each trial took less than 20 seconds to complete and was performed over a visually cluttered table with the collaborating person nearby. A

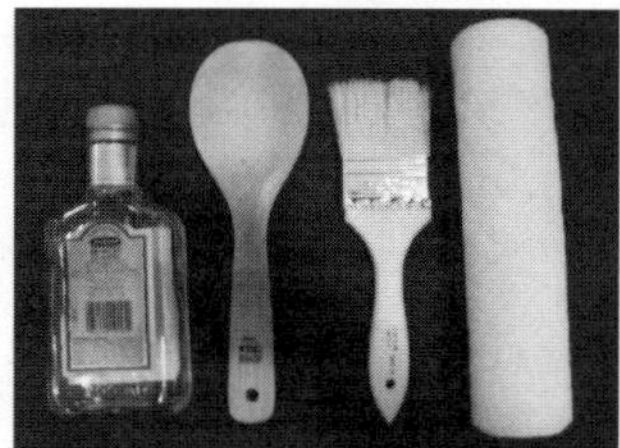

	Paper cup	Bowl	Box	Coffee mug	Jar
Mixing spoon	7/7	7/7	7/7	6/7	7/7
Bottle	6/7				
Paint brush	6/7				
Paint roller	5/7				
Spoon (open loop)	1/7				

Fig. 4. Task success for *BimanualInsert*. In a successful trial, Domo inserted the tool (rows, top left) into the container (columns, top right). For comparison, the last row shows results where the visual detection of the tip was disabled. Trials for the blank entries were not attempted.

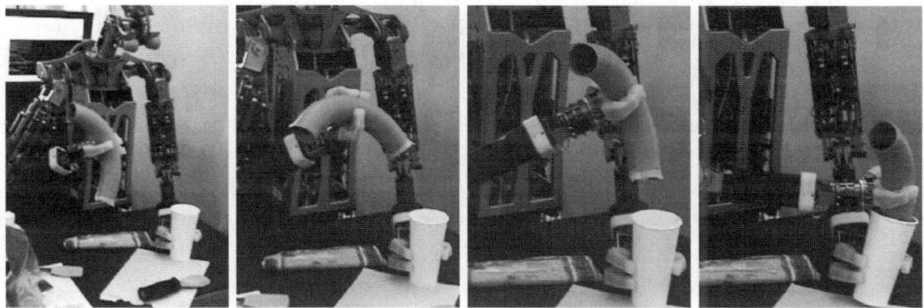

Fig. 5. Execution of *BimanualInsert* using a flexible hose. The unknown bend in the hose requires the active perception of its distal tip and realignment prior to insertion.

trial was successful if the object was fully inserted into the container. The success rates for both experiments are shown in Figure 4. As the results show, *BimanualInsert* was successful in roughly 90% of the trials. When the visual detection of the tip was disabled, the success rate fell to about 15%.

As a final example, we tested *BimanualInsert* using a flexible hose. The hose has an unknown bend, making it essential that Domo actively sense its distal tip in order to orient the hose prior to insertion. The execution of this test is shown in Figure 5. While *BimanualInsert* can handle the flexible hose in many cases, the single point representation for the tip lacks the orientation information required to reorient the hose and successfully perform the insertion task when the hose has a very large bend. Extending the tip detection system with estimation of the tip's orientation would be useful for these situations.

6 Discussion

With bimanual manipulation, a robot can simultaneously control two handheld objects in order to better perceive key features, control the objects with respect to one another, and interact with the user. Within this paper, we have presented evidence that these advantages can dramatically simplify manipulation tasks that involve two handheld objects. The control system we have presented relies on both arms, and would not succeed otherwise.

Maintaining rigid grasps on the objects throughout the manipulation task enables the robot to reliably maintain pose estimates for object features, and actively control the objects in order to facilitate new perceptual detections and reestimations. Rigidly grasping the two objects enables the robot to attach the objects to its body and the accompanying coordinate system. Although the world in which the robot is operating is uncontrolled and unmodeled, the robot's body is controlled and well-modeled. Once the robot is holding the two objects, it effectively brings them into a controlled environment.

Within this controlled environment, the robot can efficiently move the objects into favorable configurations for sensing and control. For example, by actively fixturing an object with one arm, the robot can ensure that the object maintains a favorable configuration in the presence of interaction forces. The ability to handle interaction forces is important to our approach, since it enables the robot to use physical interactions between the objects that help with the task, such as compliance during the insertion. By maintaining contact with the fixtured object, the robot also has the opportunity to physically sense whether or not the fixtured object's state has changed, and provides another channel with which to measure the influence of the interactions between the objects.

With respect to human-robot interaction, the use of two arms enables the robot to directly indicate the objects with which it is working. If the robot is only holding one object, this will be readily apparent to the human. For example, if the task is to pour a drink and the robot is only holding an empty cup, the user can readily infer that the robot should be handed a bottle. Likewise, if the robot is holding a spoon and a mixing bowl, the user can determine an appropriate task for the robot to perform, such as stirring, or decide that the objects are inappropriate.

In the long run, we suspect that these advantages, and others, may outweigh the costs and complexity associated with two armed robots that manipulate in human environments.

References

1. Bluethmann, W., Ambrose, R., Fagg, A., Rosenstein, M., Platt, R., Grupen, R., Brezeal, C., Brooks, A., Lockerd, A., Peters, R., Jenkins, O., Mataric, M., Bugajska, M.: Building an autonomous humanoid tool user. In: Proceedings of the 2004 IEEE International Conference on Humanoid Robots, Santa Monica, Los Angeles, CA (2004)

2. Brooks, R.: Cambrian Intelligence. MIT Press, Cambridge (1999)
3. Connell, J.: A behavior-based arm controller. IEEE Transactions on Robotics and Automation 5(5), 784–791 (1989)
4. Edsinger, A.: Robot Manipulation in Human Environments. PhD thesis, Massachusetts Institute of Technology, Cambridge, MA (2007)
5. Edsinger, A., Kemp, C.: Manipulation in human environments. In: Proceedings of the 2006 IEEE International Conference on Humanoid Robots, Genoa, Italy (2006)
6. Edsinger, A., Kemp, C.: Human-robot interaction for cooperative manipulation: Handing objects to one another. In: ROMAN 2007. Proceedings of the IEEE International Workshop on Robot and Human Interactive Communication, Jeju, Korea (2007)
7. Edsinger-Gonzales, A., Weber, J.: Domo: A Force Sensing Humanoid Robot for Manipulation Research. In: Proceedings of the 2004 IEEE International Conference on Humanoid Robots, Santa Monica, Los Angeles, CA (2004)
8. Inoue, H.: Force feedback in precise assembly tasks. In: Winston, P., Brown, R. (eds.) Artificial Intelligence: An MIT Perspective, The MIT Press, Cambridge (1979)
9. Kemp, C., Edsinger, A.: Visual tool tip detection and position estimation for robotic manipulation of unknown human tools. Technical Report AIM-2005-037, MIT Computer Science and Artificial Intelligence Laboratory (2005)
10. Kemp, C., Edsinger, A.: Robot manipulation of human tools: Autonomous detection and control of task relevant features. In: ICDL 2006. Proceedings of the 5th IEEE International Conference on Development and Learning, Bloomington, Indiana (2006)
11. Khatib, O., Yokoi, K., Brock, O., Chang, K., Casal, A.: Robots in human environments: Basic autonomous capabilities. International Journal of Robotics Research 18(684) (1999)
12. Lozano-Perez, L., Mason, M., Taylor, R.: Automatic synthesis of fine-motion strategies for robots. International Journal of Robotics Research 3(1) (1984)
13. Norman, D.: The Design of Everyday Things. Doubleday, New York (1990)
14. Pfeifer, R., Iida, F.: Morphological computation: Connecting body, brain and environment. Japanese Scientific Monthly 58(2), 48–54 (2005)
15. Pratt, G., Williamson, M.: Series Elastic Actuators. In: IROS 1995. Proceedings of the IEEE/RSJ International Conference on Intelligent Robots and Systems, Pittsburg, PA, vol. 1, pp. 399–406 (1995)
16. Saxena, A., Driemeyer, J., Kearns, J., Osondu, C., Ng, A.: Learning to grasp novel objects using vision. In: ISER. Proceedings of the International Symposium on Experimental Robotics (2006)
17. Shimoga, K.: Robot grasp synthesis algorithms: a survey. International Journal of Robotics Research 15(3), 230–266 (1996)
18. Sian, N., Yoki, K., Kawai, Y., Muruyama, K.: Operating humanoid robots in human environments. In: Proceedings of the Robotics, Science & Systems Workshop on Manipulation for Human Environments, Phillidelphia, Pennsylvania (2006)
19. Yang, Y., Brock, O.: Elastic roadmaps: Globally task-consistent motion for autonomous mobile manipulation. In: Proceedings of Robotics: Science and Systems, Philadelphia, USA (2006)
20. Zinn, M., Khatib, O., Roth, B., Salisbury, J.: Playing it safe: human-friendly robots. IEEE Robotics & Automation Magazine 11(2), 12–21 (2004)
21. Zöllner, R., Asfour, T., Dillmann, R.: Programming by demonstration: Dual-arm manipulation tasks for humanoid robots. In: IROS 2004. IEEE/RSJ International Conference on Intelligent Robots and Systems, Sendai, Japan (2004)

A Customizable, Multi-host Simulation and Visualization Framework for Robot Applications

Tim Braun, Jens Wettach, and Karsten Berns

Robotics Research Lab
University of Kaiserslautern
Kaiserslautern, Germany
{braun, wettach, berns}@informatik.uni-kl.de

Summary. A highly flexible framework for visualization and sensor simulation in three-dimensional environments is presented. By allowing the insertion of freely programmable elements for online scene modification, a programmer can customize the framework to fulfill the exact simulation or visualization needs of an application of interest. Furthermore, the framework provides simple external interfaces so that multiple clients can be attached to it with ease. The frameworks' capabilities are demonstrated with two complex robotic applications that require both a high quality simulation of cameras and lasers scanners and an intuitive 3D visualization.

1 Introduction

Due to the high complexity of modern robotic systems, almost any research conducted in the area of robotics can benefit from a *simulation* of the system behavior before experiments on a real platform take place. Aside from reduced development time, a simulation allows to validate safety properties and test new algorithms more objectively with increased control over sensor/actor noise and more repeatable conditions. On the other hand, the control and understanding of a robotic system in operation can often be substantially improved by offering the user a good *visualization* of the current robot situation.

Both of these aspects can (and commonly are) be approached using 3D-models of the involved objects. For example, in order to simulate a camera mounted on a mobile robot, a three-dimensional scene needs to be modeled and an image of it must be rendered from the camera's current viewpoint. For visualization, the same approach can be applied: Modeling a scene after the actual area where the robot is deployed and displaying it to the user from a virtual camera he or she controls. However, in both cases it is crucial to use a framework that is powerful enough to support all simulation and visualization requirements that arise for the application at hand. For example, the visualization of a robot arm might require to allow the parametrization of all joint angles and the highlighting of internal collisions or invalid configurations, while the visualization of a mobile robot might need to display navigational points at varying positions.

Commercially available toolkits typically provide several pre-made robots and a variety of standard sensors like cameras, bumpers and laser scanners. With

S. Lee, I.H. Suh, and M.S. Kim (Eds.): Recent Progress in Robotics, LNCIS 370, pp. 357–369, 2008.
springerlink.com   © Springer-Verlag Berlin Heidelberg 2008

these, an user can simulate or visualize a scene as long as the desired functionality is provided by the toolkit. However, these toolkits are inadequate for highly specific requirements not foreseen by the developers. For instance, the specialized outdoor simulation presented later in section 4 needs to control the visual transparency of foliage to test image processing algorithms in thick vegetation and this effect cannot be realized easily by standard out-of-the-box software.

The **SimVis3D** framework presented in this paper aims at providing a general approach to such complex and highly specific adaptation requirements. It is a modular framework usable both for the simulation of optical sensor-systems like cameras, laser scanners or PSDs and the visualization of spatial information such as 3D-environments, maps or topological graphs. Besides providing basic functionality to construct and parametrize three-dimensional scenarios, SimVis3D offers strong extensibility by providing a mechanism to easily include new, manually coded components. Since its main field of application are complex real time robotic systems which normally contain more than one controlling computer, an additional feature of SimVis3D is the ability to support situations where its input is generated by multiple client computers.

In the next section, a short overview of related robot simulation toolkits is given. Section 3 describes the SimVis3D framework architecture itself, while section 4 elaborates how the framework performed in two real applications. Based on these results, a conclusion is given.

2   Related Work

There already exists a variety of toolkits for realistic robot simulation. In *SimRobot*[1], arbitrary robots can be defined based on an XML description using predefined generic bodies, sensors and actuators. Available sensor types are cameras, laser scanners and bumpers. Robot dynamics are simulated via ODE[1]. *Gazebo*[2] is a 3D multi-robot simulator that contains predefined models of real robots as Pioneer2DX and SegwayRMP. Provided sensors are sonar, range scanner, GPS, inertial system and cameras. Robot and sensors are created as plug-ins and the simulation environment is described via XML. *Webots*[3] is a commercial simulation tool containing several models of real robots as Aibo, Khepera and Koala. It provides a virtual time to speed up simulation processes. Physics are implemented by ODE and sensor classes comprehend light, touch, force and distance sensors, scanners, GPS and cameras. *USARSim*[4] is a simulation tool for urban search and rescue robots based on the *Unreal Tournament* game engine. Virtual environments can be modelled via the *Unreal Editor* and dynamics is based on the *Karma*[2] engine. Sensor types are sonar, laser scanner and forward looking infrared (FLIR). *EyeSim*[5] is a simulator for *EyeBot* robots (a development platform for vehicle, walking and flying robots). Sensor classes contain virtual cameras and position sensitive devices which can provide realistic, noisy data.

---

[1] http://www.ode.org
[2] http://wiki.beyondunreal.com/wiki/Karma

Although these tools provide excellent support for standard mobile robot setups, highly specialized scenarios having special scene manipulation requirements are not adequately supported due to the toolkits' limited customizability and extensibility. Also, most do only support mobile robot settings; other robot types like stationary manipulators are not available.

3   Framework Architecture

SimVis3D was designed to fulfill four goals:

- Allow users to easily create custom setups by combining basic building blocks to meaningful scenarios.
- Allow users to parametrize exactly the scene aspects they want to control.
- Provide strong support for people wanting to add new functionality or alter previously static aspects.
- Use very simple external interfaces to facilitate data transport to and from remote processes.

To achieve the desired level of customizability, it was decided to use only open source libraries and eventually make SimVis3D available as a GPLed open source project. Therefore, the framework is built on top of the widely used 3D rendering library *Coin3D*[3], which is API-compatible to Open Inventor. Both rely on OpenGL for the actual rendering process and use a graph data structure called **scene graph** to store and render graphics.

In the next section, the main ideas behind the scene graph data structure will be shortly revisited in order to prepare the presentation of how a scene is composed from basic blocks (Section 3.3) and how specific scenario aspects can be made parameterizable (Section 3.4).

3.1   Scene Graph

A three-dimensional scene in Coin3D is created from nodes of the scene graph (see Fig.2). Information that defines actual 3D shapes, attributes, cameras and light sources is stored in *leaf nodes*, which are children of a hierarchy of *group nodes*. The group nodes serve to provide a logical structure by grouping nodes together that for example make up the same object.

To render an image, the graph is recursively traversed starting with the scene graph root (group) node. When a group node is traversed, all children of it are traversed in a fixed order. The order is indicated in pictures by the child nodes relative position: The left-most node comes first, the right-most node last. Each traversed graph node can manipulate the current OpenGL state by changing parameters like the active model transformation matrix or transmit geometrical primitives to the graphics hardware. Geometry is then rendered using the current OpenGL state; its appearance therefore depends on all nodes that have been traversed earlier. To limit the scope of a state change, there exists

---

[3] http://www.coin3d.org

```
<part file="world.iv" name="WORLD" insert_at="ROOT"
 pose_offset="0 0 0 0 0 0"/>
<part file="robot.iv" name="ROBOT" insert_at="WORLD"
 pose_offset="0 0 0 0 0 0"/>
<element type="pose" name="R_POSE" insert_at="ROBOT"
 x="1" y="2" z="3" roll="0" pitch="0" yaw="-90"/>
<sensor type="camera" name="ROBOT_CAM" insert_at="ROBOT"
 pose_offset="3 0 5 0 0 0"/>
```

Fig. 1. Example XML based scene description

a special variant of group nodes called *separator nodes*. Separator nodes behave like ordinary group nodes but also save the OpenGL state at the beginning of their traversal and restore this state upon its end. They are generally indicated by a horizontal line inside the node.

3.2   Insertion Points

SimVis3D extensively uses the hierarchical structure of the scene graph to express dependencies between components. Since coordinate transformations accumulate during the rendering traversal, scene components placed as a child of another component automatically inherit any transformations that influence the parent. Thus, a sensor connected to a mobile robot is automatically moved together with it if the sensor node is inserted as a child of the nodes that make up the robot.

Because of this, it is especially important to control the exact place of insertion when inserting new nodes into an existing scene graph. For this, SimVis3D uses named placeholder leaf nodes called **insertion points**, which can be located in a scene graph by their name but have no influence on rendering. Adding such insertion points in a scene allows to mark semantically meaningful locations, for example the attachment point of a camera sensor or places where other parts could be attached to a robot. To prime scene construction, SimVis3D initially provides a group node with an insertion point called ROOT.

3.3   Scene Construction

The SimVis3D framework allows users to specify the construction of a scene with a XML based description file (Fig.1 shows an example). Each line contains either a **part**, **element** or **sensor** command.

The **part** command simply adds another scene graph stored in a separate file (using a standard format like Open Inventor or VRML) to the current scene. Parts are the basic building blocks that constitute the scene and generally make up individual objects like robots, chairs or the environment. New parts can be added to the existing scene only at insertion points, but it is possible to supply a static offset which is interpreted relative to the coordinate frame active at the specified insertion point. Each part command also creates a new insertion point in this local frame named like the corresponding part itself, leading to the possibility to create nested parts without manually adding insertion points. More exactly, a part command inserts a separator node containing the static offset transformation, the new insertion point and the actual part right *before* the specified parent insertion point. Fig.2 shows a schematic scene graph after the example scene description has been processed. The dashed node is added by the **pose** element discussed in section 3.4.

With this mechanism, it is very easy to create a hierarchical scene by combining several parts. By using nested insertion points, the user can specify relations between objects, i.e. parts that are attached to another (for example, a scanner attached to a robot sitting upon a table). In this case, the pose offset and all subsequent transformations of a nested part are automatically interpreted in the coordinate frame of the parent part, agreeing with intuition.

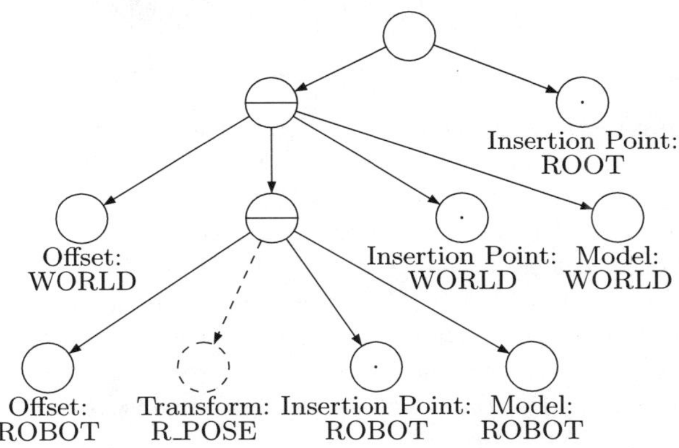

Fig. 2. Scene graph after parsing command file from Fig.1

3.4   Scene Parametrization

The **element** command is the most versatile instruction to SimVis3D and lies at the core of the frameworks flexibility and extensibility. It causes the instantiation of a class object identified by the element name using a factory pattern. The class constructor receives a reference to the elements insertion point and the whole XML command that led to its construction. It is then free to extract any needed extra configuration data from the command and modify the scene graph in any way it sees fit (it generally adds nodes at the position indicated by the insertion point). It is important to stress that all modifications are encapsulated inside the element; the SimVis3D framework is unaware of the elements' internal workings. The only interface between them are named **parameters** exported by the element, which are floating-point scalar values that can be modified by the framework. Parameter modifications are propagated through the element to effect changes in the scene graph. The exact nature of the changes again depends solely on the instantiated element. Coming back to the example in Fig.1, the third line adds an element to the scene that allows to modify the pose of the robot relative to the environment - a natural requirement for mobile robots. In the actual implementation of SimVis3D, the **pose** element reads its initial position from the XML command, exports the six parameters (x, y, z, roll, pitch, yaw) to the framework and propagates them to a new coordinate transformation (SoTransform) node inserted right before the specified insertion point.

This architecture allows users to extend SimVis3D with almost any effect or interaction capability he or she desires by implementing a custom element class and adding it to the element factory. The encapsulation guarantees that no changes to other components of SimVis3D are required. It is easy to envision elements that change the shape or color of parts, alter scene lighting, add geometrical data like point clouds visualizing 3D scanner data etc. By customizing elements, the SimVis3D framework can be tuned to the special modeling requirements for the problem at hand with ease and efficiency.

3.5   Sensors

The **sensor** command of SimVis3D is required for the inclusion of simulated sensors in the scene, for example a camera attached to a robot. Employing the same mechanism as the instantiation of elements, the sensor command effects the creation of a sensor object of the requested type at a given insertion point. Line 4 in Fig.1 creates a camera attached to the robot with a static offset relative to the robots local frame center. Differing from elements, sensors do not offer parameters to modify their behavior online. Instead, they offer sensor data to the user - plain images in the case of camera sensors, distance data in the case of laser scanners or PSDs. However, parametrized sensors can be emulated by inserting elements right before the sensor.

Although currently only cameras and laser scanners are supported by SimVis3D, the encapsulation and insertion mechanism of sensors is closely modeled after the one used for elements, so the inclusion of custom sensor types can be implemented with effort comparable to a new element class. The encapsulation principle is especially important here, since for example the inclusion of a new camera requires modifications of the scene graph not only at the insertion point, but also at its root. However, since the actual modifications are hidden for the framework anyway, this implementation detail does not bear any architectural consequences and is thus not further discussed.

3.6   Additional Framework Services

For increased ease of use, the SimVis3D framework allows to embed XML commands directly into Open Inventor files. When a part is created using this file, the embedded commands are retrieved and executed. It is thus possible to package a single file describing a complete robot with all degrees of freedoms, sensors etc. using element and sensor commands inside the file.

Furthermore, the framework optimizes the scene graph after all commands have been executed by replacing all scene graph nodes having identical content with references to one of these nodes. Identical nodes can easily appear in the scene if for example two identical part commands are given (two tables need to be placed inside a room). The optimization ensures that no performance penalty is incurred, regardless how a scenario is constructed.

3.7 Accessing SimVis3D from Multiple Hosts

Aside from easy customizability and extensibility, SimVis3D explicitly supports that parts of the simulation and visualization input data come from different processes or even different physical machines. The actual job of transporting data between computers is not in the scope of the framework and needs to be performed by standard interprocess communication (IPC) using shared memory, named pipes etc. or libraries like ACE. However, SimVis3D was designed to support a variety of these techniques by offering a very simple interface to the outside world.

This interface consists of *four data arrays* located at given memory locations and *stored compactly*. The arrays contain

- structs of element descriptors
- floating-point parameter values
- strings forming a 'scene change request log'
- strings forming a 'scene change log'

The element descriptors are structures containing all information required to identify a specific element present in the current scene and locate their parameter values in the second array. For each element, they hold the elements' name and insertion point, the number and names of its parameters and most importantly, the starting index of the parameter values in the parameter values interface array. This second array is a simple vector of floating point scalars containing, as the name suggests, the actual values of all element parameters stored consecutively. With access to these two interfaces, any process can analyze an existing scene and update the parameters of any element that it requires to access. For example, a process that calculates a robot position based on various sensors could locate the R_POSE element of the running example by scanning the element descriptors, extract the index of its parameter values and write the calculated robot pose into the value array at this location. The reason to split the element descriptors and the actual values into two arrays is plain efficiency: After looking up the exact array index of a parameter value once, the client process can from this point on directly manipulate the float values, reducing data transfer to the SimVis3D host process enormously. Fig.3 shows the interfaces that the example used in the previous section would produce.

While the first two arrays allow the manipulation of existing elements, the last two interfaces permit external processes to add new components or track these structural changes. In order to add a new scene component, a client can write XML commands similar to those in Fig.1 to the scene change request log. When

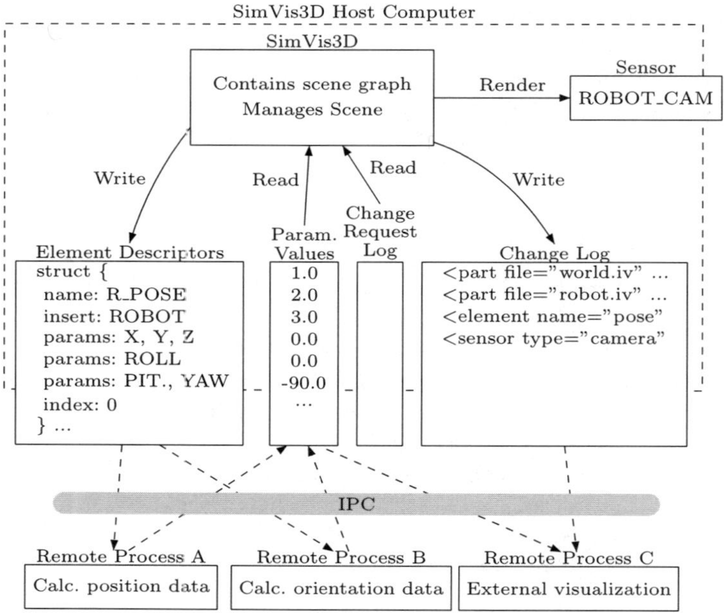

Fig. 3. Remote access interfaces provided by SimVis3D. Two use cases are depicted: a) Remote processes A and B manipulate part of the robot pose by setting the parameters exported by the R_POSE element. b) An external visualization process C copies the scene setup by repeating the commands in the change log and copying all parameter values. The external visualization could then for example render the scene on screen from a remote viewpoint without any performance penalty for the SimVis3D host computer.

SimVis3D notes a new command, the framework executes it and confirms the execution by adding it to the scene change log. This mechanism allows clients to request the addition of new elements or parts during runtime. It also permits the construction of an identical scene by simply repeating all commands stored in the change log on a local scene graph. Thereby, an external client could render the same scene as is used for simulation from a different viewpoint for visualization purposes, or even exclusively render a scene that is only set up but not rendered on a robotic platform. With this approach, SimVis3D supports both simulation *and* visualization tasks (and even mixtures of both) seamlessly within the same basic framework. Process C in Fig.3 shows an example of such a use, and Fig.6 has been created using this setup.

4   Applications

The SimVis3D framework has been used to simulate sensors and visualize complex 3D scenes for two research projects involving a mobile indoor robot named

MARVIN and an outdoor robot termed RAVON[4]. Although both projects focus on different topics and therefore have differing simulation and visualization needs, the framework was easily adapted to their requirements by implementing custom elements as described in 3.4.

For both projects, SimVis3D has been embedded into already existing robotic control systems based on MCA2[6]. The core functionality has been wrapped into two MCA2-modules (see [6] for details), with one handling all scene graph related tasks and another providing frame grabber and scanner interfaces for the virtual sensor data. Both control systems are split across two PCs and control data needed to be passed to SimVis3D from multiple hosts using the interfaces described in section 3.7. The IPC component was implemented based on the 'Blackboard' mechanism available in MCA2, which essentially provides blocks of shared memory that are transparently passed to different hosts via TCP/IP if needed. For this, it proved especially beneficial that SimVis3D exhibits simple, compact array interfaces and no complicated data exchange protocols were used.

4.1   MARVIN Scenario

MARVIN is an autonomous mobile robot for exploration and mapping of indoor environments. Its obstacle avoidance and navigation capabilities primarily depend on two planar laser range scanners at its front and rear with horizontal measurement planes 10 cm above the floor. Furthermore, a third scanner is mounted on a frontal tilt unit 1 m above ground to take 3D scans of environmental features (walls, doors, etc). Thus in order to test algorithms like line and plane extraction from distance data, the simulation for the MARVIN project mainly required a realistic 3D scanner simulation. This has been realized by adding a distance sensor class to the SimVis3D framework that provides a settable number of distance values calculated from a given sensor center to nearest objects in the scene. The calculation had been based initially on ray tracing operations which are directly supported by *Coin3D*. Due to the bad runtime performance of this method (about 2s for one scan with 361 values), a more sophisticated method has been developed that evaluates the depth buffer of a standard camera sensor which is virtually attached to the scanner. With this, all three scanners on MARVIN can be simulated at a rate of 15 Hz providing distance data in a half-circle measurement plane with 0.5° resolution (about 360 values per scanner) which is fast enough to test obstacle avoidance strategies in real time.

In Fig.4 the visualization result of the 3D scanner simulation is shown. The simulation uses the same base scene graph as the visualization as input for calculating the intersection of the laser beams with the scene objects (floor, walls, tables). The only difference to the visualization is that in the simulation scene the laser beams and data points are not visible as they might disturb the ray tracing operation.

Consequently this setup results in three different programs with four scene graphs that might transparently be spread on three computers if a suitable

---

[4] http://agrosy.informatik.uni-kl.de/en/robot_gallery/

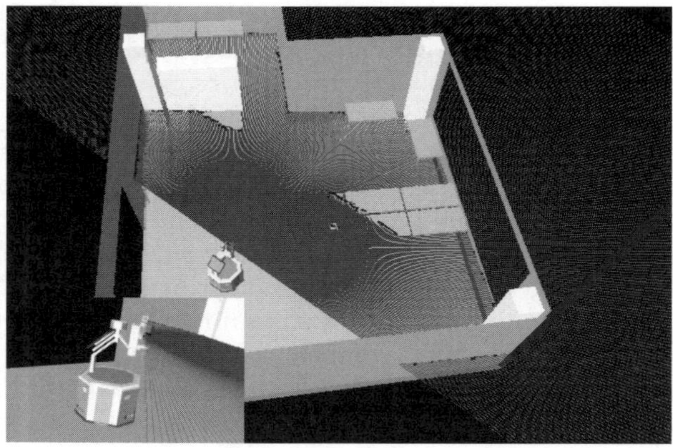

Fig. 4. Visualization of laser beams and simulated distance measurements for tiltable 3D scanner mounted on mobile robot MARVIN in a typical indoor scene. In the lower left corner a detailed view of the robot and the tilted scanner is given.

network layer is provided, as for example in the *MCA* framework that is used in this project:

- simulation node with scene graph composed of robot and environment (lab) as basis for scanner data calculation
- visualization node with same scene graph as simulation, enhanced with plotted laser beams and intersection points
- GUI node for visualizing the simulation and visualization scene, reflecting these two scene graphs in two *SoQt* viewer plugins (for ease of illustration only the visualization scene is shown in Fig.4)

The only precondition for this strategy is that simulation and visualization node get the same basic scene description file as input. The GUI node takes all information for the scene graph setup from data structures provided by the other nodes via the network link.

For easy evaluation of the simulation results a new element has been added to the SimVis3D toolkit that allows to plot an arbitrary 3D point cloud with definable number, color and size of points. To create a realistic data set the points are affected by Gaussian noise so they do not exactly fit the intersection between ray and object. This is for example the cause that some points on the wall diagonally opposite to the robot are hidden and do not appear in the visualization. This procedure allows feeding the data filtering and feature extraction algorithms with realistic distance measurements and facilitates the offline test of navigation and mapping strategies before downloading them on the real robot.

4.2 RAVON Scenario

The RAVON outdoor project focuses on autonomous obstacle avoidance and navigation and the respective algorithms rely heavily on visual stereo reconstruction and visual odometry calculation. Furthermore, color and lighting invariance algorithms are used for preprocessing.

To create an useful simulation for this research project, SimVis3D was required to deliver simulated camera images with highly realistic textures and allow the variation of lighting color and direction. In order to meet these requirements, an outdoor scene similar to the standard testing area of RAVON has been modeled with detailed obstacles and high resolution textures. Furthermore, elements allowing the on-line adjustment of foliage transparency and lighting color settings were created.

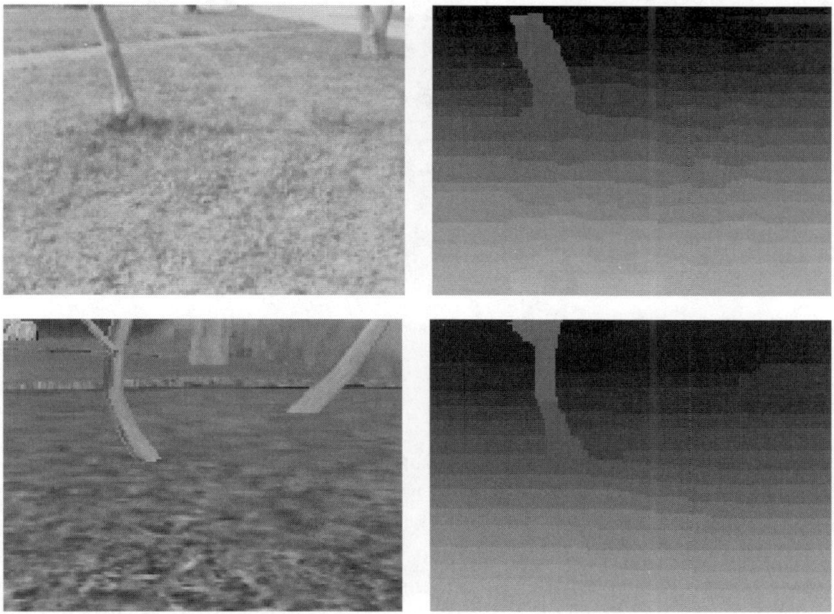

Fig. 5. Real (upper row) and simulated (lower row) camera images and disparity maps

To test the framework performance, images taken with the real robots stereo camera system have been compared with images generated with virtual camera sensors in simulation. Fig.5 exemplarily shows a pair of such real and simulated images along with disparity maps calculated from them (the images depict the typical performance of the stereo algorithm). In both cases, the actual stereo matching algorithm used in the RAVON project was used without any adjustments. The results demonstrate that the simulation clearly suffices for stereo reconstruction. Similar experiments performed for the visual odometry computation also led to realistic results, although a texture with higher contrast had to be used in the simulation.

Asides from simulation, the framework has also been used for visualization of mapping and route planning. For this, elements that overlay markers for map nodes and node connections at parametrized locations have been implemented (Fig.6). By using two separate instances of SimVis3D for simulation and visualization, both could be used simultaneously but contained different elements - therefore, the overlaid map does not appear in the simulated camera images, although the other environment and the robot pose are equal in both Fig.5 and Fig.6.

Fig. 6. Visualization of RAVON scene with overlaid navigation graph

5   Conclusion

A framework for simulation and visualization of three-dimensional scenes has been presented that is geared towards robotic applications with highly specific scene parametrization needs. By supporting manually coded extensions, a programmer can easily customize all control possibilities that an application of interest requires. The use in two complex mobile robot scenarios has provided examples for the frameworks extensibility and has shown that realistic sensor data can be produced. Although currently not as feature-rich as commercially available toolkits, the capability for extensions and the multi-host friendly architecture make SimVis3D a useful addition to available simulation/visualization frameworks that adds several new aspects.

6   Future Work

Future work includes the application of SimVis3D to stationary robots (arms, robotic heads) to explore their special requirements and the integration of a dynamics simulation based on the element inclusion methodology. Preliminary experiments with the *newton*[5] library are very promising. As already mentioned, the publication of SimVis3D as an open-source project is also one of the next steps.

---

[5] http://www.newtondynamics.com/

References

1. Laue, T., Spiess, K., Refer, T.: A general physical robot simulator and its application in robocup. In: Proc. of RoboCup Symp., Bremen, Germany (2005)
2. Koening, N., Howard, A.: Gazebo-3d multiple robot simulator with dynamics (2006), `http://playerstage.sourceforge.net/gazebo/gazebo.html`
3. Michel, O.: Webots: Professional mobile robot simulation. Int. Journal Of Advanced Robotic Systems 1(1), 39–42 (2004)
4. Wang, J., Lewis, M., Gennari, J.: A game engine based simulation of the nist urban search & rescue arenas. In: Proc. of the 2003 Winter Simulation Conference (2003)
5. Koestler, A., Braeunl, T.: Mobile robot simulation with realistic error models. In: ICARA (December 2004)
6. Scholl, K.-U., Albiez, J., Gassmann, B., Zöllner, J.: Mca - modular controller architecture. In: Robotik 2002. VDI-Bericht 1679 (2002)

Imitation of Walking Paths with a Low-Cost Humanoid Robot

Jacopo Aleotti[1] and Stefano Caselli[1]

RIMLab - Robotics and Intelligent Machines Laboratory
Dipartimento di Ingegneria dell'Informazione
University of Parma, Italy
{aleotti,caselli}@ce.unipr.it

Summary. The ability to follow specific and task-relevant paths is an essential feature for legged humanoid robots. In this paper, we describe a technique for programming walking paths for humanoid robots based on imitation. Demonstrated paths are synthesized as NURBS (Non Uniform Rational B-Spline), and can be adapted by the robot based on local and dynamic information. We report simple experiments performed with a Robosapien V2 low-cost humanoid toy which has been suitably enhanced to support imitation-based programming. We show that despite robot limitations rather complex navigational tasks can be achieved through visual guidance. The system combines inputs from multiple sensory devices, including a motion tracker and a monocular vision system for landmark recognition.

1 Introduction

This work is aimed at investigating the capabilities of a low-cost humanoid robot which imitates human demonstrated walking paths. We have explored the idea of augmenting an entertainment humanoid with multiple sensors to accomplish rather complex navigational tasks with partially autonomous behavior. The system combines an electromagnetic sensor and a monocular vision system. The main contribution is the experimental evaluation of vision-guided path following tasks with dynamic obstacle avoidance. We have also experimented autonomous vision-guided grasping tasks of objects on the ground, which can serve as basic skills for more complex imitation tasks.

Our research on imitation and human following is motivated by the fact that robot programming by demonstration [7] is an effective way to speed up the process of robot learning and automatic transfer of knowledge from a human to a robot. As it is well known that robot programming using traditional techniques is often difficult for untrained users, especially in the context of service robotics, programming by demonstration provides an intuitive solution by letting the user act as a teacher and the robot act as a follower. The need for a high-level approach for robot programming is even more apparent with humanoid robots, where conventional programming of a high number of degrees of freedom would be clearly unacceptable. The development of humanoid systems is a tremendous challenge for robotics research. Many academic laboratories as

S. Lee, I.H. Suh, and M.S. Kim (Eds.): Recent Progress in Robotics, LNCIS 370, pp. 371–383, 2008.
springerlink.com

well as industrial companies are devoting substantial efforts and resources in building advanced humanoid robots. Several authors [14, 8, 3] have pointed out that the principal challenges in building effective humanoid robots involve different aspects such as motor actuation, perception, cognition and their integration. Typically, complex mechanical systems must be regulated and a high number of degrees of freedom must be controlled, which requires the study of appropriate materials, motors, power supplies and sensors. Moreover, humanoid systems require advanced algorithms and software solutions to achieve learning skills for autonomous operations. The development of natural user interfaces for human robot interaction is also required, since one of the main goals is to overcome the traditional approaches of robot design by creating robots able to interact with humans in everyday life. However, the technological requirements and the costs to develop effective humanoid robots still prevent such advanced systems to become available to the public.

Few low-cost humanoids have been designed for both entertainment and research purposes. One of the first prototypes is Robota [2], a small humanoid doll which has been used as educational toy for robotic classes playing vision-based imitation games. Recently, several low-cost humanoid platforms have appeared in the consumer market. Besides being considered as high quality toys for personal entertainment, these devices can also be exploited for research projects which can lead to promising results. One of such robots is Robosapien V2 (RSV2) developed by WowWee and released at the end of 2005. The earlier model Robosapien V1 has been used by Behnke et al. [1] for research in soccer skills development. A Pocket PC and a color camera were added to the robot to make it autonomous. RSV2 has been chosen for the experimental evaluation reported in this paper as it is one of the most promising low-cost robots available to the market.

The rest of the paper is organized as follows. Section 2 presents an overview of works regarding trajectory imitation by teaching and playing-back. Section 3 describes the experimental set-up, including a description of the augmented RSV2, the input devices and the software architecture. Section 4 presents the walking path imitation technique, along with an experimental evaluation of the system. The paper closes in section 5 discussing the obtained results.

2 Related Work

Many authors have investigated the problem of tracking and following demonstrated navigational routes with mobile wheeled robots. Dixon and Khosla [5] presented a system for learning motor-skill tasks by observing user demonstrations with laser range finder. The system is able to extract subgoals and associate them with objects in the environment generalizing across multiple demonstrations. Hwang et al. [6] proposed a touch interface method to control a mobile robot in a supervised manner. The algorithm extracts a set of significant points from the user-specified trajectory, produces a smooth trajectory using Bezier curves and allows on-line modification of the planned trajectory in a dynamic environment. In [4] a reactive sensor-motor mapping system is described where a mobile robot learns navigational routes and obstacle avoidance. Morioka et al. [13] focused on the problem of human-following for a mobile robot in an intelligent environment with distributed sensors. The control law is based on a virtual spring model. Some interesting research projects have considered the use of humanoid robots for navigation and autonomous mapping of indoor environments. Michel et al. [12] presented an approach to autonomous humanoid walking for the Honda ASIMO in the presence of dynamically moving obstacles. The system combines vision processing for real-time environment mapping and footsteps planning for obstacle avoidance. A similar approach was presented by Kagami et al. [9]. In [11] an online environment reconstruction system is presented. The system utilizes both external sensors for global localization, and on-body sensors for detailed local mapping for the HRP-2 humanoid robot.

3   Experimental Set-Up

3.1   Robosapien V2

Robosapien V2, shown in figure 1, has $60cm$ height, $7Kg$ weight, 12 degrees of the freedom and is driven by 12 DC motors. RSV2 is fully controllable by a remote infrared controller and has many capabilities which make it suitable for entertainment but also for research purposes. The main functionalities of the robot include true bi-pedal walking with multiple gaits, turning, bending, sitting and getting up. RSV2 can also pick up, drop and throw small objects with articulated fingers. The robot is equipped with an infrared vision system, a color camera, stereo sonic sensors and touch-sensitive sensors in his hands and feet. The locomotion of RSV2 is achieved by alternatively tilting the upper body and moving the leg motors in opposite directions. The lateral swinging movement of the upper body generates a periodic displacement of the center of mass between the two feet. To achieve fully autonomous capabilities RSV2 must be augmented with external computing power. There are two possible ways for augmenting RSV2. The first method is hacking the onboard electronics to get access to the motor and sensor signals. The second method is bypassing the remote controller with an infrared transmitter connected to a processing unit. This strategy is less intrusive but it only allows to send motor commands to the robot, therefore it also requires external sensors like a camera, as proposed in [1]. In this work we chose a similar approach. We adopted a USB-UIRT (Universal Infrared Receiver Transmitter) device, which is connected to a host PC. The device is able to transmit and receive infrared signals. Reception has been used to learn the IR codes of the RSV2 remote controller. As a result, the IR codes of the motor commands for the RSV2 have been decoded and stored in a database.

3.2   Sensor Devices

The experimental set-up of the proposed system comprises a Fastrak 3D motion tracking device (by Polhemus, Inc.) and a monocular vision system. The Fastrak is a six degrees of freedom electro-magnetic sensor that tracks the position and

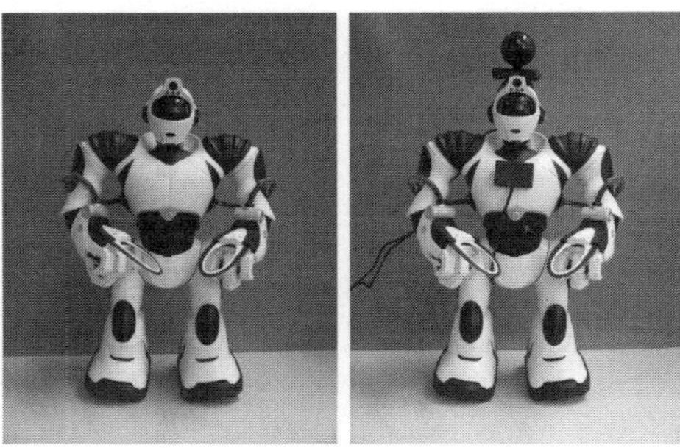

Fig. 1. Original RSV2 (left image) and augmented RSV2 (right image)

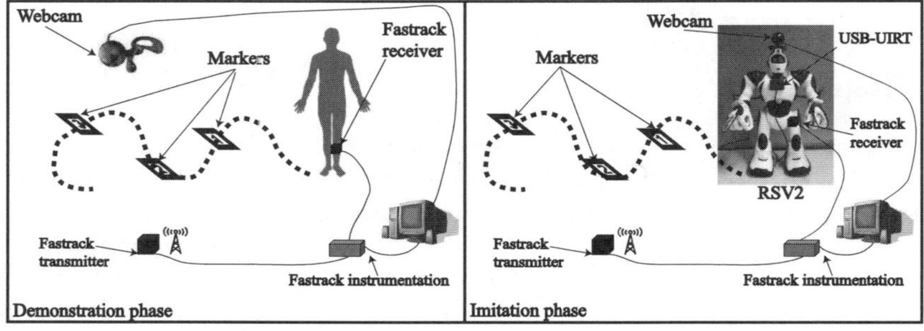

Fig. 2. System Architecture for imitation of walking paths with RSV2

orientation of a small receiver relative to a fixed transmitter. The vision system exploits a Webcam with a CMOS image sensor with 640x480 resolution and 30 frame per second refresh rate. In the current hardware set-up the actual measured frame rate is $12\,fps$. Figure 1 (right image) shows the augmented RSV2. The Webcam has been mounted on the head of the robot to minimize negative effects on its balance. This arrangement provides an additional advantage as the pan-tilt head motors can be exploited to rotate the camera. Camera rotation along two axes is crucial for enhancing the robot capabilities of marker recognition, as will pointed out in the following section. The Fastrak receiver has been mounted on the back of RSV2 and is used for global localization. The USB-UIRT device has been mounted on RSV2 as well. The device is located on the torso and, being very close to the RSV2 infrared receiver, it helps in avoiding packet loss in transmission of motor commands. Figure 2 shows the structure of the system and highlights the hardware components along with the sensors and their connections. The system comprises a demonstration phase, where the user demonstrates a walking path, and an imitation phase, where the robot follows the provided trajectory. The setup also includes a fixed ceiling camera for initial recognition of the topology of the markers on the ground.

3.3   Software

The application which runs the imitation system has a multithreaded architecture. The main thread is devoted to sensor data acquisition from the devices, visual processing and computation of the motor commands for the Robosapien. A second thread is devoted to sending the motor commands to the humanoid robot through the IR transmitter. The ARToolKit software library was adopted for the development of the vision system. ARToolKit [10] is an open-source multiplatform software library explicitly designed for augmented reality applications. The library exploits a pattern recognition algorithm and allows overlaying of computer graphics images on the video stream captured by a camera in real-time. In this paper we used the toolkit to recognize both markers on the ground, acting as via points, and markers attached to static obstacles. The overlay option was enabled for testing the effectiveness of the recognition algorithm. In particular, ARToolKit uses computer vision techniques to compute the real camera position and orientation relative to markers. Markers are squares of known size ($8 \times 8cm$ in the proposed experiments) with inner symbols. One snapshot of each marker must be provided to the system in advance as training pattern. The recognition algorithm is a loop which consists of different steps. First the live video image is binarized using a lighting threshold. Then the algorithm searches for square regions. The squares containing real markers are then identified and the corresponding transformation matrices are computed. OpenGL is used for setting the virtual camera coordinates and drawing the virtual images.

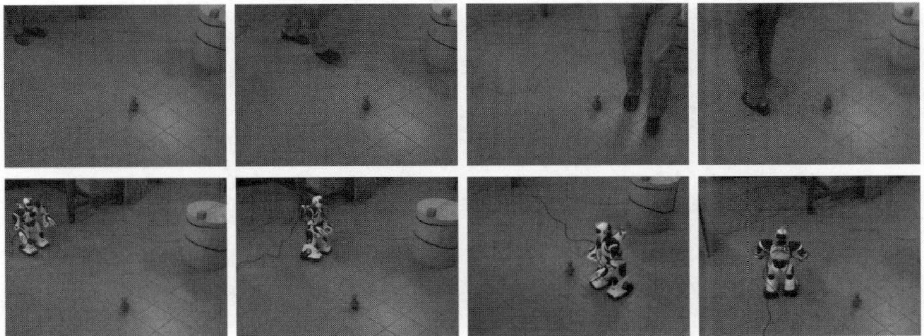

Fig. 3. Experiment 1: demonstration phase (top row) and imitation phase (bottom row)

4   Imitation of Walking Paths

We adopted a hybrid deliberative-reactive control architecture for imitation of walking paths. At the highest level the robot mission is to follow a demonstrated trajectory with a priori information about landmark models and their meaning. Each landmark is a semantic object which lends information to the system. Landmarks are used for representing obstacles, via-points or ground regions with objects to be grasped. A collection of intelligent reactive motor behaviors has also been developed. Such basic behaviors are triggered once the robot perceives a landmark. Experiments involving reactive obstacle avoidance, marker following and grasping will be described next.

4.1   Trajectory Following

The first experiment is a simple walking imitation task where the robot has to follow a taught route. In the demonstration phase a human performs a walking path while his movements are tracked by the Fastrak receiver, which is attached to one of the legs of the demonstrator. After the demonstration phase RSV2 imitates the demonstrated path relying only on the tracker sensor without visual guidance. The transformation matrix from the Fastrak receiver frame (rx) to the world reference frame (W) is given by

$$T_{W \leftarrow rx}(t) = T_{W \leftarrow tx} T_{tx \leftarrow rx}(t) \tag{1}$$

where $T_{W \leftarrow tx}$ is the transformation between the Fastrak transmitter frame to the world frame. Figure 3 shows a sequence of images taken from both the demonstration and the imitation phases. The demonstrated path is a smooth

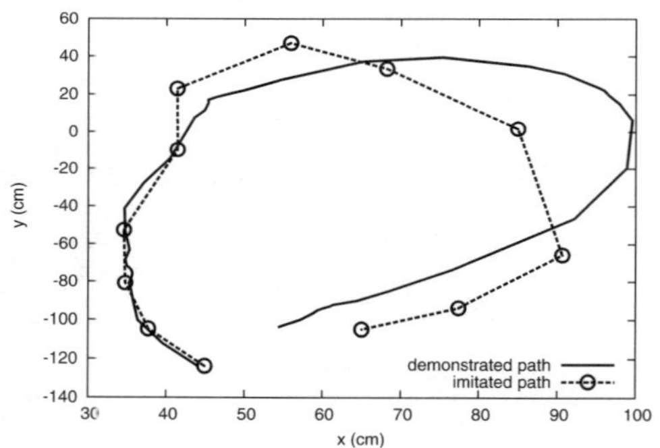

Fig. 4. Experiment 1: demonstrated path and imitated path

U-turn around a static obstacle. Figure 4 reports the evolution of RSV2 walking trajectory compared to the demonstrated trajectory. The demonstrated trajectory is approximated as a NURBS (Non Uniform Rational B-Spline) curve. The systems computes a piecewise linear approximation of the NURBS generating a set of via-points and RSV2 follows the approximated path through a sequence of rotations and walking steps. The computed mean error of the path following task is about $15cm$. The demonstrated path shown in figure 4 appears deformed on the right side due to sensor inaccuracies of the tracker, which increase as the distance between the transmitter and the receiver increases. The same inaccuracies are the main cause of the worsening of path following performance in the last stage of the task after the U-turn.

4.2   Visual Dynamic Obstacle Avoidance

In the second experiment, a vision-guided imitation task is performed which requires reactive obstacle avoidance. After the demonstration phase an obstacle is positioned on the original path. RSV2 is programmed to follow the path and to scan the environment for possible obstacles by turning its head whenever it reaches a via-point. A marker is attached to the obstacle for detection.

Fig. 5. Experiment 2: imitation phase with dynamic vision-guided obstacle avoidance

The algorithm used by ARToolkit for estimating the transformation matrix of markers finds, for each marker, the transformation T_{cm} which relates the coordinates (X_m, Y_m, Z_m) in the marker reference frame to the coordinates (X_c, Y_c, Z_c) in the camera reference frame. The translation component of T_{cm} is obtained by the correspondence between the coordinates of the four vertices of the marker in the marker coordinate frame and the coordinates of the vertices in the camera coordinate frame. The rotation component is estimated by projecting the two pairs of parallel lines contouring the marker. Figure 5 shows a sequence of images taken from the imitation phase of the task. Figure 6 reports the demonstrated path together with the replanned path and the actual imitated path (the obstacle is detected when the robot reaches the second via-point). Figure 7 shows the output of the onboard camera when the marker is recognized with a colored square superimposed to the recognized marker. The algorithm used for replanning the trajectory performs a local deformation of the original NURBS around the obstacle. Initially the NURBS is resampled and a deformation is applied to each sample point $(x(u), y(u))$ which falls within a certain distance from the obstacle according to the following equation

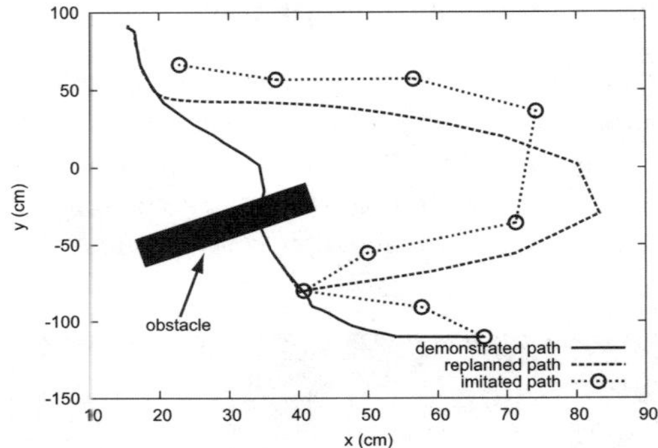

Fig. 6. Experiment 2: demonstrated path, replanned path and imitated path

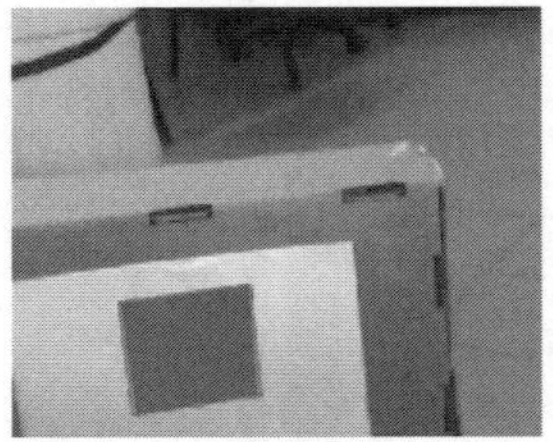

Fig. 7. Experiment 2: obstacle detection

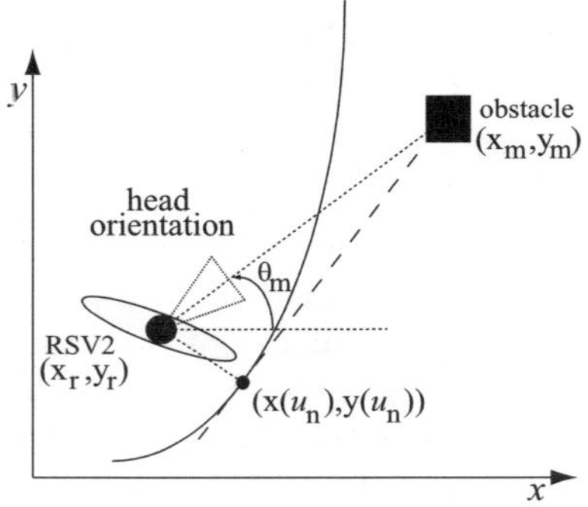

Fig. 8. Experiment 2: approximation of obstacle coordinates

$$\begin{cases} \tilde{x}(u) = x(u) + \alpha e^{-\frac{d^2}{\beta}} \left(x(u) - x_m \right) \\ \tilde{y}(u) = y(u) + \alpha e^{-\frac{d^2}{\beta}} \left(y(u) - y_m \right) \end{cases} \tag{2}$$

where α and β are constant. The deformation stretches the samples with a gaussian modulation factor which depends on the distance d between the robot and the obstacle. The value of d is approximated by the measured distance d_{cm}

between the camera and the marker which are approximately orthogonal. The obstacle coordinates (x_m, y_m) are set equal to the coordinates of the marker in the world reference frame and are given by

$$\begin{cases} x_m = x_r + d\cos\theta_m \\ y_m = y_r + d\sin\theta_m \end{cases} \tag{3}$$

where (x_r, y_r) is the current position of the robot and θ_m is the angle between the x-axis and the obstacle, which depends on both RSV2 body and head rotation. The angle θ_m can not be easily computed, as the actual head rotation is hard to estimate due to the non ideal alignment between the head and the body when the robot stops. Therefore, the obstacle coordinates are approximated as follows

$$\begin{cases} x_m \approx x(u_n) + d_{cm}\dfrac{\dot{x}(u_n)}{\sqrt{\dot{x}^2(u_n)+\dot{y}^2(u_n)}} \\ y_m \approx y(u_n) + d_{cm}\dfrac{\dot{y}(u_n)}{\sqrt{\dot{x}^2(u_n)+\dot{y}^2(u_n)}} \end{cases} \tag{4}$$

where $(x(u_n), y(u_n))$ is the nearest point on the original NURBS to the current position of the robot. This means that the obstacle is assumed to lie on the tangent to the curve at the closest point to the robot (Figure 8).

4.3 Visual Landmark Navigation

In the third experiment another vision-guided imitation task has been investigated. A set of markers is positioned on the ground of the workspace and a fixed ceiling camera detects the initial configuration of the markers (with an estimation error approximately equal to $7cm$). The demonstration consists of a walking task where a human reaches a sequence of markers. The correct sequence of reached markers is computed from the measured trajectory and from the topology of the markers. After the demonstration phase, the markers are

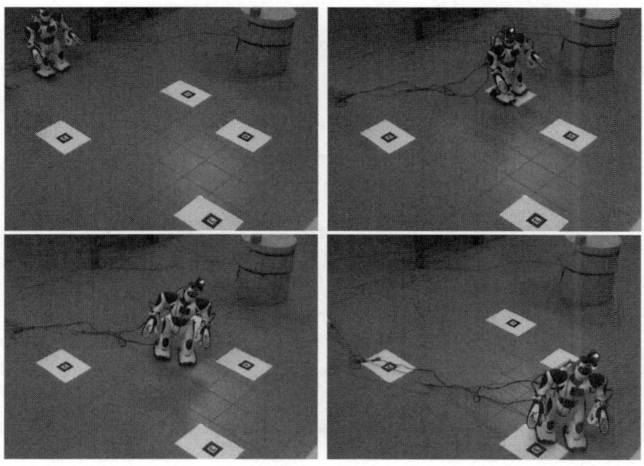

Fig. 9. Experiment 3: imitation phase with vision guided marker following

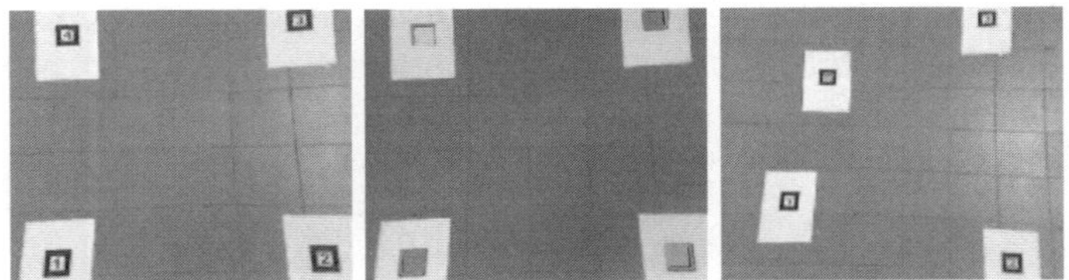

Fig. 10. Experiment 3: initial marker configuration (left), marker detection from ceiling camera (middle) and modified marker configuration (right)

slightly moved into a new configuration and RSV2 is programmed to reach the demonstrated sequence of markers. In the proposed experiment the task is to reach marker 1 and then marker 3. Figure 9 reports images of the imitation phase, while figure 10 shows the initial marker configuration and the modified marker configuration. Figure 11 shows the demonstrated path, the configuration of the markers and the imitated path. RSV2 imitates the reaching task by following the demonstrated path. Once the robot detects a marker on the floor it tries to approach the marker through a sequence of small movements by estimating the position and the distance of the target with the onboard camera with equation (3). The estimation error is less critical than in experiment 2 since the robot performs small movements while approaching the marker, as shown in figure 11.

4.4 Vision-Guided Grasping

RSV2 autonomous grasping skills have been evaluated in the last experiment, where the goal is to reach a landmark on the floor indicating the position of a set of graspable objects. The robot starts approximately $80cm$ away from the landmark and in its initial configuration the head and the body are properly

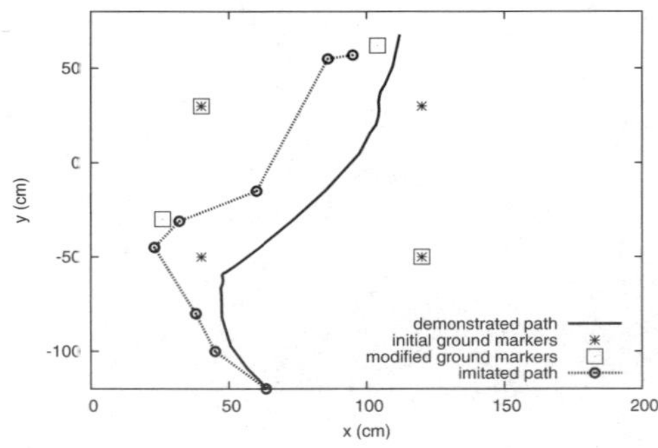

Fig. 11. Experiment 3: demonstrated path, marker positions and imitated path

aligned. Hence RSV2 is able to detect the landmark with tolerable error using the onboard camera and moves towards the objects. Once the robot gets closer to the landmark it tries to reach a proper alignment with the target with small rotations and walking steps. Figure 12 shows the execution of the grasping task while

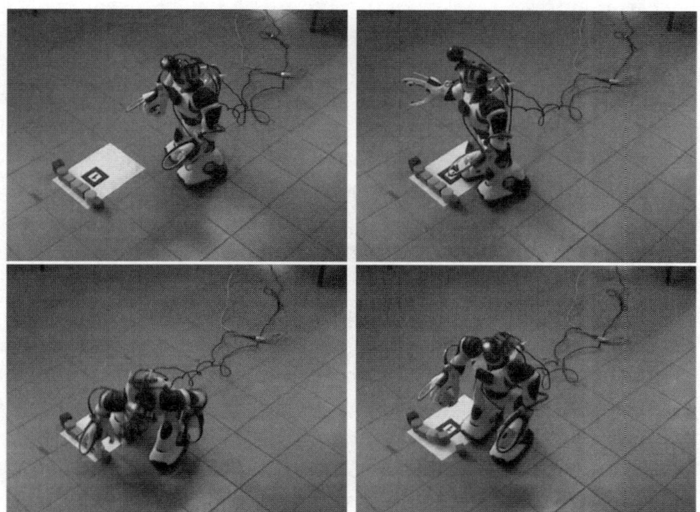

Fig. 12. Experiment 4: vision guided grasping task

figure 13 shows the marker detection phase at two different configurations. Due to the limitations of RSV2 the grasping procedure is not controllable and allows only to lower the body and one arm of the robot trying to pick up objects which are located near the foot of the robot. To overcome the kinematics limitations of the toy and the sensor errors, multiple objects have been placed near the landmark for redundancy. Unfortunately, the rate of successful grasping is still low (approximately 40%).

Fig. 13. Experiment 4: marker detection

5 Discussion and Conclusion

In this paper, a path specification system based on imitation of human walking path for a humanoid robot has been presented. The experimental system investigated comprises a Robosapien V2 humanoid and includes multiple sensor devices such as an electromagnetic tracker and a monocular vision system. The novelty of the approach lies in the investigation of a high level programming paradigm such as imitation with a low cost humanoid toy robot. Vision-guided trajectory imitation and dynamic obstacle avoidance have been successfully experimented, along with autonomous grasping skills. The experiments show that the ability to teach motion paths enables a toy robot to achieve complex navigation tasks. The proposed imitation approach is rather general, even though its implementation is constrained by the limitations of RSV2 and by sensor inaccuracies. The usable workspace is currently restricted to a square of approximately $2m^2$ due to the range resolution of the tracking device. Moreover, the tilting motion of the upper body while the robot is walking prevents the use of a continuous visual feedback. Therefore the vision system can be exploited only when the robot is not moving. Finally, the non ideal alignment between the head and the body of the robot when it stops strongly affects the estimation of landmarks pose used for visual guidance. The kinematics limitations of RSV2 also affect the imitation accuracy. Such limitations include the difficulty of performing sharp turns, the absence of side steps commands and the reduced grasping abilities. Nonetheless, we believe that low-cost humanoid platforms such as RSV2 provide an exciting and affordable opportunity for research in humanoid programming based on imitation.

Acknowledgment

This research is partially supported by Laboratory LARER of Regione Emilia-Romagna, Italy.

References

1. Behnke, S., Müller, J., Schreiber, M.: Playing soccer with robosapien. In: 9th RoboCup Int'l Symposium, Osaka, Japan (July 2005)
2. Billard, A.: Robota: Clever toy and educational tool. Robotics and Autonomous Systems 42(3-4), 259–269 (2003)
3. Breazeal, C., Scassellati, B.: Challenges in building robots that imitate people. In: Dautenhahn, K., Nehaniv, C. (eds.) Imitation in Animals and Artifacts, MIT Press, Cambridge (2001)
4. Chow, H.N., Xu, Y., Tso, S.K.: Learning human navigational skill for smart wheelchair. In: IROS. IEEE/RSJ Int'l Conference on Intelligent Robots and Systems (September 2002)
5. Dixon, K.R., Khosla, P.K.: Learning by Observation with Mobile Robots: A Computational Approach. In: ICRA. IEEE Int'l Conference on Robotics and Automation, New Orleans, LA (April 2004)

6. Hwang, J.H., Arkin, R.C., Kwon, D.S.: Mobile robots at your fingertip: Bezier curve on-line trajectory generation for supervisory control. In: IROS. IEEE/RSJ Int'l Conference on Intelligent Robots and Systems, Las Vegas, Nevada (October 2003)
7. Ikeuchi, K., Suehiro, T.: Toward an assembly plan from observation, Part I: Task recognition with polyhedral objects. IEEE Trans. Robot. Automat. 10(3), 368–385 (1994)
8. Mataric, M.J.: Getting humanoids to move and imitate. IEEE Intelligent Systems 15(4), 18–24 (2000)
9. Kagami, S., Nishiwaki, K., Kuffner, J.J., Okada, K., Inaba, M., Inoue, H.: Vision-Based 2.5D Terrain Modeling for Humanoid Locomotion. In: ICRA. IEEE Int'l Conference on Robotics and Automation, Taipei, Taiwan (September 2003)
10. Kato, H., Billinghurst, M.: Marker Tracking and HMD Calibration for a video-based Augmented Reality Conferencing System. In: IWAR. 2nd Int'l Workshop on Augmented Reality (October 1999)
11. Michel, P., Chestnutt, J., Kagami, S., Nishiwaki, K., Kuffner, J., Kanade, T.: Online Environment Reconstruction for Biped Navigation. In: ICRA. IEEE Int'l Conference on Robotics and Automation, Orlando, Florida (May 2006)
12. Michel, P., Chestnutt, J., Kuffner, J., Kanade, T.: Vision-Guided Humanoid Footstep Planning for Dynamic Environments. In: 5th IEEE-RAS Int'l Conference on Humanoid Robots (2005)
13. Morioka, K., Lee, J.-H., Hashimoto, H.: Human-Following Mobile Robot in a Distributed Intelligent Sensor Network. IEEE Trans. Ind. Electron. 51(1), 229–237 (2004)
14. Schaal, S.: Is imitation learning the route to humanoid robots? Trends in Cognitive Sciences 3(6), 233–242 (1999)

Intelligent Robot Software Architecture

Jonghoon Kim[1], Mun-Taek Choi[1], Munsang Kim[1], Suntae Kim[2],
Minseong Kim[2], Sooyong Park[2], Jaeho Lee[3], and ByungKook Kim[4]

[1] Center for Intelligent Robotics, Frontier 21 Program at Korea Institute of Science
and Technology, Seoul, Korea
`jhkim, mtchoi, munsang@kist.re.kr`
[2] Department of Computer Science, Sogang University, Seoul, Korea
`jipsin08, minskim, sypark@sogang.ac.kr`
[3] Department of Electrical and Computer Engineering, The University of Seoul,
Seoul, Korea
`jaeho@uos.ac.kr`
[4] Real-Time Control Laboratory, Department of Electrical Engineering and
Computer Science, Korea Advanced Institute of Science and Technology, Daejeon,
Korea
`bkkim@eeg.kaist.ac.kr`

Summary. This paper describes the software architecture of intelligent robots, developed by Center for Intelligent Robotics (CIR). The architecture employs a hybrid type consisting of deliberate, sequencing and reactive behaviors. The primary contributions of the architecture are the followings: 1) reusability and extensibility for the rapid development of robot systems with different requirements and 2) adaptability to distributed computing environments, different OS's and various programming languages. The architecture has successfully been applied to the development of CIR's many robot platforms such as T-Rot, Kibo and Easy Interaction Room. T-Rot and Kibo have been exhibited at the 2005 Asia-Pacific Economic Cooperation (APEC) in Korea.

1 Introduction

Modern intelligent robots to assist people in the home environments necessitate advancement in robotic services. Such services may include the following:

- Physical supports such as fetch-and-carry, meal assistance, mobility aids, etc
- Mental/Emotional supports such as being a companion for conversation and entertainments, educational assistance, etc
- Informational supports such as providing weather information or schedule information, health care, etc.

In order to provide those services, it is required to have sophisticated intelligent software in the robot systems. The robot software has to be able to recognize uncertain surrounding environments, plan for tasks based on knowledge from the recognition, and execute behaviors such as navigation or manipulation.

S. Lee, I.H. Suh, and M.S. Kim (Eds.): Recent Progress in Robotics, LNCIS 370, pp. 385–397, 2008.
springerlink.com

The components of the software, subject to systematic integration may consist of the following:

- Interfacing components for hardware sensors such as range sensors, vision cameras, microphones, etc
- Intellectual components that recognize objects, human's face and voice from sensory information
- Task managing components that plan for tasks through reasoning drawn by interacting with human and knowledge from recognition
- Behavior components to execute the task planning, navigating to a destination, manipulating a target object to carry, etc
- Interfacing components for hardware actuators such as wheels for mobile base, joint actuators for manipulators, pan/tilt actuators for a head, etc.

To make intelligent robot software, a well-defined architecture is essential. This well-defined software architecture can separate the internal logic of components from the overall structure. Based on this software architecture, components can be deployed efficiently and added easily. Researches have been continued to design effective architectures up to date and a hybrid-type has frequently been used to develop intelligent robots in recent years. The hybrid architecture is intensively discussed by Gat [1]. This hybrid architecture composed of deliberative layer and reactive layer. The architecture of MINERVA [2] is composed of four layers, high-level control and learning, human interaction modules, navigation modules and hardware interface modules. MINERVA is controlled by a generic software approach for robot navigation and human robot interaction. The architecture of Care-O-bot II [3, 4] is a hybrid control architecture for task planning and execution that controls concurrent, reactive, and deliberative activities. The architecture of PSR [5, 6] also employs a three-layered architecture in conjunction with a tripodal schematic design to incorporate extensibility.

This paper is organized as follows. Chapter 2 introduces robot platforms. Chapters 3 and 4 describe the design requirements and the software architecture of the robots, respectively. The results of experiments are discussed in Chapter 5. Finally, some concluding remarks and future works are represented in Chapter 6.

2   Robot Platforms

CIR developed several hardware platforms including T-Rot, Kibo and Easy Interaction Room. T-Rot (Thinking Robot) is an intelligent service robot with manipulation capability as shown in Figure 1. T-Rot consists of 4 Single Board Computers (SBCs), two-differential-wheel mobile base, two 7-DOF manipulators, two 3-fingered hands, 2-DOF pantilt, speech unit, battery and several sensors(e.g., laser range finder, infrared ray, sonar, encoder, camera, microphone, etc).

Kibo is a little humanoid robot for friendly interaction with humans and Easy Interaction Room is a test bed to research service robot's role in ubiquitous computing environment as shown in Figure 2.

Fig. 1. T-Rot's body, pantilt and arm with hand

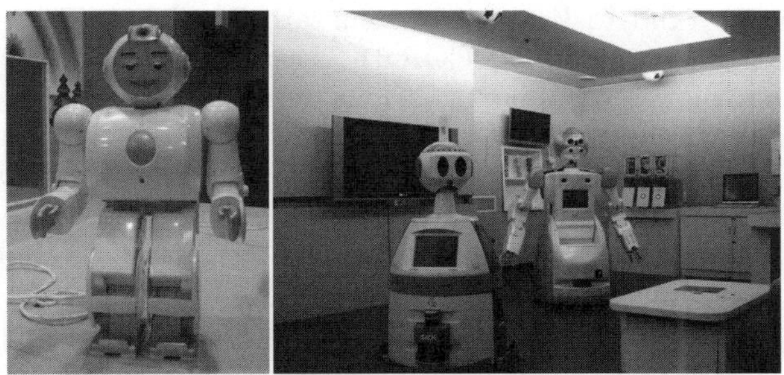

Fig. 2. Kibo and Easy Interaction Room

3 Design Requirements on Architecture

The architectural requirements of the robots are described as follows:

- Hybrid architecture: Robot software architecture consists of three layers as shown in Figure 3. The lowest level is the reactive layer for real-time services. Above the reactive layer is the sequencing layer which supports primitive behavior. On top of the sequencing layer, deliberate layer for task managing exists.
- Component Based Development (CBD): Each robot application module is componentized and then turned into "services" that are controllable by task manager. There are two types of Control Component (CC), as we explain later, C++ and Java component. Because CC is encapsulated by abstract class that our framework offers, CC can easily communicate with task manager through exchanging standardized XML messages in distributed environment. Therefore, our software architecture provides scalability and flexibility. Actually, many teams took part in the development of the robot component and T-Rot's system was integrated by component based style. We use XML messages over TCP/IP to communicate data conforming to a defined interface specification between

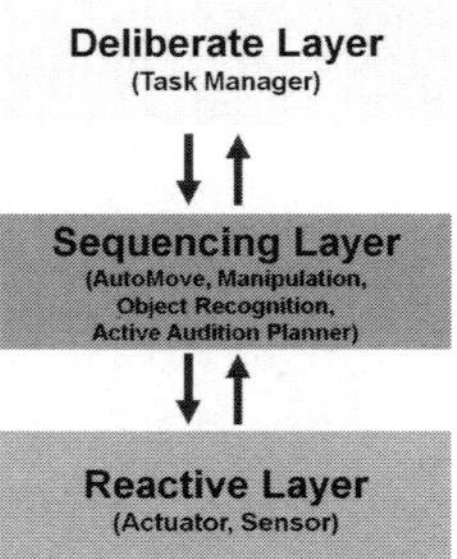

Fig. 3. Layered software architecture

components and task manager. Our architecture is not tied to a specific technology including CORBA, DCOM or Web Services. Because our robot platform supports various development languages, we choose RPC-style method call to exchange messages.

- Real-Time Application Interface (RTAI): At the reactive layer, in order to access sensors and actuators in real-time, components are implemented using RTAI which makes hard real-time functionality available within the Linux environment.

- Transparent and asynchronous control of distributed component: The robot system consists of many distributed components that communicate with task manager for performing their behavior. So those components must be controlled by task manager transparently over the network. Actually, four computers are embedded in the robot and components deployed in each PC are connected to task manager. Only task manager knows the location of components at execution time by XML description and components don't have to know anything about location. If some component's location changes, what we have only to do is changing XML description. Messages between task manager and components are transferred asynchronously so that task manager can send commands to several components at once and several components are executed at the same time.

- Multiple OS and multiple languages: Linux is main operating system and Windows is used for robot monitoring. For more flexibility, languages like C, C++ or Java are used to develop components. Java is used to develop task manager, control component manager and some components. Navigation, manipulation, vision and voice components are implemented by C++.

- UML-based design: One of the best known object oriented modeling languages is the Unified Modeling Language (UML [10]) that is a standard notation for software analysis and design. Previously our navigation pilot project [7] was carried out using COMET [11]. In this paper we design our software architecture and components using UML. At the analysis phase, activity diagram helps us to identify components clearly. At the design phase, we use sequence diagram and class diagram to design components in detail. Also collaboration diagram is very helpful to find out the relation between components and framework modules.

4 Component Based Architecture

4.1 Conceptual View

While the robots are interacting with people, to make the robots smart, i.e., being able to cope with exceptional situations, high-level control software modules are required. Task manager plays a key role in the deliberate layer. At the sequencing layer, many control components receive commands from task manager and execute their work at a given time. At the reactive layer, various RTAI modules interface directly to the robot's sensors and actuators. Each layer is explained in detail in the following sections.

Our application architecture contains various constituent in each layer as shown in Figure 4. Through the monitor module, states of task manager and control components are observed. Also monitor can send command messages to task manager like fact change as an event trigger. Task Manager (TM) is an agent with goals, plans, and world model that commands and controls the control components based on the contexts and deliberative planning and reasoning [8]. TM which resides in the deliberate layer performs tasks sequentially or concurrently according to the current conditions. TM also controls the operations of the modules responsible for deliberative computations such as time-consuming planning and high level reasoning.

Control Component (CC) is an independent software module in the sequencing layer that receives TM's command messages and executes its own process. CC's state is reported automatically to TM. CC can notify its event to TM asynchronously, and then the TM deals with that event. CC can be implemented as Java or C++. To transfer command to distributed CC, each distributed SBC must have a Control Component Manager (CCM). Control Component Manager (CCM) is a daemon process on each SBC to manage CCs to be deployed. Device wrappers are device interface modules in the sequencing layer that can access to sensors and actuators. CC can use these device wrappers regardless of the reactive layer's construction. Various real-time components acquire sensor's data periodically, store them in shared memory to service real-time as well as upper layer's software components and also manage actuator operation.

4.2 Deliberate Layer

Task Manager works as an integration middleware and provides a consistent and unified control view for the functional components which may be distributed over a network. Basically, the framework is designed following the agent model. An

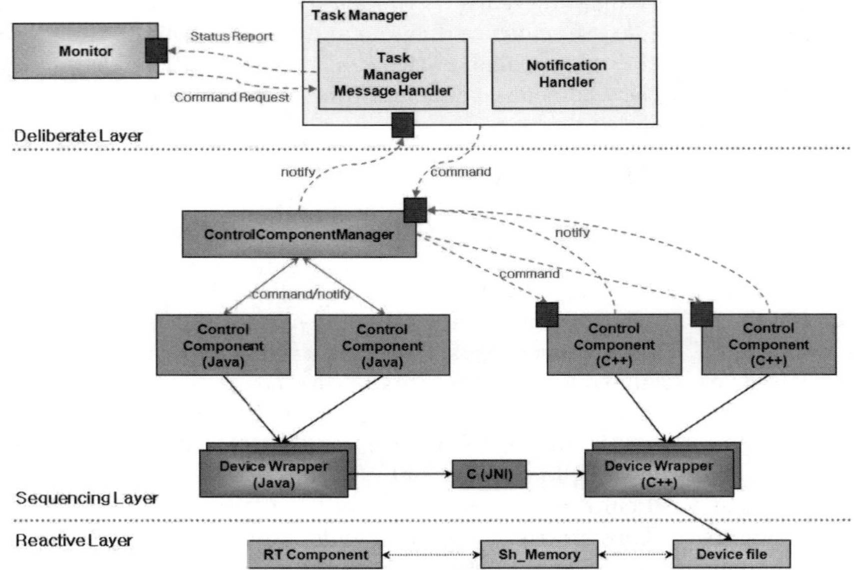

Fig. 4. Application Architecture

agent is a software system that is situated in an environment and that continuously perceives, reasons, and acts [13]. Task Manager, the top-most controller of the system, as shown in Figure 5 is especially built using the BDI model. BDI agents are modeled in terms of beliefs, desires and intentions and believed to provide sufficient level of abstraction to overcome the complexity of intelligent robots. TM is implemented as a JAM. It comprises a belief base, a set of current desires or goals, a plan library and an intention structure. The plan library provides plans describing how certain sequences of actions and tests may be performed to achieve given goals or to react to particular situations. The intention structure contains those plans that have been chosen for execution. An interpreter manipulates these components, selecting appropriate plans based on the system's beliefs and goals, placing those selected on the intention structure and executing them.

Using JAM agent, we could write out the following plans in advance.

- Fetch and carry beverage from dispenser
- Go to the kitchen and identify object
- Talking about TV program

With the TM, robot can achieve its goal such as above scenarios. Figure 6 shows a collaboration diagram to achieve plan such as fetching and carrying beverage from dispenser.

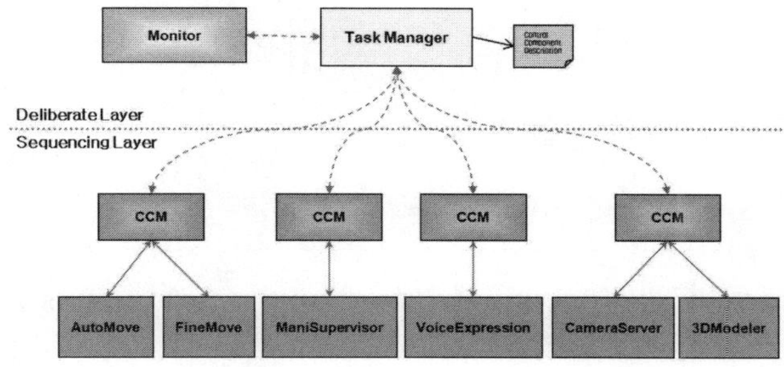

Fig. 5. Task Manager Architecture

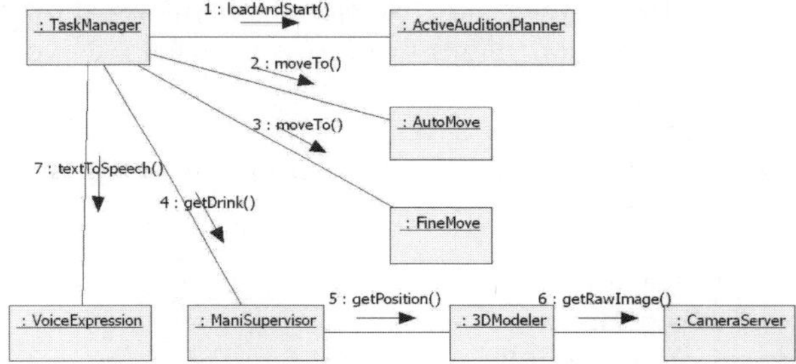

Fig. 6. High-level Collaboration Diagram

4.3 Sequencing Layer

In the sequencing layer, command message from task manager is parsed into
task and the task is executed by control component. Sequencing layer comprises
CCM, CC and Device Wrapper. CCM works as a proxy between the CCs and
the TM, that is, dynamically loads or unloads CCs on request of the TM, and
delivers TM's commands to CC and CC's notification of events to the TM.
It uses agent communication language (ACL) to communicate with TM [8].
Because the format of command message is XML, CC must parse and interpret
the XML message. After that, CC executes its own process and the state of the
CC is changed to *RUNNING*. The state of the CC is one of the three states,
which are *STOPPED*, *RUNNING* or *SUSPENDED*. A more detailed description
of state definition and transition of CC can be found in [8].

In our software framework, communication logic is optimized independently
of CCs so that CC developers can concentrate only on their core logic. This
facilitates the separation of concerns in developing robots, which is an important
design principle in software engineering.

Control Component is composed of several sub-components. Sub-component
is a subordinate and cooperative software module that performs its function in
a CC. For example, AutoMove is a CC with 4 sub-components, namely, Map

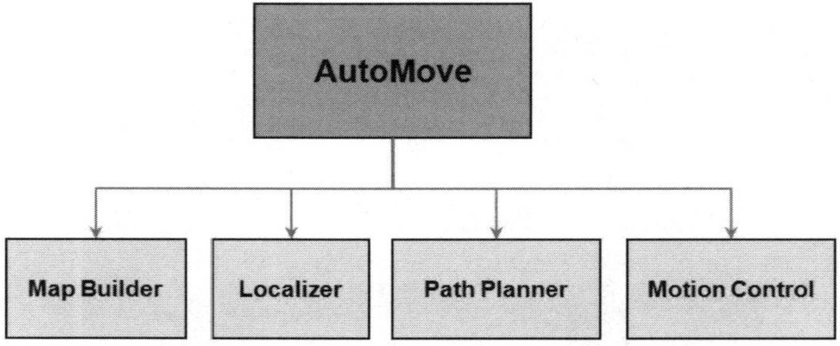

Fig. 7. Control Component and Sub-Component

Builder, Localizer, Path Planner and Motion Control as shown in Figure 7. Figure 8 shows a collaboration diagram of AutoMove to move to destination position.

In the sequencing layer there are several distributed CCs which communicate with task manager asynchronously. The followings are CCs in the sequencing layer.

- AutoMove, FineMove
- Cognitive Interaction
- ManiSupervisor (for handling arm and hand)
- 3DModeler, Camera Server
- Active Audition Planner, Voice Expression
- Multi-Modal Input

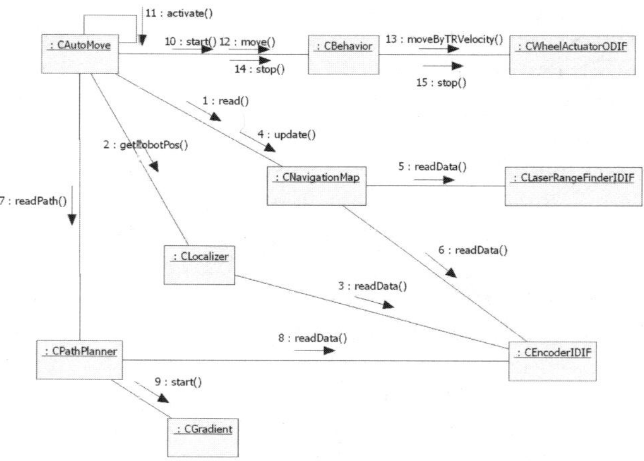

Fig. 8. CC-level Collaboration Diagram

4.4   Reactive Layer

Software components comprising the proposed real-time reactive control layer on
Main SBC are implemented as RTAI modules as shown in Figure 9. The reactive
layer is a behavior-based type with Resource, Actuator, Behavior, and Behavior
Coordinator components as described by Jeon [14]. Each Resource components
configures a given sensor hardware for periodic data acquisition, manages hardware operation, and also stores the sensor data in shared memory for other
software components. An Actuator component configures and manages actuator
hardware and sends generated control outputs. A Behavior component is a basis
of reactive action which uses the sensory data to compute the control output.
Finally, a Behavior Coordinator (BC) component collects outputs from a set of
Behavior components and fuses them for Actuator component. Currently, there
are ten Resource components, one Actuator component for the actuator on the
platform, one BC component, and two Behavior components. To communicate
with components in sequencing layer, RT Task Supervisor (RTTS) component is
used to manage all real-time tasks and handle service requests issued to reactive
layer [9].

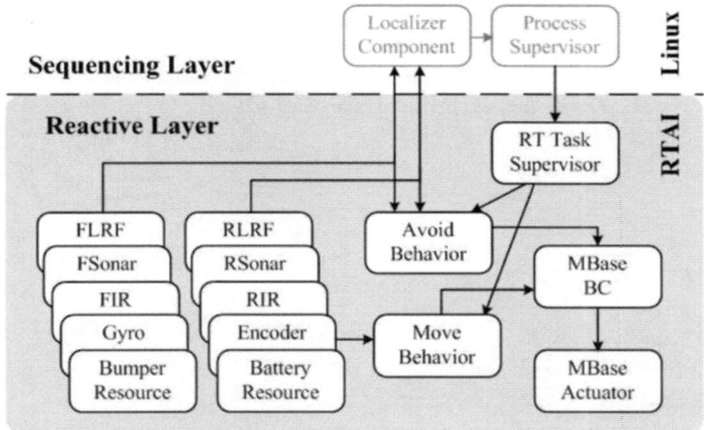

Fig. 9. Reactive Layer Software Components

4.5   Communicating with XML

To communicate with distributed object, reliable transfer mechanism and standard message structure are necessary. We use XML message structure and
TCP/IP framework for middleware instead of other middleware like CORBA
or UPnP. That is to say, format of command and notification messages is XML.
It is fully accepted as the standard for representing data over the internet. Therefore messages between the TM and the CCs are easily exchanged by our software
framework. Interface designer must previously define XML message structures

between TM and CCs. Figure 10 shows a XML format of command message which was transmitted from TM to CC. For instance, if AutoMove CC which is responsible for auto-navigation receives a XML parameter that contains information of destination position, it parses the XML and move to the destination position.

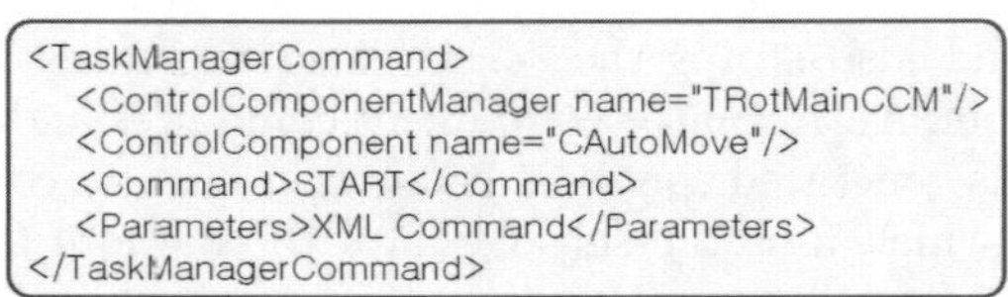

Fig. 10. XML Command Message

4.6 Device Wrapper

Component located in the sequencing layer should access to the reactive layer. Device wrappers are modules that are able to read sensor's information and command to actuator. Those are designed by object oriented method [10]. Each device interface must have restricted number of object, because of the nature of the sensors and actuators. Therefore every Input Device Interface (IDIF) and Output Device Interface (ODIF) is applied to singleton design pattern [11, 12]. Figure 11 and 12 show the design of IDIF and ODIF.

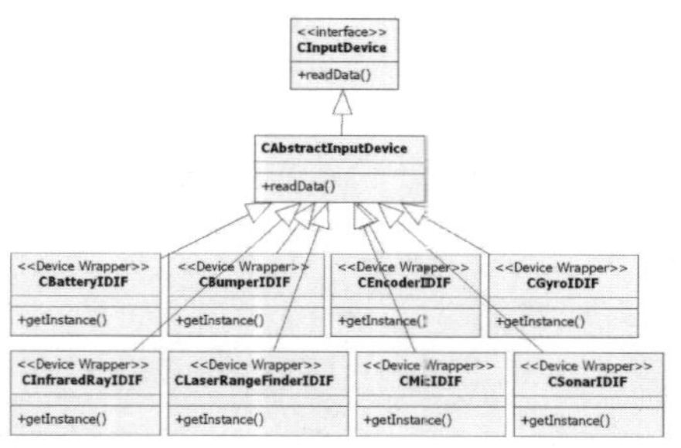

Fig. 11. Input Device Interface

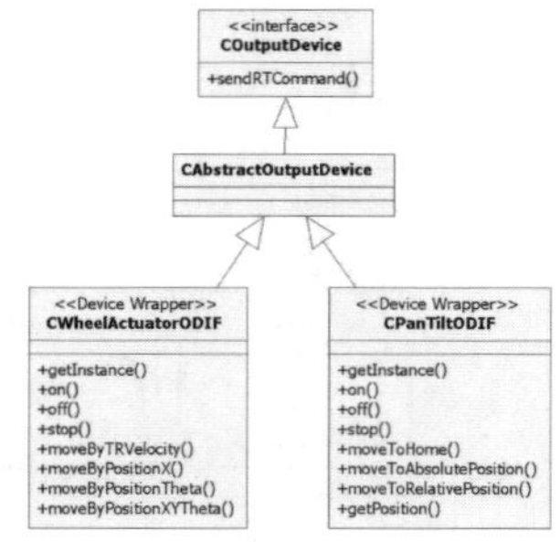

Fig. 12. Output Device Interface

5 Experiments and Results

Our software architecture is implemented on the service robot "T-Rot", humanoid robot "Kibo" and Easy Interaction Room. T-Rot and Kibo have been exhibited at the 2005 Asia-Pacific Economic Cooperation (APEC) in Korea. During the exhibition, T-Rot has successfully provided drinking services to many summit leaders in a robot cafe as shown in Figure 13.

Fig. 13. T-Rot in Robot Cafe

T-Rot can do its job making an order by voice recognition and getting drink from a dispenser. A guest talks to T-Rot using a microphone to ask something to drink, and then T-Rot recognizes the meaning of speech through components such as Active Audition Planner and Cognitive Interaction. Although the inner space of the robot cafe is smaller than 4 m $\times$ 3 m, T-Rot must know it's map for navigation beforehand. Using AutoMove CC which knows the grid map of robot cafe, T-Rot can navigates roughly in the robot cafe. Because getting drink from the dispenser and putting drink in a cup are very accurate operations, it's essential to have functions such as exact navigation, object recognition and manipulation. Using two laser range finders, T-Rot can move to the exact position within the range of $\pm$3cm and $\pm$1°. Vision CCs which use two cameras recognize the target object such as the dispenser and the cup. Manipulator CC which receives the position of a target object calculates the optimal trajectory and moves T-Rot's waist, arm and hand to deliver the ordered drink.

The software architecture was also applied to Kibo and Easy Interaction Room. Figure 14 shows dancing Kibo at the APEC and Easy Interaction Room which is a ubiquitous computing environment.

Fig. 14. Dancing Kibo and Easy Interaction Room

The software architecture enabled us to integrate distributed heterogeneous components relatively easy. Therefore component developers could concentrate on developing their functional logic.

6 Conclusion and Future Works

The proposed software architecture has proven its effectiveness on reusability and extensibility for the rapid developments of robot systems by successfully developing T-Rot, Kibo and Easy Interaction Room. To cope with distributed and heterogeneous computing environments, the architecture guarantees adaptability to distributed systems, Linux as well as Windows and C/C++ and Java. Future research issues include the improvement on robot intelligence by developing an ontological knowledge system based on blackboard architecture. In addition, further work is planned for developing service oriented architecture and frameworks for repurposing and reconfiguring of services.

Acknowledgments

This research was performed for the Intelligent Robotics Development Program, one of the 21st Century Frontier R&D Programs funded by the Ministry of Commerce, Industry and Energy of Republic of Korea.

References

1. Gat, E.: On Three-Layer Architectures. In: Kortenkamp, D., et al. (eds.) Artificial Intelligence and Mobile Robots, AAAI Press, Stanford (1998)
2. Thrun, S., Bennewitz2, M., Burgard2, W., Cremers2, A.B., Dellaert1, F., Fox, D.: MINERVA: A Second-Generation Museum Tour-Guide Robot. In: Proc. of the IEEE Int. Conf. on Robotics and Automation (1999)
3. Hans, M., Baum, W.: Concept of a Hybrid Architecture for Care-O-bot. In: Proceedings of ROMAN-2001, pp. 407–411 (2001)
4. Graf, B., Hans, M., Schraft, R.D.: Care-O-bot II - Development of a Next Generation Robotic Home Assistant. Autonomous Robot Journal 16(2), 193–205 (2004)
5. Kim, G., Chung, W., Kim, M., Lee, C.: Tripodal Schematic Design of the Control Architecture for the Service Robot PSR. In: Proceeding of the IEEE Conference on Robotics and Automation, Taipei, Taiwan, pp. 2792–2797 (2003)
6. Kim, G., Chung, W., Kim, M., Lee, C.: Implementation of Multi-Functional Service Robots Using Tripodal Schematic Control Architecture. In: Proceedings of the International Conference on Robotics and Automation, New Orleans, LA, USA (2004)
7. Kim, M., Kim, S., Park, S., Choi, M.-T., Kim, M., Gomaa, H.: UML-Based Service Robot Software Development: A Case Study. In: 28th International Conference on Software Engineering, Shanghai, China 5 (2006)
8. Lee, J., Kwak, B.: A Task Management Architecture for Control of Intelligent Robots. In: Shi, Z.-Z., Sadananda, R. (eds.) PRIMA 2006. LNCS (LNAI), vol. 4088, pp. 59–70. Springer, Heidelberg (2006)
9. Lee, H.S., Choi, S.W., Kim, B.K.: Real-time reactive control layer design for intelligent silver-mate robot on RTAI. In: Proc. of the 7th Real-Time Linux Workshop, pp. 9–15 (November 2005)

10. Booch, G., Rumbaugh, J., Jacobson, I.: The Unified Modeling Language User Guide. Addison-Wesley, Reading (1999)
11. Gomaa, H.: Designing Concurrent, Distributed, and Real-Time Application with UML. Addison-Wesley, Reading (2000)
12. Gamma, E.: Design Patterns: Elements of Reusable Object-Oriented Software. Addison-Wesley, Reading (1995)
13. Russell, S., Norvig, P.: Artificial Intelligence: A Modern Approach. Prentice Hall, Englewood Cliffs (1995)
14. Jeon, S.Y., Kim, H.J., Hong, K.S., Kim, B.K.: Reactive Layer Control Architecture for Autonomous Mobile Robots. In: ICMIT 2005. Proceedings of the 3rd International Conference on Mechatronics and information Technology, Chongqing, China (2005)

Ontology-Based Semantic Context Modeling for Object Recognition of Intelligent Mobile Robots

Jung Hwa Choi[1], Young Tack Park[1], Il Hong Suh[2], Gi Hyun Lim[2], and Sanghoon Lee[2]

[1] School of Computing, Soongsil University, 511, Sangdo-Dong, Dongjak-Gu, Seoul, Korea
cjh79@ailab.ssu.ac.kr, park@ssu.ac.kr
[2] College of Information and Communications, Hanyang University, 17 Haengdang-dong, Seongdong-gu, Seoul, Korea
ihsuh@hanyang.ac.kr, {hmetal, shlee}@incorl.hanyang.ac.kr

Abstract. Object recognitions are challenging tasks, especially partially or fully occluded object recognition in changing and unpredictable robot environments. We propose a novel approach to construct semantic contexts using ontology inference for mobile robots to recognize objects in real-world situations. By semantic contexts we mean characteristic information abstracted from robot sensors. In addition, ontology has been used for better recognizing objects using knowledge represented in the ontology where OWL (Web Ontology Language) has been used for representing object ontologies and contexts. We employ a four-layered robot-centered ontology schema to represent perception, model, context, and activity for intelligent robots. And, axiomatic rules have been used for generating semantic contexts using OWL ontologies. Experiments are successfully performed for recognizing partially occluded objects based on our ontology-based semantic context model without contradictions in real applications.

1 Introduction

Object recognition in the intelligent mobile robot domain requires coping with a number of challenges, starting with difficulties intrinsic to the robot. These difficulties are caused by incomplete recognition such as imprecise perception and action, and by the limited knowledge [1, 2]. These affect the robot judgment to represent the environment and to predicate the behavior variation by user. The context modeling may lie in the imposition by the designer of an organization of the physical-world data into logical structure. We dismiss the obvious recognition of contexts in spite of hidden and partial data in the scene, preferring to promote semantic context model as the necessary basic knowledge for intelligent mobile robots.

Knowledge for robot [3] can be used to resolve the ambiguity problem when the observed positions of objects are ambiguous. Several approaches have been proposed to enhance the performance of location information [1, 2, 6].

In this paper, we represent a semantic context model on the similar principle. Context is any information that can be used to characterize the situation of an entity. An entity may be a person, place, or object that is considered relevant to the interaction between a user and an application, including the user and application themselves [4]. In general, contexts consist of location, identity, time and activity. The context is essential for dealing with complex vision-information for intelligent robots in the dynamic world.

S. Lee, I.H. Suh, and M.S. Kim (Eds.): Recent Progress in Robotics, LNCIS 370, pp. 399–408, 2008.
springerlink.com

Recently, ontology has been used to represent sharable schema knowledge using classes, properties, and individuals. Ontology is useful for representing background knowledge for knowledge processing. In this paper, ontology has been used for describing various schema knowledge using taxonomy and relations between classes. Moreover, we represent all the relevant knowledge in the form of ontology for real time processing of observed objects. The idea comes from storing background knowledge in advance and using the stored knowledge for resolving ambiguities in real applications.

Contexts are represented by vocabulary defined in the ontology. Classes and properties play important roles in representing contexts. We use OWL ontology [5] to present common concepts and inference of axiomatic rules. In this way, we model the different context hierarchies more adaptively and effectively. In order to maintain coherences and consistent inferred contexts, we propose a method to automate the maintenance of visible information. The semantic context model is applicable in dynamic situation where certain sets of observed contexts are considered to be contradictory and the system must make sure that the semantic context model remains free of contradiction.

2 Related Work

Recently, many related academic research efforts and commercial reality are made for intelligent robot with vision sensor. Many knowledge-based vision systems have been designed for the application domain.

[6] proposed a framework for building ontology to provide semantic interpretations for image contents. Their ontology is focused on visual information such as texture, color and edge features in ontology. However the ontology is developed only for visual concepts. [1] proposed OWL as context ontology for modeling context and supporting logic based context reasoning. The context ontology includes person, place, entity and activity. Furthermore they developed context reasoning process. Using the mechanism, implicit context of activities such as cooking, sleeping can be deduced from explicit context such as person, device, place and so on. However they only modeled objects, places and activities, and defined rules for reasoning between them. Thus, their approach does not provide sensor-driven image or geometry information. [2] studied how to model context information in a robot environment. They developed the context model using rules and ontology. The rules are used for modeling dynamic information such as user's current location, current robot dictator and so on. Their ontology is mainly used for describing static information about most parts of device, space, person and artifact. Their research focuses on modeling objects, space and activities. They do not concern low level context which is generally used to recognize their context of interest in practical application.

Our goal is to acquire all the domain knowledge without requiring image processing skills. We propose a semantic context model based on ontology. The context ontology uses visual concept knowledge in order to hide the low-level vision layer complexity and to guide the rich knowledge in everyday environment. Then, we want to build knowledge bases of intelligent robots relying on the visual concept.

3 Semantic Context Model

A robot cannot process and understand information like humans do. The robot that acts within an environment needs machine-understandable representation of objects, in terms of their shapes, feature and usage. Our input to robots has to be very explicit so that they can handle objects and determine what to do with these. We have developed a context in everyday environment using OWL ontology for everyday physical objects to support context understanding.

The need for context is even greater when we solve problem in a dynamic environments. The autonomous service robot has given the users the convenience that they can access whatever information and services human wants, whenever they want and wherever they are. Having the informative and accurate context, we will be able to obtain different information from the same services in different situations. Context can be used to determine what information or services to make available or to bring forward to the users.

> *Context is any information that can be used to enhance robot vision related with an entity situation. An entity is a nearby object, person, their relation, and changes to those objects.*

The context $p(s, v)$ is expressed using predicate logic, that is, functions over situation, that return true or false, depending on whether or not the property holds. S is a subject of context; p is an attributes of s; and v is value of s. For example, the predicate $left(x, y)$ is defined to return true if an object x is on the left of an object y. As this example shows, the state argument is implicit in the notation of a predicate.

A context model $CM = P(S, V)$ represents a set of contexts related to one object that invisible contexts may support. $S = \{s_1, s_2, ..., s_n\}$ is a set of context's subject; $P=\{p_1, p_2, ..., p_n\}$ is a set of attributes of S; and $V = \{v_1, v_2, v_n\}$ is a set of values of S. These model the ability to exploit resources relevant to the robot position and its ability to infer relevant information to an object. This model represents basic knowledge for robots. We work with a fixed set atoms formulas (such as $encloses(cup, kitchen)$), These atoms stand for atomic facts which may hold of a robot, like "the cup is enclosed by the kitchen." Verbs like "encloses," "on," "left," "before," "meets," all denote predicate symbols, sometimes called propositional symbols, relationships between robots and propositions. Every variable is a term such as subject "cup" of p and value "kitchen" of p.

There are two categories: (i) spatial context is an extension of previous research [7]. We have added context categories of presenting context to the robot. The context can be categorized into three types (with a robot, with objects and with space). An example of a context with objects might be expressed using objects relation "right", where an object x is on the right of object y. (ii) temporal context is time information when defining a scene as an instance (i.e., begins, ends). For example, if we define the beginning time with the sensor information of a particular actor as the value of begins and the coordinates of the actor stay unchanged for a given time, the finishing time of the actor action is stored as the value of ending. At this time, the relations of time among actions are not considered [7]. Our context and context category for a context model are summarized in Table 1.

Semantic context is proposed to represent richer contexts using relevant ontologies. The contexts can provide richer high-level contextual information to intelligent robots. We employ a context inference engine in which axioms are used to infer contexts based on ontology as well as observed context predicates. The observed context predicates need to be removed and all the inferred semantic contexts from then need to be maintained as well. For example, when a person brings an object in a room, semantic contexts related to the object are renewed based on ontological inferences. When a person takes the object out of the room, it is necessary to retract all the semantic contexts generated. Thus, we need a system to maintain the consistencies of semantic contexts. Contexts are labeled "old_datum" or "new_datum" to indicate whether the context is observed in the situation. For instance, when a object with person enters a kitchen, the observed context node, "encloses(*object, kitchen*)," becomes "new_datum" in blackboard. Later, if the object leaves the kitchen, the observed context, "encloses(*object, kitchen*)" becomes "old_datum".

Table 1. Context and context category

Context Type		Predicate of Context	Description
Spatial Context	with robot	near	A robot is near an object.
		far	A robot is far an object.
		narrow	The road narrows for robot.
		wide	It is a wide road for robot.
	with objects	above	An object is above another one, it is directly over it or higher than it.
		below	It is inverse of "above."
		over	An object is over another object.
		under	It is inverse of "over."
		left	An object is on the left of another object.
		right	An object is on the right of another object.
		beside	It is "left" or "right."
		between	Something is between two objects or is in between them.
	with place	encloses	A place or an object is enclosed by something.
		overlap	A thing overlaps another.
Temporal Context			before, after, meets, met-by, overlaps, overlapped-by, starts, started-by, during, contains, finishes, finished-by, equals

4 Robot-Centered Ontology and Reasoning for Semantic Context

4.1 Robot-Centered Ontology

Context predicates are generated based on vision-sensor of mobile robots. For instance, when an object movement is detected by vision sensors, a context model related to the object is regenerated. By contexts, robot refers to any information from ubiquitous sensors and infers changed information based on associated ontologies. The ontology has characteristics such as definitions of representational vocabulary, a well-defined syntax, well-understood semantics, efficient reasoning and sufficient expressive capabilities that make robot intelligent.

As in [12], we define major ontology as some ontology of our full robot-centered ontology, related to vision into object ontology, space ontology and context ontology. Object ontology and space ontology are comprised in model ontology. Figure 1 shows architecture of robot-centered ontology.

The object ontology defines objects in an environment where a robot lives. For instance, it classifies objects named container (i.e., can, cup, pot, etc), furniture (i.e., table, chair, cabinet), and food (i.e., rice, ice-cream) in space and defines their features (i.e. size, color). A sensing data is low-level information (i.e., object features) of an image including pixel information and visual features such as hue, saturation, brightness, size, and shape.

Contexts are then inferred from the data by using an axiomatic context inference engine. The inference engine analyzes the locations of objects with respect to location of a robot. Here, robot class is subsumed to the object class in order to refer to robot data such as size and height of the robot.

The space ontology represents spatial information. As it is important to infer the movements of an actor and objects in a particular space, the space must be characterized and classified by using symbols such as path, table zone in the kitchen and corridor.

The context ontology defines various components that represent situation. It can come from the vision-data of a robot, like a human is looking around. It is the schema for context model. The ontology represents context model of objects. The context predicate is the properties for the ontology. Context ontology describes spatial context and temporal context. (i) spatial context ontology: All property values of spatial context ontology are inferred as sensing coordination of the resources in ubiquitous environment. The properties are spatial predicates of context (as shown in Table 1).

Humans and objects can move in and out of a space or change their position in same space. For instance, when a cup moves to another spot, context inference engine generates all the relevant contextual information by triggering context inference axioms. When it moves out the space, all of the inferred contextual information must be removed from the model. Likewise, when it moves in again, the same process will be repeated. Hence, we need to keep context dependency structure to enhance the

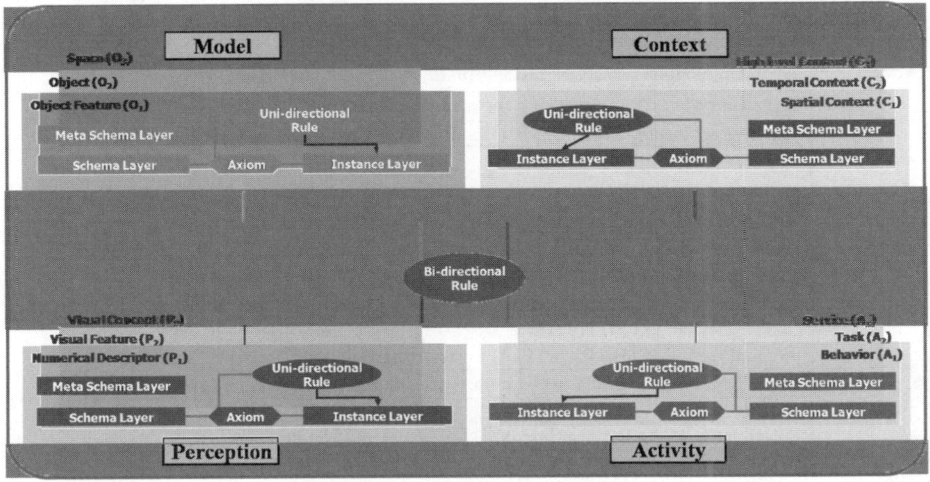

Fig. 1. Architecture of robot-centered ontology

performance of context inference engines. (ii) Temporal context ontology: We defined time ontology as OWL-time ontology [8], where the time information can be obtained according to the location of user or object. The OWL-time ontology formalizes time by expressing it with concepts and properties.

For this ontology, we reference thirteen interval relations which are proposed to formalize time by James Allen with defining relations among scenes. Temporal ontology makes it possible to obtain knowledge which is not explicitly represented in ontology by inferring relations among scenes using Allen's interval relation [9].

4.2 Semantic Context Inference Engine Using Axiomatic Rules

In this paper, these robot-centered ontology instances are used by an ontology inference engine to generate a context model. The ontology inference engine reasons about the subsumption relations between contextual resources and infers new contextual information based on heuristic knowledge using axiomatic inference rules.

Our ontology inference engine is used to identify the taxonomy relations between humans and objects. This infers new knowledge using axioms provided by ontology. For instance, if we define that the domain of the property named "encloses" is "Object" class and that "Robot" class must have "encloses" property with at least one "Space" class as a range, the relation of "Robot" class's inclusion in "Object" class is defined as (Robot ⊆ Object). We therefore can infer that the instance defined in "Robot" class is subsumed by "Object" class.

We classify rules into four categories, namely "create," "retrieve," "update" and "retract" rules, as discussed in Table 2. These APIs are implemented using a Prolog [10] with context inference engine using the rules. When retrieve-API of ontology inference engine calls, the engine infers semantic contexts using coordinate information of resources that are generated from ubiquitous sensors.

Table 2. Axiomatic rules for ontological inference

Category	API Name and Rule
Create	createOntologyInstance(CN) → findInstance(CN, Triple) ∧ assert(new_datum(Triple)).
	createOntologyInstance(CN, Inst) → findInstance(CN, Inst) ∧ assert(new_datum(Inst)).
	setProperty(Inst, Prop, Val) → assert(new_datum(Prop(Inst, Val))).
Retrieve	getOntologyInstance(CN, [Inst]) → findInstance(CN, [Inst]) □ ssert(new_datum([Inst])).
	getProperty(CN, class, [Prop]) → findProperty(CN, class, [Prop]) ∧ assert(new_datum([Prop])).
	getProperty(Inst, instance, [Prop]) → x_position(Inst, Xpos) ∧ y_position(Inst, Ypos) ∧ theta(Inst, Theta), spatialContext(context(Inst, Xpos, Ypos, Theta, [SpatialCont])) ∧ temporalContext(context(Inst, Xpos, Ypos, Theta, [TemporalCont])) ∧ assert(new_datum([SpatialCont])) ∧ assert(new_datum([Temporalcont])).
	getPropertyValue(Inst, Prop, [Val]) → findPropValue(Inst, Prop, [Val]) ∧ assert(new_datum([Prop(Inst, Val)])).
Retract	retractOntologyInstance(Inst) → retract(new_datum(Inst)) ∧ assert(old_datum(Inst)).
	retractPropertyValue(Inst, Prop, Val) → retract(new_datum(Prop(Inst, Val))) ∧ assert(old_datum(Prop(Inst, Val))).
Update	update(Inst, Prop, Val) → retract(new_datum(Prop(Inst, [beforeVal]))) ∧ assert(old_datum(Prop(Inst, beforeVal))) ∧ assert(new_datum(Prop(Inst, Val))).

4.3 Semantic Context Acquisition Process

As illustrated in Figure 2, the proposed semantic context acquisition process is guided by the semantic context model. It models time as a sequence of states, extending infinitely into the future. This sequence of states is simply a path of request from a robot. The model allows robot to find contexts in spite of hidden and partial data. Moreover, our semantic context inference engine using axiomatic rules (see Table 2) enables robot to query through any directional reasoning between each scene.

Our inference engine is used to infer objects or some necessary information for object recognition when basic visual features (e.g., hue, saturation, brightness, size, shape, etc) are given. A simple example of Figure 2 presents how we can recognize an unknown object as a cup with objects relationship that is recognize previous scenes. In general, we wish to determine whether the object exists at the current space or not.

We can use the following procedure:

1. Construct a list L of triples p(s, v), where s and v is any variable mentioned in the KB (Knowledge Base: working memory) and any instance in the ontology, and p is any constant mentioned in the KB and the property of the robot-centered ontology. Our inference engine translates OWL documents into triples as facts. They are the simplest form of predicate, and held in a working memory.
2. If no such triple of visual features can be found in current space, we can use the semantic context inference engine included in ontology manager. The reasoner use collected facts in KB. In spite of unknown in current scenes, our robot inferences hidden object via fact (of the true value) in KB.

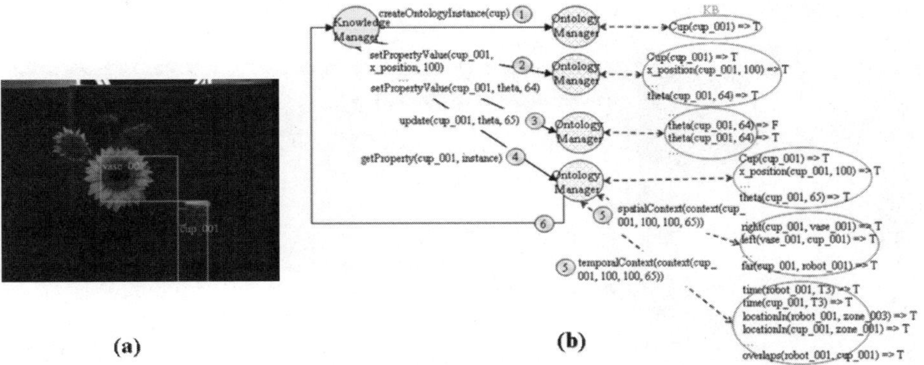

Fig. 2. An example of semantic context acquisition process; (a) Snapshot of object recognition: cup and vase (b) Semantic context acquisition process in the case of (a)

We may assume that the query of the robot to be sent is divided according to the inference API based on ontology, which are sent sequentially. Its role is to connect the requested API, and giving them initial values of their parameters.

At a time, the cup is on the right of the vase, knowledge manager requests ontology manager to fetch ontology instance of a cup and vase, and their relation (see Figure 2).

The ontology manager translates ontology instances into predicate forms. When a visual recognition process is completed, the context inference engine embedded in ontology manager partially instantiates related ontology instances in real time.

In the example, the context inference engine generates the following semantic contexts which become "new_datum" in blackboard:

```
beside(cup, vase).            beside(vase, cup).
left(vase, cup).              right(cup, vase).
far(cup, robot).              above(cup, table).
encloses(cup, kitchen).
```

A few seconds later, temporal contexts occur when the context inference engine generates new datum. Three temporal predicates; "time(robot, T1)," "time(cup, T1)" and "begin(T1)" are generated [7]. Even if some objects are not recognized due to their movement or occlusion, our system can estimate the object position by using all previous contexts telling that their propositions are true. The predicate "new_datum(far(cup, robot))" is generated by sensing data (see Figure 3). The followings are predicate forms remained "new_dautm" by previous contexts;

```
beside(cup, vase).            beside(vase, cup).
above(cup, table).            encloses(cup,
kitchen).
```

Our inference engine can infer "behind(*cup*, *vase*)" using following a rule.

```
behind(obj1, obj2) :-
    \+ left(obj1, obj2),        \+ left(obj2, obj1),
    above(obj1, x),         above(obj2, x),
    beside(obj1, obj2).
```

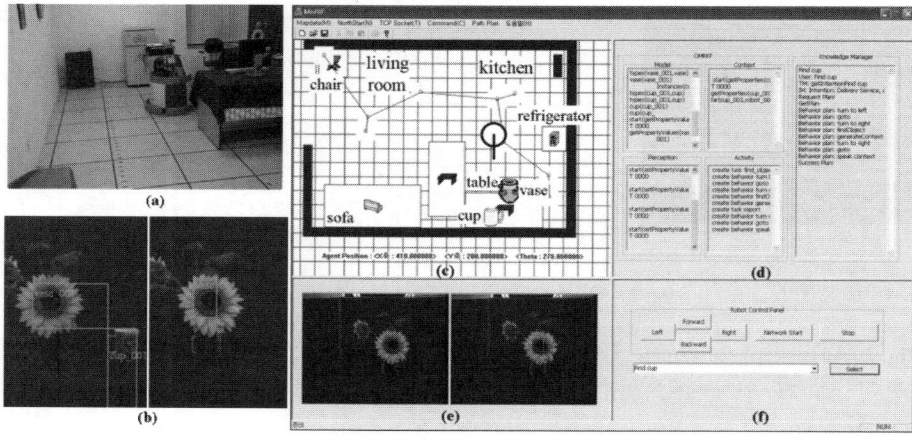

Fig. 3. Snapshots of our experiments; (a) View of external camera (b) The left image is a result scene when cup and vase are recognized. The right image is a result scene when cup is occluded by vase. (c) Top view of our experimental environment (d) Logs and messages about robot-centered ontology (e) Input and result images for object recognition (f) Robot control panel and user command box.

4.4 Empirical Results

In this paper, a mobile robot is used in which a laptop and a camera are mounted. The camera gets the state of human and objects with the robot as the central position [11]. Figure 3 shows the snapshot of our experiments. Images are continuously captured by the stereo camera. The objects are firstly recognized by SIFT features. And objects and their location can be inferred by our proposed approach and compensated by our proposed axiomatic rules even if they are occluded. Note that the proposed system can extract the relationship between objects which are not directly recognizable but inferred from relationships among the instances of spatial contexts and temporal contexts created previously (see also Figure 3). In this experiment, the cup is on the right of the vase, while the robot is far from the object. In the next, although the cup hides behind the vase, the robot is still able to recognize the cup because of the contexts in previous scenes.

5 Conclusion and Future Directions

Semantic context modeling points out the necessity of situation recognition in terms of what is happening in a particular place at a particular time, or what is happening to the user. In this paper, we have presented different degrees of awareness in respect to their influences on the search space, where the context model is proven to be of tremendous importance in the recognition ability.

Acknowledgement

This work was performed for the Intelligent Robotics Development Program, one of the 21st Century Frontier R&D Programs funded by Korea Ministry of Commerce, Industry and Energy.

References

1. Wang, X.H., Zhang, D.Q., Gu, T., Pung, H.K.: Ontology Based Context Modeling and Reasoning using OWL. In: Proc. of the Second IEEE Annual Conference on Pervasive Computing and Communication Workshops, pp. 18–22 (2004)
2. Go, Y.C., Shon, J.-C.: Context Modeling for Intelligent Robot Services using Rule and Ontology. In: The 7th international conference, vol. 2, pp. 813–816 (2005)
3. Hector, J.L., Ronald, J.B.: Expressiveness and tractability in knowledge representation and reasoning. Computational Intelligence 3, 78–93 (1987)
4. Anind, K.: Providing Architectural Support for Building Context-Aware Applications. PhD thesis, Georgia Institute of Technology (2000)
5. Smith, M., Welty, C., Deborah, L., McGuinness, D.: OWL Web Ontology Language Guide. W3C Recommendation (2004)
6. Liu, S., Chia, L.-T., Chan, S.: Ontology for Nature-Scene Image Retrieval. In: Meersman, R., Tari, Z. (eds.) On the Move to Meaningful Internet Systems 2004: CoopIS, DOA, and ODBASE. LNCS, vol. 3291, pp. 1050–1061. Springer, Heidelberg (2004)

7. Choi, J.H., Suh, I.H., Park, Y.T.: Ontology-based User Context Modeling for Ubiquitous Robots. In: URAI 2006. The 3rd International Conference on Ubiquitous Robots and Ambient Intelligence (2006)
8. Hobbs, J.R., Pan, F.: An Ontology of Time for the Semantic Web. ACM Transactions on Asian Language Processing (TALIP): Special issue on Temporal Information Processing 3, 66–85 (2004)
9. Allen, J.F.: Planning as Temporal Reasoning. In: Proc. Of the Second Int'1 Conf. On Principles of Knowledge Representation and Reasoning, Cambridge, MA (April 1991)
10. Bratko, I.: Prolog Programming for Artificial Intelligence, 3rd edn. Pearson education, London (2001)
11. Lim, G.H., Chung, J.L., et al.: Ontological Representation of Vision-based 3D Spatio-Temporal Context for Mobile Robot Applications. Artifical Life and Robotics (AROB) (2007)
12. Suh, I.H, Lim, G.H., Hwang, W., Suh, H., Choi, J.H., Park, Y.T.: Ontology-based Multi-layered Robot Knowledge Framework (OMRKF) for Robot Intelligence. In: Proc. of IROS 2007 (2007)

Lecture Notes in Control and Information Sciences

Edited by M. Thoma, M. Morari

Further volumes of this series can be found on our homepage:
springer.com

Vol. 370: Lee S.; Suh I.H.;
Kim M.S. (Eds.)
Recent Progress in Robotics:
Viable Robotic Service to Human
410 p. 2008 [978-3-540-76728-2]

Vol. 369: Hirsch M.J.; Pardalos P.M.;
Murphey R.; Grundel D.
Advances in Cooperative Control and
Optimization
423 p. 2007 [978-3-540-74354-5]

Vol. 368: Chee F.; Fernando T.
Closed-Loop Control of Blood Glucose
157 p. 2007 [978-3-540-74030-8]

Vol. 367: Turner M.C.; Bates D.G. (Eds.)
Mathematical Methods for Robust and Nonlinear
Control
444 p. 2007 [978-1-84800-024-7]

Vol. 366: Bullo F.; Fujimoto K. (Eds.)
Lagrangian and Hamiltonian Methods for
Nonlinear Control 2006
398 p. 2007 [978-3-540-73889-3]

Vol. 365: Bates D.; Hagström M. (Eds.)
Nonlinear Analysis and Synthesis Techniques for
Aircraft Control
360 p. 2007 [978-3-540-73718-6]

Vol. 364: Chiuso A.; Ferrante A.;
Pinzoni S. (Eds.)
Modeling, Estimation and Control
356 p. 2007 [978-3-540-73569-4]

Vol. 363: Besançon G. (Ed.)
Nonlinear Observers and Applications
224 p. 2007 [978-3-540-73502-1]

Vol. 362: Tarn T.-J.; Chen S.-B.;
Zhou C. (Eds.)
Robotic Welding, Intelligence and Automation
562 p. 2007 [978-3-540-73373-7]

Vol. 361: Méndez-Acosta H.O.; Femat R.;
González-Álvarez V. (Eds.):
Selected Topics in Dynamics and Control of
Chemical and Biological Processes
320 p. 2007 [978-3-540-73187-0]

Vol. 360: Kozlowski K. (Ed.)
Robot Motion and Control 2007
452 p. 2007 [978-1-84628-973-6]

Vol. 359: Christophersen F.J.
Optimal Control of Constrained
Piecewise Affine Systems
190 p. 2007 [978-3-540-72700-2]

Vol. 358: Findeisen R.; Allgöwer
F.; Biegler L.T. (Eds.): Assessment and Future Directions of Nonlinear
Model Predictive Control
642 p. 2007 [978-3-540-72698-2]

Vol. 357: Queinnec I.; Tarbouriech
S.; Garcia G.; Niculescu S.-I. (Eds.):
Biology and Control Theory: Current Challenges
589 p. 2007 [978-3-540-71987-8]

Vol. 356: Karatkevich A.:
Dynamic Analysis of Petri Net-Based Discrete
Systems
166 p. 2007 [978-3-540-71464-4]

Vol. 355: Zhang H.; Xie L.:
Control and Estimation of Systems with
Input/Output Delays
213 p. 2007 [978-3-540-71118-6]

Vol. 354: Witczak M.:
Modelling and Estimation Strategies for Fault
Diagnosis of Non-Linear Systems
215 p. 2007 [978-3-540-71114-8]

Vol. 353: Bonivento C.; Isidori A.; Marconi L.;
Rossi C. (Eds.)
Advances in Control Theory and Applications
305 p. 2007 [978-3-540-70700-4]

Vol. 352: Chiasson, J.; Loiseau, J.J. (Eds.)
Applications of Time Delay Systems
358 p. 2007 [978-3-540-49555-0]

Vol. 351: Lin, C.; Wang, Q.-G.; Lee, T.H., He, Y.
LMI Approach to Analysis and Control of
Takagi-Sugeno Fuzzy Systems with Time Delay
204 p. 2007 [978-3-540-49552-9]

Vol. 350: Bandyopadhyay, B.; Manjunath, T.C.;
Umapathy, M.
Modeling, Control and Implementation of Smart
Structures 250 p. 2007 [978-3-540-48393-9]

Vol. 349: Rogers, E.T.A.; Galkowski, K.;
Owens, D.H.
Control Systems Theory
and Applications for Linear
Repetitive Processes 482
p. 2007 [978-3-540-42663-9]

Vol. 347: Assawinchaichote, W.; Nguang, K.S.;
Shi P.
Fuzzy Control and Filter Design
for Uncertain Fuzzy Systems
188 p. 2006 [978-3-540-37011-6]

Vol. 346: Tarbouriech, S.; Garcia, G.; Glattfelder,
A.H. (Eds.)
Advanced Strategies in Control Systems
with Input and Output Constraints
480 p. 2006 [978-3-540-37009-3]

Vol. 345: Huang, D.-S.; Li, K.; Irwin, G.W. (Eds.)
Intelligent Computing in Signal Processing
and Pattern Recognition
1179 p. 2006 [978-3-540-37257-8]

Vol. 344: Huang, D.-S.; Li, K.; Irwin, G.W. (Eds.)
Intelligent Control and Automation
1121 p. 2006 [978-3-540-37255-4]

Vol. 341: Commault, C.; Marchand, N. (Eds.)
Positive Systems
448 p. 2006 [978-3-540-34771-2]

Vol. 340: Diehl, M.; Mombaur, K. (Eds.)
Fast Motions in Biomechanics and Robotics
500 p. 2006 [978-3-540-36118-3]

Vol. 339: Alamir, M.
Stabilization of Nonlinear Systems Using
Receding-horizon Control Schemes
325 p. 2006 [978-1-84628-470-0]

Vol. 338: Tokarzewski, J.
Finite Zeros in Discrete Time Control Systems
325 p. 2006 [978-3-540-33464-4]

Vol. 337: Blom, H.; Lygeros, J. (Eds.)
Stochastic Hybrid Systems
395 p. 2006 [978-3-540-33466-8]

Vol. 336: Pettersen, K.Y.; Gravdahl, J.T.;
Nijmeijer, H. (Eds.)
Group Coordination and Cooperative Control
310 p. 2006 [978-3-540-33468-2]

Vol. 335: Kozłowski, K. (Ed.)
Robot Motion and Control
424 p. 2006 [978-1-84628-404-5]

Vol. 334: Edwards, C.; Fossas Colet, E.;
Fridman, L. (Eds.)
Advances in Variable Structure and Sliding Mode
Control
504 p. 2006 [978-3-540-32800-1]

Vol. 333: Banavar, R.N.; Sankaranarayanan, V.
Switched Finite Time Control of a Class of
Underactuated Systems
99 p. 2006 [978-3-540-32799-8]

Vol. 332: Xu, S.; Lam, J.
Robust Control and Filtering of Singular Systems
234 p. 2006 [978-3-540-32797-4]

Vol. 331: Antsaklis, P.J.; Tabuada, P. (Eds.)
Networked Embedded Sensing and Control
367 p. 2006 [978-3-540-32794-3]

Vol. 330: Koumoutsakos, P.; Mezic, I. (Eds.)
Control of Fluid Flow
200 p. 2006 [978-3-540-25140-8]

Vol. 329: Francis, B.A.; Smith, M.C.; Willems,
J.C. (Eds.)
Control of Uncertain Systems: Modelling,
Approximation, and Design
429 p. 2006 [978-3-540-31754-8]

Vol. 328: Loría, A.; Lamnabhi-Lagarrigue, F.;
Panteley, E. (Eds.)
Advanced Topics in Control Systems Theory
305 p. 2006 [978-1-84628-313-0]

Vol. 327: Fournier, J.-D.; Grimm, J.; Leblond, J.;
Partington, J.R. (Eds.)
Harmonic Analysis and Rational Approximation
301 p. 2006 [978-3-540-30922-2]

Vol. 326: Wang, H.-S.; Yung, C.-F.; Chang, F.-R.
H_∞ Control for Nonlinear Descriptor Systems
164 p. 2006 [978-1-84628-289-8]

Vol. 325: Amato, F.
Robust Control of Linear Systems Subject to
Uncertain
Time-Varying Parameters
180 p. 2006 [978-3-540-23950-5]

Vol. 324: Christofides, P.; El-Farra, N.
Control of Nonlinear and Hybrid Process Systems
446 p. 2005 [978-3-540-28456-7]

Vol. 323: Bandyopadhyay, B.; Janardhanan, S.
Discrete-time Sliding Mode Control
147 p. 2005 [978-3-540-28140-5]

Vol. 322: Meurer, T.; Graichen, K.; Gilles, E.D.
(Eds.)
Control and Observer Design for Nonlinear Finite
and Infinite Dimensional Systems
422 p. 2005 [978-3-540-27938-9]

Vol. 321: Dayawansa, W.P.; Lindquist, A.;
Zhou, Y. (Eds.)
New Directions and Applications in Control
Theory
400 p. 2005 [978-3-540-23953-6]

Vol. 320: Steffen, T.
Control Reconfiguration of Dynamical Systems
290 p. 2005 [978-3-540-25730-1]

Vol. 319: Hofbaur, M.W.
Hybrid Estimation of Complex Systems
148 p. 2005 [978-3-540-25727-1]

Vol. 318: Gershon, E.; Shaked, U.; Yaesh, I.
H_∞ Control and Estimation of State-multiplicative
Linear Systems
256 p. 2005 [978-1-85233-997-5]

Vol. 317: Ma, C.; Wonham, M.
Nonblocking Supervisory Control of State Tree
Structures
208 p. 2005 [978-3-540-25069-2]

Vol. 316: Patel, R.V.; Shadpey, F.
Control of Redundant Robot Manipulators
224 p. 2005 [978-3-540-25071-5]